LONDON MATHEMATICAL SOCIETY LECTURE NOTE SERIES

Managing Editor: Professor I.M. James,
Mathematical Institute, 24-29 St Giles, Oxford

1. General cohomology theory and K-theory, P.HI
4. Algebraic topology, J.F.ADAMS
5. Commutative algebra, J.T.KNIGHT
8. Integration and harmonic analysis on compact groups, R.E.EDWARDS
9. Elliptic functions and elliptic curves, P.DU VAL
10. Numerical ranges II, F.F.BONSALL & J.DUNCAN
11. New developments in topology, G.SEGAL (ed.)
12. Symposium on complex analysis, Canterbury, 1973, J.CLUNIE & W.K.HAYMAN (eds.)
13. Combinatorics: Proceedings of the British Combinatorial Conference 1973, T.P.McDONOUGH & V.C.MAVRON (eds.)
15. An introduction to topological groups, P.J.HIGGINS
16. Topics in finite groups, T.M.GAGEN
17. Differential germs and catastrophes, Th.BROCKER & L.LANDER
18. A geometric approach to homology theory, S.BUONCRISTIANO, C.P. BOURKE & B.J.SANDERSON
20. Sheaf theory, B.R.TENNISON
21. Automatic continuity of linear operators, A.M.SINCLAIR
23. Parallelisms of complete designs, P.J.CAMERON
24. The topology of Stiefel manifolds, I.M.JAMES
25. Lie groups and compact groups, J.F.PRICE
26. Transformation groups: Proceedings of the conference in the University of Newcastle-upon-Tyne, August 1976, C.KOSNIOWSKI
27. Skew field constructions, P.M.COHN
28. Brownian motion, Hardy spaces and bounded mean oscillations, K.E.PETERSEN
29. Pontryagin duality and the structure of locally compact Abelian groups, S.A.MORRIS
30. Interaction models, N.L.BIGGS
31. Continuous crossed products and type III von Neumann algebras, A.VAN DAELE
32. Uniform algebras and Jensen measures, T.W.GAMELIN
33. Permutation groups and combinatorial structures, N.L.BIGGS & A.T.WHITE
34. Representation theory of Lie groups, M.F. ATIYAH et al.
35. Trace ideals and their applications, B.SIMON
36. Homological group theory, C.T.C.WALL (ed.)
37. Partially ordered rings and semi-algebraic geometry, G.W.BRUMFIEL
38. Surveys in combinatorics, B.BOLLOBAS (ed.)
39. Affine sets and affine groups, D.G.NORTHCOTT
40. Introduction to Hp spaces, P.J.KOOSIS
41. Theory and applications of Hopf bifurcation, B.D.HASSARD, N.D.KAZARINOFF & Y-H.WAN
42. Topics in the theory of group presentations, D.L.JOHNSON
43. Graphs, codes and designs, P.J.CAMERON & J.H.VAN LINT
44. Z/2-homotopy theory, M.C.CRABB
45. Recursion theory: its generalisations and applications, F.R.DRAKE & S.S.WAINER (eds.)
46. p-adic analysis: a short course on recent work, N.KOBLITZ
47. Coding the Universe, A.BELLER, R.JENSEN & P.WELCH
48. Low-dimensional topology, R.BROWN & T.L.THICKSTUN (eds.)

49. Finite geometries and designs, P.CAMERON, J.W.P.HIRSCHFELD & D.R.HUGHES (eds.)
50. Commutator calculus and groups of homotopy classes, H.J.BAUES
51. Synthetic differential geometry, A. KOCK
52. Combinatorics, H.N.V.TEMPERLEY (ed.)
53. Singularity theory, V.I.ARNOLD
54. Markov processes and related problems of analysis, E.B.DYNKIN
55. Ordered permutation groups, A.M.W.GLASS
56. Journees arithmetiques 1980, J.V.ARMITAGE (ed.)
57. Techniques of geometric topology, R.A.FENN
58. Singularities of smooth functions and maps, J.MARTINET
59. Applicable differential geometry, F.A.E.PIRANI & M.CRAMPIN
60. Integrable systems, S.P.NOVIKOV et al.
61. The core model, A.DODD
62. Economics for mathematicians, J.W.S.CASSELS
63. Continuous semigroups in Banach algebras, A.M.SINCLAIR
64. Basic concepts of enriched category theory, G.M.KELLY
65. Several complex variables and complex manifolds I, M.J.FIELD
66. Several complex variables and complex manifolds II, M.J.FIELD
67. Classification problems in ergodic theory, W.PARRY & S.TUNCEL

London Mathematical Society Lecture Note Series. 56

Journées Arithmétiques 1980

Edited by
J.V. ARMITAGE
Principal of the College of St. Hild and St. Bede
University of Durham

CAMBRIDGE UNIVERSITY PRESS

CAMBRIDGE

LONDON NEW YORK NEW ROCHELLE

MELBOURNE SYDNEY

CAMBRIDGE UNIVERSITY PRESS
Cambridge, New York, Melbourne, Madrid, Cape Town, Singapore, São Paulo

Cambridge University Press
The Edinburgh Building, Cambridge CB2 8RU, UK

Published in the United States of America by Cambridge University Press, New York

www.cambridge.org
Information on this title: www.cambridge.org/9780521285131

First published 1982
Re-issued in this digitally printed version 2007

A catalogue record for this publication is available from the British Library

Library of Congress Catalogue Card Number: 81–18032

ISBN 978-0-521-28513-1 paperback

CONTENTS

* B.J. Birch and J. Coates also gave invited addresses, but preferred not to publish them in the proceedings.

**We are able to publish only a selection of splinter group contributions.

PREFACE

For a number of years, French mathematicians, supported by the Société Mathématique de France, have organized colloquia in the theory of numbers, to which they have always invited British mathematicians (and others). The London Mathematical Society returned their hospitality during the 'Journées Arithmétiques' held at the University of Exeter from 13th April to 19th April 1980.

This volume contains the texts of the invited addresses, in some cases considerably expanded, with the exception of expository papers by Dr B.J. Birch (a progress report on elliptic curves) and Professor J. Coates (on work of Mazur and Wiles on cyclotomic fields), who preferred not to publish their talks. In addition there are selected contributions, also sometimes in expanded versions, from the several splinter groups organized during the colloquium. Our gratitude is due to the lecturers for making the publication of this volume possible, and indeed for their original contribution to the success of the Journées'.

The first part of Professor Chudnovsky's article is a report of his talk. We are grateful to him for making available for publication a proof of his results, as presented here in the second part of his article. Because of the importance of the results, it was an editorial decision to publish them here, even though there was no time to submit the manuscript to the usual refereeing procedure.

We should like to express our thanks to the London Mathematical Society for supporting these 'Journées', to the organisers and especially to the secretary, Dr R.W.K. Odoni of the University of Exeter. Our thanks are due too to the Cambridge University Press for help with the preparation of this volume and, finally, to the Press and the London Mathematical Society for agreeing to publish it in their Lecture Notes Series.

PREFACE

[illegible]

ADDRESSES OF CONTRIBUTORS

MAIN SPEAKERS

D. Bertrand, Institut de Mathématiques et Sciences Physiques, Parc Valrose, 06034 Nice, Cedex, France.

G.V. Chudnovsky, Department of Mathematics, Columbia University in the City of New York, New York, N.Y. 10027, U.S.A.

R. Heath-Brown, Mathematical Institute, 24-29 St. Giles, Oxford.

C. Hooley, Department of Pure Mathematics, University College, Cardiff, P.O. Box 78, Cardiff, CF1 1XL.

H.W. Lenstra, Jr., Mathematisch Institut, Universiteit van Amsterdam, Roetersstraat 15, 1018 WB, Amsterdam, Netherlands.

J. Martinet, UER de Mathématique et d'Informatique, 351 Cours de la Libération, 33405 Talence, Cedex, France.

L. McCulloch, Department of Mathematics, University of Illinois, Urbana, Ill. 61801, U.S.A.

W. Narkiewicz, Uniwersytet Wrockawski im Boleslawa Bieruta, Wydzial Matematyki, Fizyki i Chemii, Instut Matematyczny, 50-384 Wroclaw, pl. Grunwaldzki 2/4, Poland.

A. Śchinzel, ul. Brzozowa 12m.27, 00286 Warszawa, Poland.

M. Taylor, Department of Pure Mathematics, Queen Mary College, London, E1 4NS.

SPLINTER GROUP SPEAKERS

R.C. Baker, Department of Mathematics, Royal Holloway College, Egham, Surrey, TW20 0EX.

W.L. Chen, Department of Mathematics, Imperial College, Princes Gate, London, SW7 2AZ.

D. Coray, Département de Mathématiques, Université de Genève, Geneva, Switzerland.

E. Dubois, Département de Mathématiques, Université de Caen, 14032 Caen, Cedex, France.

J. Elstrodt, Mathematisches Institut der Universität Münster, Noxeler Str. 64, 4400 Münster, W. Germany.

H. Faure, Val de Riou, Pont de l'Etoile, 13360 Roquevaire, France.

G. Gras, Département de Mathématiques, Faculté des Sciences de Besancon, Route de Gray, La Bouloie, 25030 Besancon, Cedex, France.

G. Grekos, Department of Mathematics, University of Crete, Iraklio, Crete, Greece.

F. Grunewald, Sonderforschungsbereich 'Theoretische Mathematik' der Universität Bonn, Beringstrasse 4, 5300 Bonn, W. Germany.

H. Iwaniec, Institut Matematyczny, Akademia Nauk, Warszawa, Poland.

E. Kani, Mathematisches Institut der Universität, 6900 Heidelberg, Im Neueuheimer Feld 288, W. Germany.

G. Lachaud, 48 rue Monsieur le Prince, 75006 Paris, France.

F.J. van der Linden, Mathematisch Institut, Universiteit van Amsterdam, P.O. Box 2023, 1000 HE Amsterdam, Netherlands.

D.W. Masser, Department of Mathematics, University of Nottingham, University Park, Nottingham, NG7 2RD.

J. Mennicke, 48 Bielefeld, Mathematisches Fakultät der Universität, Postfach 8640, W. Germany.

M. Mignotte, Département de Mathématique, 7 rue R. Descartes, 67084 Strasbourg, France.

G.J. Rieger, Mathematisches Institut, Universitat Hannover, Hannover, W. Germany.

K. Rubin, Department of Mathematics, Harvard University, Cambridge, MA 02138, U.S.A.

C-G. Schmidt, Universität Saarbrücken, Fachbereich 9 - Mathematik, D6600, Saarbrücken, W. Germany.

G. Tenenbaum, UER de Math. et d'Informatique, 351 Cours de la Libération, 33405 Talence, Cedex, France.

G. Wüstholz, Gesamthochschule Wuppertal, Gaussstrasse 20, D5600, Wuppertal, W. Germany.

PARTICIPANTS

J-P. Allouch	Bordeaux
Amara Hedi	Tunis
R. Apery	Caen
J.V. Armitage	Durham
C. Ayoub	Penn State/Warwick
R. Ayoub	Penn State/Warwick
R.C. Baker	London
B. Baktavatsalov	Abidjan
K. Barner	Kassel
D. Barsky	Paris
S. Basarab	Heidelberg/Bucharest
C. Batut	Bordeaux
H-J. Bentz	Osnabruck
B. Benzaghou	Bouzarea
A-M. Bergé	Bordeaux
D. Bernardi	Paris
M-J. Bertin	Paris
D. Bertrand	Nice
B.J. Birch	Oxford
S. Böge	Heidelberg
J-P. Borel	Limoges
J.L. Boxall	Oxford
D.W. Boyd	Vancouver
J. Brinkhuis	London
P. Bundschuh	Köln
D.A. Burgess	Nottingham
C.J. Bushnell	London
M. Car	Marseille
D. Chatelain	Besancon
W.L. Chen	London
Chen-Jing-Rung	Nottingham
G.V. Chudnovsky	Columbia
J. Coates	Paris
S.D. Cohen	Glasgow
J.L. Colliot-Thélène	Paris
B. Conrey	Michigan/Cambridge
S. Cooper	Oxford
D. Coray	Geneva
J. Cosgrave	Dublin
J. Cougnard	Besancon
J. Cremona	Oxford
J-M. Deshouillers	Bordeaux

Dr. Ding Ping	Nottingham
F. Dress	Bordeaux
E. Dubois	Caen
A. Durand	Limoges
R. Dvornicich	Pisa
L.C. Eggan	Michigan
M. Emsalem	Paris
P. Erdös	Budapest/Limoges
H. Faure	Marseille
F. Frei	Quebec
Dr. Fresnel	Bordeaux
A. Fröhlich	London
A. Ghosh	Nottingham
C. Goldstein	Paris
F. Gramain	Paris
M. Grandet	Caen
G. Gras	Besancon
M-N. Gras	Besancon
G.R.H. Greaves	Cardiff
G. Grekos	Crete
M.R. Gupta	Cambridge
S.K. Gupta	Delhi
J.A. Haight	London
H. Halberstam	Nottingham
R.R. Hall	York
F. Halter-Koch	Essen
G. Harman	London
R. Heath-Brown	Oxford
Y. Hellegouarch	Caen
G. Henniart	Paris
J. Hoffstein	Cambridge
C. Hooley	Cardiff
M.N. Huxley	Cardiff
K.H. Indlekofer	Paderborn
H. Iwaniec	Warsaw
T.H. Jackson	York
H. Jager	Amsterdam
M. Jarden	Heidelberg/Tel Aviv
J-F. Jaulent	Besancon
E. Kani	Heidelberg
P. Kaplan	Nancy
M. Keating	London
A. Kerkour	Rabat

G. Lachaud	Paris
M. Laborde	Paris
H. Langevin	St. Cloud
F. Laubi	Paris
F. Lazami	Tunis
O. Lecacheux	Paris
R.C. Ledgard	Manchester
H.W. Lenstra, Jr.	Amsterdam
C. Levesque	Quebec
D.J. Lewis	Michigan/Heidelberg
P. Liardet	Marseille
F.J. van der Linden	Amsterdam
L.R. McCulloch	Urbana
J. Martinet	Bordeaux
D.W. Masser	Nottingham
R.C. Mason	Cambridge
R. Massy	Brest
Prof. de Mathan	Bordeaux
M. Matignon	Bordeaux
C.R. Matthews	Cambridge
M. Mendès-France	Bordeaux
K. Mennicke	Bielefeld
J. Meyer	Reims
M. Mignotte	Strasbourg
H. Montgomery	Michigan/Cambridge
B. Moroz	Jerusalem
N. Moser	Grenoble
W. Narkiewicz	Wroclaw
A.M. Nelson	London
J-L. Nicolas	Limoges
R. Odoni	Exeter
M. Olivier	Bordeaux
B. Oriat	Besancon
R. Paysant-le-Roux	Caen
A. Perelli	Pisa
B. Perrin-Riou	Paris
M. Peters	Münster
P. Philippon	Paris
R. Pinch	Oxford
J. Pintz	Budapest
P.A.B. Pleasants	Cardiff
G. Poitou	Paris
M. Polzin	Bordeaux

A. Prince	Edinburgh
M. van der Put	Bordeaux
J. Queyrut	Bordeaux
F. Rayner	Liverpool
Dr. Remmal	Paris
M. Reversat	Bordeaux
E. Reyssat	Paris
G. Rhin	Metz
G.J. Rieger	Hannover
K. Rubin	Harvard
B. Saffari	Paris
J.J. Sansuc	Paris
P. Satgé	Caen
N. Schappacher	Paris
R. Scharlau	Bielefeld
A. Schinzel	Warsaw
C-G. Schmidt	Saarbrücken
A.J. Scholl	Oxford
R.J. Schoof	Leiden
E.J. Scourfield	London
N. Sebti	Paris
J-P. Serre	Paris
I. Schiokawa	Keio
P. Shiu	Loughborough
A. Silverberg	Cambridge
J.B. Slater	Salford
R.A. Smith	Toronto
C.J. Smyth	Vancouver
N.M. Stephens	Cardiff
C. Stewart	Canada
W.W. Stothers	Glasgow
R.J. Stroeker	Rotterdam
M. Taylor	Besancon/London
G. Tenenbaum	Bordeaux
R. Tijdeman	Leiden
P. Toffin	Caen
C. Touibi	Tunis
G.W.M. Turk	Amsterdam
R.C. Vaughan	London
J. Velu	Caen
C. Viola	Pisa
G. Wagner	Stuttgart
M. Waldschmidt	Paris

P.L. Walker	Lancaster
C.D. Walter	Dublin
S.M.J. Wilson	Durham
C.F. Woodcock	Kent
G. Wuestholz	Wuppertal
R.I. Yager	Paris
Kun-Rui Yu	Cambridge
H. Zantema	Amsterdam
M. Zitouni	Alger

ABELIAN FUNCTIONS AND TRANSCENDENCE

Daniel Bertrand
Department of Mathematics
University of Nice
Nice, France

After the first results of Siegel [23], the study of abelian functions from the point of view of transcendence theory was developed by Schneider in his article [17] on the values of Euler's beta function. This area of research has been much revived in the past few years, as reflected in a recent meeting [1] on these topics. One of its main applications, the study of diophantine equations, has been discussed on several occasions in previous 'Journées arithmétiques'. I have therefore left this matter aside, preferring to concentrate on the arithmetic nature of certain constants connected with abelian integrals. Three aspects have been selected: Part 1 describes a result of Laurent on elliptic integrals; abelian varieties proper are treated in Part 2, but the main application of this study (Corollary 2) concerns elliptic curves; Part 3 deals with an important theorem of Waldschmidt related to the values of the Grössencharacters of CM fields, and concludes with a discussion on zero estimates.

The results described in this survey all refer in a sense to the list of problems raised by Schneider in [18]. We have specified below some of the questions that remain to be settled in this direction.

If V is an algebraic variety defined over a field k, we write $V(k)$ for the set of k-rational points on V. We denote by $\overline{\mathbb{Q}}$ the field of all complex algebraic numbers.

1. PERIODS OF ELLIPTIC INTEGRALS

The third problem in Schneider's list consists in extending his results on elliptic integrals of the 2nd kind to those of the 3rd kind. An important advance has recently been made in this direction by Laurent [9], in the case of periods.

Let E be an elliptic curve, given by a Weierstrass equation

$$y^2 = 4x^3 - g_2x - g_3 \quad , \qquad (*)$$

where g_2 and g_3 are algebraic, and let ξ be a non-exact differential form on E, defined over $\overline{\mathbb{Q}}$. Denote by $P_1, \ldots, P_m$ the poles of ξ at finite distance, and assume that the corresponding residues $c_1, \ldots, c_m$ are rational integers. If all the c_i's are 0, i.e. if ξ is of the 2nd kind, Schneider's theorems [17] imply that its non-trivial periods are transcendental. More generally, if an integral multiple of the point

$$P = \sum_{i=1}^{m} c_i P_i$$

is the origin O of E, the work of Masser (see, e.g., [2], Chapter 6) provides a complete description of the arithmetic nature of the different determinations of the periods of ξ. Assume from now on that P is any non-zero point on E. Then, ξ is equal to the sum of a differential of the 2nd kind, the logarithmic differential of a rational function on E, and the unique form

$$\xi_P = \frac{1}{2} \frac{y + y(P)}{x - x(P)} \frac{dx}{y}$$

with poles at O and P, and corresponding residues $-1, 1$ respectively.

In order to compute the different determinations of the period of ξ_P along a fundamental cycle γ, we consider the complex parametrization p of $E(\mathbb{C})$ by means of the Weierstrass elliptic functions associated to (*). We shall denote by $M(p)$ the set of complex numbers which are mapped by p into $E(\overline{\mathbb{Q}})$. Let ω be the period of p corresponding to γ, and let v be an element of $p^{-1}(P)$. Introducing the standard function $\zeta(z)$, $\sigma(z)$ related to p, we set

$$f_v(z) = \sigma(z-v)(\sigma(z)\sigma(v))^{-1}\exp(\zeta(v)z) ,$$

$$\eta(\omega) = \zeta(z+\omega) - \zeta(z) ,$$

$$\lambda_v(\omega) = \omega\zeta(v) - \eta(\omega)v \equiv \mathrm{Log}(f_v(z+\omega)/f_v(z)) \pmod{2\pi i\, \mathbb{Z}} .$$

Since the pull-back of ξ_P under p is equal to the logarithmic differential of $f_v(z)$, we have

$$\int_\gamma \xi_P = \lambda_v(\omega) \pmod{2\pi i\, \mathbb{Z}} ,$$

and, in view of the Legendre relation, any determination of the period of ξ along γ can be expressed as a linear combination Λ of ω, $\eta(\omega)$ and $\lambda_u(\omega)$, for some element u in $p^{-1}(P)$.

The fact that ξ is defined over $\overline{\mathbb{Q}}$ implies that the coefficients of Λ are algebraic, and that u belongs to $M(p)$. Now,

Laurent's result states that ω , $\eta(\omega)$, $\lambda_u(\omega)$ and 1 are linearly independent over $\overline{\mathbb{Q}}$ for any element u of $M(p)$ such that the order of $p(u)$ in E is not finite. Combining this with Masser's theorem, we have, in particular:

THEOREM 1 ([9], Théorème 3). *Let* ξ *be a differential form on* E *, defined over* $\overline{\mathbb{Q}}$ *, with rational residues. Then, the non-zero periods of* ξ *are transcendental.*

The proof of Laurent's theorem relies on Baker's techniques. Its concluding argument requires an upper bound for the number of zeros of polynomials in the functions introduced above. We shall come back to these zero estimates at the end of Part 3.

It would of course be very desirable to remove the assumption on the residues in the hypothesis of Theorem 1, in other words, to solve the following

PROBLEM 1. *Let* $u_1,\ldots,u_n$ *be elements of* $M(p)$ *such that* $p(u_1),\ldots,p(u_n)$ *are linearly independent over* $\mathbb{Q}$ *. What is the dimension over* $\overline{\mathbb{Q}}$ *of the* $\overline{\mathbb{Q}}$*-vector space generated by* $\lambda_{u_1}(\omega),\ldots,\lambda_{u_n}(\omega)$, ω , $\eta(\omega)$, $2\pi i$ *and* 1 ?

(When E has complex multiplications by an order $\mathbb{Z}[\tau]$ of an imaginary quadratic field, this problem includes the study of numbers of the form $\lambda_u(\omega)$, $\lambda_{\tau u}(\omega)$. It may be pointed out that in this case, $\lambda_u(\omega)$, $\lambda_{\tau u}(\tau\omega)$ and ω *are* linearly dependent over $\overline{\mathbb{Q}}$.)

The values of $f_u(z)$ are also of interest (see [16],[24]). For instance, on studying the analytic subgroups of the extension of E by the mulplicative group parametrized by ξ_p , Waldschmidt has proved that if $P = p(u)$ is a non-torsion point of $E(\overline{\mathbb{Q}})$, the number $\sigma(u)\exp(-u\zeta(u)/2)$ is transcendental ([24], §3.2.e). A p-adic version of this result can be obtained in a similar fashion. In this context, we mention the problem of the irrationality of the integral powers of $\sigma(u)\exp((-(\eta(\omega)/2\omega)u^2+\alpha)$ for algebraic α's ; under the further hypothesis that g_2 , g_3 and $p(u)$ are rational, this would imply the transcendency of the Néron-Tate height of the point $p(u)$.

2. ABELIAN INTEGRALS

Abelian integrals of the 2nd kind form the theme of Schneider's fourth problem. The gaps in our knowledge are here much broader. The theory discussed below, which was elaborated in a joint work with Masser [5,6], concerns only a special class of abelian varieties. Our main tool will be a transcendence criterion, due to Schneider and Lang (and considerably improved since then by Bombieri, cf. [24]), according to which $n+1$ algebraically independent functions, meromorphic in $\mathbb{C}^n$, which satisfy certain systems of algebraic partial differential equations, cannot simultaneously assume values in a fixed number field at all the points of a $\mathbb{Z}$-module Γ of rank n over $\mathbb{C}$ (see [8], Chapter 4). The special class we consider enables us to construct such modules Γ in a natural way. Before describing it, we mention that in view of the p-adic version of the Schneider-Lang theorem given in [24], Appendix I, §4.2, all our results admit p-adic analogues.

From now on, we fix an arbitrary number field F of degree n over $\mathbb{Q}$, and we denote by Σ the regular representation of F.

a) Abelian varieties of type (F,Σ): let A be an abelian variety, defined over $\overline{\mathbb{Q}}$, and let j be an embedding of F into the algebra $\mathrm{End}_0 A$ obtained from the ring of endomorphisms of A by extending the scalars to $\mathbb{Q}$. We shall say that (A,j) is an abelian variety of type (F,Σ) if the complex representation of F induced by j on the tangent space TA of A at the origin is equivalent to Σ. Under this assumption, the dimension of A is equal to n, and we can associate to every embedding σ of F into $\mathbb{C}$ a $\overline{\mathbb{Q}}$-vector space $H_\sigma = H_\sigma(A,j)$ of dimension 2, consisting of the classes of meromorphic functions on $TA(\mathbb{C})$ modulo abelian functions, whose differential is the pull-back of a differential of the 2nd kind η on A, defined over $\overline{\mathbb{Q}}$, such that, for any element α in F, the form $\eta \circ j(\alpha) - \sigma(\alpha)\eta$ is exact. Finally, we say that (A,j) is irreducible if, for any proper abelian subvariety B of A, there does not exist any injective homomorphism from F into $\mathrm{End}_0 B$.

Consider an irreducible abelian variety (A,j) of type (F,Σ), and let M_A be the set of points of $TA(\mathbb{C})$ whose image under the exponential map on $A(\mathbb{C})$ lie in $A(\overline{\mathbb{Q}})$. A first application of the Schneider-Lang theorem yields the following fact: the orbit Γ of any non zero element of M_A under an order of $j(F)$ generates $TA(\mathbb{C})$ over $\mathbb{C}$. On applying the Schneider-Lang theorem a second time, we obtain:

THEOREM 2 ([4], théorème 2.2). *Let* (A,j) *be an irreducible abelian variety of type* (F,Σ)*, and let* σ *be an embedding of* F *into* $\mathbb{C}$*. Consider a non-zero element* u *of* M_A*, and a representative* H_σ *of a non-zero element of* H_σ*, analytic at* 0 *and* u*,* *and vanishing at* 0 . *Then,* $H_\sigma(u)$ *is transcendental.*

Note that we do not assume the simplicity of the abelian variety A itself. This will be fundamental for the applications we have in mind. As for the hypothesis of irreducibility, it suffices here to say that if (A,j) is not irreducible, then A is isogenous to the square of abelian variety of CM type in the sense of Shimura. Sharper results can in fact be obtained in this case (see [4], théorème 3, and Masser's earlier work [11]).

A typical example of an abelian variety of type (F,Σ) is provided by certain factors of the jacobians of modular curves. Let f be a primitive cusp form of weight 2 and conductor N : in particular, f is an eigen-form for the action of the Hecke operators; the coefficients of its Fourier expansion at infinity generate a number field $F = K_f$; and the vector-space T_f spanned by the 'companion forms' of f can be viewed as the tangent space of an abelian variety J_f of type (F,Σ) (see, e.g., [21], §2). Denoting by $h(J)$ the union of $\mathbb{Q}$ and of the set of points at which the modular invariant J assume algebraic values, we deduce from Theorem 2:

COROLLARY 1 ([3], Theorem 2). *Let* τ *be an element of* $h(J)$. *Any non-zero determination* $u(f,\tau)$ *of the integral* $2\pi i \int_{i\infty}^{\tau} f(z)dz$ *is transcendental.*

In particular, the value at $s = 1$ of the L-function $L(f,s,\chi)$ associated to f and to a Dirichlet character χ is either zero or transcendental. It would be very interesting to generalize this result to all the elements of $T_f(\overline{\mathbb{Q}})$. In the general situation described above, this can be formulated as follows:

PROBLEM 2. *Let* $\sigma_1,\dots,\sigma_n$ *be the different embeddings of* F *into* $\mathbb{C}$. *With the notations of Theorem 2, when are the numbers* $H_{\sigma_1}(u),\dots,H_{\sigma_n}(u)$ *and* 1 *linearly independent over* $\overline{\mathbb{Q}}$?

(In the case of periods, a complete solution of this problem has been given by Masser for general abelian varieties of dimension 2 ; see [12], and his article in [1].)

On a different level, one may inquire about the arithmetic nature of the special values of the L-series associated to cusp forms of weight >2 , as, e.g., $L(\Delta,6)$ for the discriminant form Δ . But this seems out of reach at the moment.

b) Back to elliptic integrals: the n-th power of a given elliptic curve E provides another fundamental example of an abelian variety of type (F,Σ). Indeed, a faithful representation $j_{(\beta)}$ of F into the ring of rational square matrices of order n can be associated to any basis (β) of F over $\mathbb{Q}$ in a canonical way. We now fix such a homomorphism $j_{(\beta)}$, and we denote by $\{\alpha_1,\ldots,\alpha_n\}$ the dual basis of (β). For simplicity, we restrict our discussion to integrals of the 1st kind.

We assume that E is defined over $\overline{\mathbb{Q}}$, and use the notations p, $M(p)$ associated in §1 to a Weierstrass model of E. Let $u_1,\ldots,u_n$ be elements of $M(p)$, not all zero. It is easily checked (see [4], §3.b) that for any embedding σ of F into $\mathbb{C}$, the number

$$u_\sigma = \sigma(\alpha_1)u_1 + \ldots + \sigma(\alpha_n)u_n$$

is the value of a holomorphic element Z_σ in $H_\sigma(E^n,j_{(\beta)})$ at a point of M_{E^n}. If $\mathrm{End}_0 E = \mathbb{Q}$, Theorem 2 then implies that u_σ is transcendental. Since F can be chosen arbitrarily, we have therefore proved:

COROLLARY 2 ([5], Theorem). *Assume* E *has no complex multiplication, and let* $u_1,\ldots,u_m$ *be elements of* $M(p)$*, linearly independent over* $\mathbb{Q}$. *Then,* $u_1,\ldots,u_m$ *and* 1 *are linearly independent over* $\overline{\mathbb{Q}}$.

The method also applies in the case of complex multiplication, which had been treated earlier by Masser with the help of Baker's method (see, e.g., [2], Chapter 7). As suggested above, stronger results can be obtained in this case (see [10],[4]). However, it remains an important open problem to obtain with our method the sharp *quantitative* results which make up the core of Baker's method.

The principle described above also applies to the pure powers of an abelian variety of type (F,Σ) (see [6] for the case of real multiplications). Considering a primitive cusp form as above, and denoting by E_f the commutant of K_f in $\mathrm{End}_0(J_f)$, we obtain:

COROLLARY 3 (see [6], théorème 2). *Let* $\tau_1,\ldots,\tau_m$ *be elements of* $h(J)$ *such that the numbers* $u_k = u(f,\tau_k)$ $(k = 1,\ldots,m)$ *are linearly independent over* E_f. *Then,* $u_1,\ldots,u_m$ *and* 1 *are linearly independent over* $\overline{\mathbb{Q}}$.

We conclude this section with a few words on the period matrix of a CM abelian variety. Consider a CM field K, a CM type Φ of K (see, e.g., [22], §1) and an element σ in Φ. For any abelian variety (B,j) of type (K,Φ), the periods of the holomorphic forms ω such that $\omega \circ j(\alpha) = \sigma(\alpha)\omega$ for any α in K can all be expressed as algebraic multiples of a complex number ω_σ. Theorem 2 and the Riemann relations (see, e.g., the 3rd article in [1]) imply that ω_σ, π/ω_σ and 1 are linearly independent over $\overline{\mathbb{Q}}$ (Schneider's theorem [17] would in fact suffice in this case). In particular, ω_σ/π, which is the number $p_K(\sigma,\Phi)$ introduced by Shimura in [22], is transcendental; so is ω_σ^2/π. On the other hand, the answer to the following problem is known only in the case of elliptic curves (see Chudnovsky [7], Theorem 8).

PROBLEM 3. *Are the numbers* $p_K(\sigma,\Phi)$ *and* π *algebraically independent over* $\mathbb{Q}$ *(for a fixed element* σ *in* Φ *)?*

For the relations between these numbers and arithmetic automorphic functions, we refer the reader to [20], [22].

3. CHARACTERS OF TYPE (A_0) AND ZERO ESTIMATES

The work of Shimura and Taniyama establishes a close connection between the study of an abelian variety B of CM type, with CM field K, and certain Grössencharacters of K, introduced by Weil [26] as characters of type (A_0). In particular, the zeta function of B over a sufficiently large field coincides with a product of the Hecke L-series attached to such Grössencharacters, and in certain cases, the special values of these Hecke L-series can be expressed in terms of the numbers $p_K(\sigma,\Phi)$ mentioned above (see [20], §3).

Let now χ be an arbitrary Grössencharacter of K (i.e. a representation of the idele class group of K in $\mathbb{C}^\times$, possibly of infinite order). Denoting by $\mathfrak{f}$ the conductor of χ, and by $K^\times(\mathfrak{f})$ the set of elements α of K congruent to $1 \bmod^\times \mathfrak{f}$, we recall that the coefficients of the Hecke L-series attached to χ can be computed by taking roots of the values of the character X of $K^\times(\mathfrak{f})$ induced by χ. In general, these values take the form

$$X(\alpha) = \prod_\sigma (\sigma(\alpha)/|\sigma(\alpha)|)^{h_\sigma} |\sigma(\alpha)|^{\tau+i\phi_\sigma} ,$$

where the product is taken over all the elements of a CM type of K, the ϕ_σ's and τ are real numbers, and the h_σ's are rational integers. The character χ is said to be of type (A) if all the ϕ_σ's are 0, and τ is rational, and of type (A_0) if, moreover,

τ is an integer, and all the h_σ's have the same parity as τ.

Thus, the coefficients of the Hecke L-series attached to a character of type (A) (resp. (A_0)) are algebraic numbers (resp. elements of a finite extension of $\mathbb{Q}$). In [26], Weil asked whether the converse statements hold. For instance, is an algebraic-valued Grössencharacter necessarily of type (A)? A positive answer to Weil's problem has recently been given by Waldschmidt [25]. The result follows from

THEOREM 3 ([25], Corollaire 1.2.a). *Let* m, n *be positive integers such that* $m \geq n^2 + n + 1$, *and let* $\ell_{\mu,\nu}$ $(1 \leq \mu \leq m;\ 1 \leq \nu \leq n)$ *be logarithms of algebraic numbers such that the* m *vectors in* $\mathbb{C}^n$ *with components* $\{\ell_{\mu,1}, \ldots, \ell_{\mu,n}\}$ $(\mu = 1, \ldots, m)$ *are linearly independent over* $\mathbb{Q}$. *Let further* $t_1, \ldots, t_n$ *be complex numbers such that* $t_1, \ldots, t_n$ *and* 1 *are linearly independent over* $\mathbb{Q}$. *Then, one at least of the* m *numbers* $\exp(\sum_{\nu=1}^{n} t_\nu \ell_{\mu,\nu})$ $(\mu = 1, \ldots, m)$ *is transcendental.*

(To justify a statement of the introduction, we mention that Schneider's first problem consists in proving that, for $n = 1$, the condition $m \geq n^2 + n + 1$ can be weakened to $m \geq 2$. In this case, Theorem 3 has been proved by Lang [8].)

Waldschmidt also establishes a p-adic analogue of this result, thus confirming a conjecture of Serre according to which a rational, semi-simple, abelian p-adic representation of the absolute Galois group of a number field should necessarily be 'locally algebraic' (see [19], Chapter III, §3).

The proof of Waldschmidt's theorem combines the construction of a new type of 'auxiliary function' with an important theorem of Masser on the zeros of exponential polynomials in several variables. In a joint work with Wüstholz [14], Masser has extended his result to general group varieties. Returning to the theme of our survey, we state the following corollary of their results.

THEOREM 4 ([13],[14]). *Let* A *be a simple abelian variety of dimension* n, *embedded in a projective space* $\mathbb{P}_d$. *There exists a positive constant* λ, *depending only on* A, *with the following property. Let* P *be a homogeneous polynomial in* $\mathbb{C}[X_0, \ldots, X_d]$ *of degree* D, *and let* Γ *be a subgroup of* A *generated by* r *elements* $\gamma_1, \ldots, \gamma_r$ *of* A *linearly independent over* $\mathbf{Z}$. *Assume that* P *vanishes at all the points of the form* $n_1\gamma_1 + \ldots + n_r\gamma_r$, *where the* n_i*'s run through the non-negative integers* $\leq N$, *and* N *is such that* $D^n \leq \lambda(N/d)^r$. *Then,* P *vanishes identically on* A.

Masser has also extended this result to torsion points on A, and this yields an interesting lower bound for the degree of the fields of rationality of such points ([13]). Another kind of application of zero estimates concerns problems of algebraic independence (see [7],[14],[15]). Thus, zero estimates are playing a more and more important role in transcendence theory, and it is likely that they will provide much further progress in this field.

REFERENCES

[1] Fonctions abéliennes et nombres transcendants, Mémoires SMF, 1980, Nouvelle serie, no. 2.

[2] A. Baker and D. Masser, Transcendence Theory: Advances and Applications, Academic Press, 1977.

[3] D. Bertrand, Transcendence properties of abelian integrals, Queen's papers in Pure and Applied Math., 1980, No. 54.

[4] D. Bertrand, Variétés abéliennes et formes linéaires d'intégrales elliptiques, Séminaire Delange - Pisot - Poitou, 1979-80, Birkhauser, Progress in Mathematics, Volume 12, 1981.

[5] D. Bertrand and D. Masser, Linear forms in elliptic integrals, Inventiones Math., 58 (1980), 283-288.

[6] D. Bertrand and D. Masser, Formes linéaires d'intégrales abéliennes, CRAS Paris, 290 (1980), 725-727.

[7] G.V. Chudnovsky, Algebraic independence of values of exponential and elliptic functions, Proc. ICM Helsinki (1978), I, 339-350.

[8] S. Lang, Introduction to transcendental numbers, Addison-Wesley, 1966.

[9] M. Laurent, Transcendance de périodes d'intégrales elliptiques, J. reine angew. Math. 316 (1980), 122-139.

[10] M. Laurent, Indépendance linéaire de valeurs de fonctions doublement quasi-périodiques, CRAS Paris, 290 (1980), 397-399.

[11] D. Masser, Linear forms in algebraic points of abelian functions III, Proc. London M.S. 33 (1976), 549-564.

[12] D. Masser, The transcendence of certain quasi-periods associated with abelian functions in two variables, Compositio Math. 35 (1977), 239-258.

[13] D. Masser, Small values of the quadratic part of the Neron-Tate height, Seminaire Delange - Pisot - Poitou, 1979-1980, Birkhauser, Progress in Mathematics, Volume 12, 1981.

[14] D. Masser and G. Wüstholz, Algebraic independence properties of elliptic functions, this volume.

[15] P. Philippon, Indépendance algébrique de valeurs de fonctions elliptiques p-adiques, Queen's papers in Pure and Applied Math., 1980, No. 54.

[16] E. Reyssat, Fonctions de Weierstrass et indépendance algébrique, CRAS Paris, 290 (1980), pp. 439-441.

[17] T. Schneider, Zur Theorie der Abelscher Funktionen und Integrale, J. reine angew. Math. 183 (1941), 110-128.

[18] T. Schneider, Einführung in die transzendenten Zahlen, Springer, 1957.

[19] J-P. Serre, Abelian ℓ-adic representations and elliptic curves, Benjamin, 1968.

[20] G. Shimura, On some arithmetic properties of modular forms of one and several variables, Ann.Math. 102 (1975), 491-515.

[21] G. Shimura, On the periods of modular forms, Math.Ann. 229 (1977), 211-221.

[22] G. Shimura, Automorphic forms and the periods of abelian varieties, J.Math.Soc. Japan, 31 (1979), 561-592.

[23] C.L. Siegel, Über die Perioden elliptischer Funktionen, J. reine angew. Math. 167 (1932), 62-69 (Gesam.Abh. I, 267-274).

[24] M. Waldschmidt, Nombres transcendants et groupes algébriques, Astérisque 69-70, 1979.

[25] M. Waldschmidt, Transcendance et exponentielle en plusieurs variables, Inventiones math. 63, 1981, pp. 97-127.

[26] A. Weil, On a certain type of characters of the idèle class-group of an algebraic number field, Proc. I.S. Alg. Number Theory, Tokyo-Nikko, 1955 (Collected papers, II, 255-261).

MEASURES OF IRRATIONALITY, TRANSCENDENCE AND ALGEBRAIC INDEPENDENCE. RECENT PROGRESS

G.V. Chudnovsky*
Department of Mathematics
Columbia University
New York, N.Y. 10027

CHAPTER 1

§0. In this paper we study arithmetic properties (such as the measure of diophantine approximations) of complex (or, perhaps, p-adic) numbers which have an 'analytic' and 'algebraic' sense. By numbers θ having 'analytic' sense we understand those that have the form $\theta = f(\alpha)$ for algebraic α and a function $f(x)$ satisfying an algebraic differential equation $P(x,f,\dots,f^{(q-1)}) \equiv 0$ with a polynomial $P(x,z_0,\dots,z_{q-1})$ having algebraic coefficients and $f(x)$ determined by algebraic initial (or boundary) conditions. The number θ has 'algebraic' or 'geometric' sense, if the corresponding function $f(x)$ arises from an algebraic variety in some canonical way. For example θ may be a period of an abelian variety or value of an abelian integral at an algebraic point.

It is natural to distinguish first of all the classes of numbers connected with exponential and with elliptic functions. For these two classes of functions we present tables of results on the measure of transcendence and algebraic independence (see page 159).

If a number α is already transcendental, then its measure of transcendence can be

i) of the general form

$$|P(\alpha)| > \psi(H,d) \quad ,$$

where $P(x) \in \mathbb{Z}[x]$, $P \not\equiv 0$ and $d(P) \le d$, $H(P) \le H$ with positive function $\psi(H,d)$. If $\log \psi(H,d)$ is a polynomial in $\log H$ and d, then we say that α is of finite type of transcendence.

More generally (following Lang) we say that n algebraically independent numbers $\theta_1,\dots,\theta_n$ have type of transcendence $\le \tau$ if

$$|P(\theta_1,\dots,\theta_n)| > \exp(-c.(\log(H+1) + d)^{\tau})$$

for $P(x_1,\dots,x_n) \in \mathbb{Z}[x_1,\dots,x_n]$, $P \not\equiv 0$ and $d(P) \le d$, $H(P) \le H$. By Dirichlet's theorem $\tau \ge n+1$.

*Supported in part by ONR and US Air Force.

ii) Measures of the 'normal' form, when $\psi(H,d)$ is $H^{-\phi(d)}$ for positive $\phi(d)$. For example we call a number α 'normal' if

$$|\alpha - \frac{p}{q}| > |q|^{-c} : \quad p,q \in \mathbb{Z},\ q \neq 0,$$

for some constant $c > 0$. In general we call a number 'normal' with respect to its measure of transcendence, if

$$|\alpha - \xi| > H(\xi)^{-c_1}$$

for all algebraic ξ of height $\leq H(\xi)$ and some exponent c_1 depending on the degree of ξ; or equivalently

$$|P(\alpha)| > H(P)^{-c_1'}$$

for all $P(x) \in \mathbb{Z}[x]$, $P \neq 0$ and some exponent c_1' depending on $d(P)$.

Almost all numbers are 'normal'; non-normal numbers are called sometimes Liouvillean.

However, if a number is 'normal' what is its exponent?

iii) Numbers with Dirichlet's type of exponent:

$$|P(\alpha)| > H(P)^{-c_2 d(P)}$$

or, for n numbers $\theta_1,\ldots,\theta_n$:

$$|P(\theta_1,\ldots,\theta_n)| > H(P)^{-c_2 d(P)^n}$$

if $P(\bar{x}) \in \mathbb{Z}[\bar{x}]$, $P \neq 0$ and constant $c_2 > 0$ depending on $\theta_1,\ldots,\theta_n$ only.

For example for the measure of irrationality, the most attractive among numbers are those that have so-called 'measure of irrationality $2 + \varepsilon$ (for any $\varepsilon > 0$)':

$$|\alpha - \frac{p}{q}| > |q|^{-2-\varepsilon}$$

for integers p, q and $|q| \geq q_0(\varepsilon)$. Among transcendental numbers having 'analytic' sense only numbers like

$$e^r,\ r \in \mathbb{Q} \quad \text{or} \quad J_0'(1)/J_0(1) \quad \text{etc.}$$

are known to have 'measure of irrationality $2 + \varepsilon$'.

The class is indeed very narrow: it can be described as values of meromorphic functions $g(z)$ satisfying a Riccati equation $g' = g^2 + V$ with $V \in \mathbb{K}(z)$ and $\mathbb{K}$ imaginary field and $\alpha = g(b)$ for $b \in \mathbb{K}$.

The paper is organised along the lines of the original lecture in the following way. In the beginning of the paper we give a short exposition of the existing results on the measure of irrationality, transcendence and algebraic independence of numbers having analytic and geometric senses. We present results, including new ones reported in this paper in the form of several summary tables. Our main attention is devoted to the normality of measures of diophantine approximations in the form $|\alpha - p/q| > |q|^{-c_1}$ (for measures of irrationality) or $|P(\alpha_1,\ldots,\alpha_n)| > H(P)^{-c_2}$ (for measures of transcendence or algebraic independence with c_2 depending on the degree of the polynomial P). In several cases we propose a general theory giving the criteria of normality of some numbers which are values of the analytic functions $f(z)$.

We take as a natural class of numbers, those for which transcendence can be proved by some version of the Siegel-Gelfond-Schneider method. In this case, functions $f(z)$ are usually meromorphic and satisfy algebraic differential equations. The most convenient general theorem applicable to a wide class of functions $f(z)$ is the Schneider-Lang theorem [29] or its multidimensional generalisations (Bombieri or Bombieri-Lang [7]). However, many numbers proved to be transcendental in this way are not known to be normal (e^{π} is the best known example). Moreover, the Schneider-Lang theorem in its traditional form apparently leads to very poor transcendence measures (cf. arguments of [8]), and in order to get better transcendence measures, people have to devise special methods for each concrete case [20]. However Siegel's E-function method [36] gives normal measures of algebraic independence for E-functions which satisfy linear differential equations. Unfortunately, E-functions should be considered to be outside the class which has an immediate algebraico-geometric interpretation, since they are confluent hypergeometric functions, satisfying linear differential equations with irregular singularities and not of the Fuchsian type. Algebraico-geometric functions (like the period or Hodge structures of an algebraic variety) lead usually to Fuchsian linear differential equations and sometimes to ${}_{p+1}F_p$-hypergeometric functions [4].

Literally the same object appears in the uniformization problem on Riemann surfaces. These functions which satisfy Fuchsian differential equations can often be represented as Taylor series with algebraic coefficients and special arithmetic properties of denominators of those coefficients. In some cases (and only in some cases) these functions are the G-functions of Siegel [35]. Hence it is natural to expect that for such functions, measures of transcendence of the normal type can be deduced using Siegel's method of G-functions [35] as in the E-function case. Unfortunately, the G-function method of Siegel, developed recently by a variety of authors [25], [24], [6], [40], does not give results on transcendence or measure of transcendence. The explanation lies in the fact that in the G-function method, in order to show that the value $f(\xi)$, of a G-function $f(z)$ at algebraic $\xi \neq 0$ satisfying a linear differential equation of order q, is not algebraic of degree D (irrational if $D = 1$), it is necessary to assume

$$|\xi| < \exp(-C(D)\ (\log H(\xi))^{\frac{q}{q+1}}) \qquad (0.1)$$

for some (not effective!) $C(D) > 0$.

Moreover, an additional arithmetic restriction of (G,C)-type must be imposed on $f(z)$. (The definition of (G,C) for Fuchsian linear differential equations is presented in this paper. For the definition of (G,C)-functions in the general case of an arbitrary linear differential equation with rational function coefficients, we refer the reader to our paper [17].) In any case, the condition of the type (0.1) does not allow us to apply the G-function method directly to concrete numbers of analytic or geometric origin such as periods of algebraic varieties. However the G-function and similar methods still look attractive since the proof of irrationality in this case is usually supplied with the normal measures of irrationality (as in the E-function case).

We are doing the most natural thing, and therefore we are trying to combine the generality of the statement of the Schneider-Lang theorem with the sharp bounds of the E or G-function methods. The main results of the present paper are based on this approach. Namely, we consider meromorphic functions satisfying algebraic differential equations that can be reduced by an appropriate change of variables to G-functions satisfying Fuchsian linear differential equations. Such an assumption which looks strange at first sight, is the most natural since it is suggested by uniformization theory. In the cases that are most interesting for applications, when we are working with Abelian varieties, the corresponding linear differential equations arise as linear differential equations satisfied by Abelian integrals of the first and second kind.

Fortunately, the (G,C)-function condition requiring the most labour reduces for Abelian integrals to the Eisenstein theorem [23] on Taylor expansion of algebraic functions. That is why most of the results of the present paper deal with algebraic points of Abelian varieties defined over $\overline{\mathbb{Q}}$. For example, one of the most longstanding conjectures in this field is finally solved: we show that a linear combination of periods and quasiperiods of Abelian varieties of CM-type with algebraic coefficients is, if non-zero, itself a transcendental number of normal type. Moreover we obtain a lower bound for a linear form of algebraic points of Abelian variety of CM-type, which is of a normal type. For example, in the elliptic case we have

$$|\beta_1 u_1 + \dots + \beta_n u_n| > H^{-C}$$

for algebraic β_i, $H(\beta_i) \le H$: $i = 1,\dots,n$ and n algebraic points $u_1,\dots,u_n$ of an elliptic function $\wp(x)$, linearly independent over $\mathbb{K}$, provided that $\wp(x)$ is defined over $\overline{\mathbb{Q}}$ and has complex multiplication in $\mathbb{K}$. These results finally put the existing bounds [38] into a 'normal' form. These results are particular cases of our general one-dimensional and multidimensional criteria of normality that incorporate meromorphic functions of finite order of growth satisfying algebraic differential equations and (G,C)-functions. We give a brief indication of all necessary steps and the outline of the proof. We do not present all proofs in all of the cases since later in the course of the paper we give the complete proof of a much more difficult theorem on the normal measure of algebraic independence of two numbers arising from elliptic function theory. This proof is given in connection with a series of results on normal measures of algebraic independence for several pairs of numbers such as π/ω, η/ω or $\zeta(u)-(\eta/\omega)u$, η/ω for which the author has previously proven algebraic independence [13],[14]. We not only present the proof of normality, but also for the first time present the complete proof of the bound for the transcendence type of π/ω, η/ω to be $\le 3 + \varepsilon$, for every $\varepsilon > 0$, as was announced before in [11a],[13]. The proof is long but straightforward and requires no fancy machinery other than algebraic properties of resultants. We decided to restrict ourselves to this proof only because other proofs of the measure of algebraic independence do not differ at all except in the change of auxiliary functions.

As an application of our result, we mention the normality of the measure of transcendence of numbers such as $\Gamma(1/3)$, $\Gamma(1/4)$, $\Gamma(1/6)$ etc. Similarly, normal measures of algebraic independence are proved for the author's generalization of the Lindemann-Weierstrass theorem [15]. Our result is as follows:

$$|P(\wp(\alpha_1),\dots,\wp(\alpha_n))| > H(P)^{-Cd(P)^n}$$

for $P(x_1,\dots,x_n) \in \mathbb{Z}[x_1,\dots,x_n]$, $\wp(x)$ being an elliptic function with algebraic invariants g_2 , g_3 and complex multiplications in the imaginary quadratic field $\mathbb{K}$ and algebraic numbers $\alpha_1,\dots,\alpha_n$ linearly independent over $\mathbb{K}$ for $C = C(d,n)$ for $d = [\mathbb{K}(g_2,g_3,\alpha_1,\dots,\alpha_n) : \mathbb{Q}]$.

In the paper we collect other results on the measure of algebraic independence, even if not of the normal type.

The final part of the paper is devoted to different questions related to (G,C)-functions. We discuss different criteria for (G,C)-functions for Fuchsian linear differential equations in complex and p-adic domains. The results of Dwork [22],[23] and new conjectures are presented along with old ones like that of Grothendieck. We use this opportunity to present results on the explicit Padé approximation, which are partially connected with (G,C)-function theory, but provide very sharp normality results for the measure of irrationality of important numbers. Estimates presented in this chapter follow the exposition of [9],[10],[17],[18].

I use this opportunity to thank the organizers of the Exeter Conference, and particularly Professor Halberstam and Dr. Odoni for their wonderful hospitality during the Conference.

Table 1. Table of transcendence measures for the exponential case; α , α_i , α_i' are non-zero algebraic numbers; β , β_i , $\gamma_i \in \overline{\mathbb{Q}}$:

		Type of transcendence	'Normality'
1)	e^{α}	3	normal; Dirichlet's type; measure of irrationality is $2+\varepsilon$ for $\alpha \in \mathbb{Q}$
2)	$\log \alpha$, $\alpha \neq 1$	3	normal
3)	π	$2+\varepsilon$	normal
4)	$\dfrac{\Sigma\beta_i \log \alpha_i}{\Sigma\gamma_i \log \alpha_i'}$	4	normal
5)	α^{β}, $\alpha \neq 1$, $\beta \notin \mathbb{Q}$	4	?
6)	$e^{\alpha_0}\alpha_1^{\beta_1}\cdots\alpha_n^{\beta_n}$	4	?

Problem 0.2. Prove $|e^{\pi} - p/q| > |q|^{-c}$, $c > 0$, i.e. normality of e^{π} .

Problem 0.3. Find an example of $\log\alpha$, $\alpha \neq 1$ with measure or irrationality $\leq 2+\varepsilon$ for all $\varepsilon > 0$.

Table 2. Transcendental numbers connected with elliptic curves. Here $\wp(x)$ has algebraic invariants g_2, g_3 ; u is algebraic point of $\wp(x)$ (i.e. $\wp(u) \in \overline{\mathbb{Q}}$); $u_1,\ldots,u_n$ are algebraic points of $\wp(x)$ l.i. over field of endomorphisms of $\wp(x)$; $\wp_1(x)$ also with algebraic invariants is a.i. with $\wp(x)$; $\zeta(x)$ and $\sigma(x)$ are Weierstrass functions ($\zeta'(x) = -\wp(x)$; $d_x \log\sigma(x) = \zeta(x)$) ; α is algebraic ($\alpha \neq 0$) .

1)*	u	2)	e^u
3)*	$\alpha_1 u + \alpha_2\zeta(u)$	4)	$e^{\alpha u}$
5)*	$\Sigma\beta_i u_i/\Sigma\gamma_i u_i'$	6)*	$\zeta(u) - \frac{\eta}{\omega}u$
7)*	$\Sigma\beta_i u_i + \Sigma\gamma_i\zeta(u_i)$ in the c.m. case	8)	$\frac{\sigma(u_1+u_2)}{\sigma(u_1)\sigma(u_2)}e^{-u_1\zeta(u_2)}$
9)*	$\wp(\alpha)$ Dirichlet's for c.m.)		($u_2 \notin L$)
10)	$\wp(\alpha u)$ (if there is no c.m. by α)	11)*	$P(u,\zeta(u))$ for c.m. case and $P \neq 0$
12)	$\wp(\Sigma\beta_i u_i)$		
13)	$\wp_1(\alpha u)$	14)	$\zeta(\alpha)$ (transcendence?)
15)*	u_1/u_2' for u_1 alg. point of $\wp(x)$ and u_2' alg. point of $\wp_1(x)$	16)	$e^{\pi i\omega_2/\omega_1}$ (transcendence?) (not c.m. case)

* means normality

Table 3. Table of algebraically independent results with measures of algebraic independence; α is algebraic, $\alpha \neq 0$, $\wp(x)$ has algebraic invariants g_2, g_3 ; u is an algebraic point of $\wp(z)$, $\wp(u) \in \overline{\mathbb{Q}}$; $\zeta(z)$ is a Weierstrass ζ-function; (ω,η) are period and quasi-period of $\wp(z)$: $\eta = 2\zeta(\omega/2)$.

	Numbers	Type of Transcendence	'Normality' etc.
1)	$(e^{\alpha_1},\ldots,e^{\alpha_n})$ for algebraic $\alpha_1,\ldots,\alpha_n$ l.i. over $\mathbb{Q}$	? (for $n \geq 2$)	normal, Dirichlet's type

2)	$(\alpha^{\beta},\alpha^{\beta^2})$ $\alpha \neq 1$; β of the third degree	?	? $\exp(-cH^{c_1 d^{1+\varepsilon}})$ is the lower bound
3)	$(\frac{\pi}{\omega}, \frac{\eta}{\omega})$	$3+\varepsilon$ for any $\varepsilon > 0$	normal (not quite Dirichlet)
4)	$(\zeta(u) - \frac{\eta}{\omega}u, \frac{\eta}{\omega})$	$9+\varepsilon$ for any $\varepsilon > 0$	normal
5)	$(\frac{\pi}{\omega}, e^{\pi i \tau})$ for $\wp(x)$ with c.m. by τ	$7+\varepsilon$ for any $\varepsilon > 0$	?
6)	$(u, \zeta(u))$ c.m. case	?	normal
7)	$(\wp(\alpha_1),\dots,\wp(\alpha_n))$ for $\wp(x)$ with c.m. in $\mathbb{K}$, $\alpha_1,\dots,\alpha_n$ l.i. over $\mathbb{K}$	? (for $n \geq 2$)	normal Dirichlet's type

§1. It is very tempting to try to represent the whole variety of measures of transcendence for different numbers through some general theorem. As we know now there exists such a general theorem that covers all (or at least almost all) transcendence results (but not those of algebraic independence). This theorem is the Schneider-Lang theorem in the one-dimensional case or the Bombieri-Lang theorem in the multidimensional case. The first qualitative version of the Schneider-Lang theorem was considered by Brownawell-Masser [8]. Here we refine these results and generalize them to include 'normal' numbers.

As in the formulation of the Schneider-Lang theorem we start with functions satisfying algebraic differential equations over $\overline{\mathbb{Q}}$.

Let $f_1(z), f_2(z)$ be two algebraically independent over $\mathbb{C}$ meromorphic functions, of orders of growth ρ_1, ρ_2, respectively.

We are interested in functions f_1, f_2 satisfying algebraic differential equations. The most general definition of an algebraic differential equation over $\overline{\mathbb{Q}}$ satisfied by $f_1(z), f_2(z)$ is the following [29],[38]. There are functions $f_3(z), \ldots, f_n(z)$ and an algebraic number field $\mathbb{K}$ such that d/dz maps $\mathbb{K}[f_1, \ldots, f_n]$ into itself. For example we may take algebraic differential equations in the form

$$Q_1(f_1', f_1, f_2) = 0, \quad Q_2(f_2', f_1, f_2) = 0 \qquad (1.1)$$

for two polynomials from $\mathbb{K}[x_0, x_1, x_2]$. When an algebraic number field $\mathbb{K}$ is given, we say that $f_1(z), f_2(z)$ satisfy an algebraic differential equation over $\mathbb{K}$.

For the system of functions satisfying an algebraic differential equation we have a very general transcendence result:

THEOREM 1.2 (Schneider-Lang [29]). ***Let*** $f_1(z), f_2(z)$ ***as above satisfy an algebraic differential equation over*** $\mathbb{K}$. ***Then there are at most*** $[\mathbb{K}:\mathbb{Q}](\rho_1 + \rho_2)$ ***complex numbers*** θ ***such that*** $f_i(\theta) \in \mathbb{K}$: $i = 1,2,3,\ldots,n$.

In particular, if $f_1(z), \ldots, f_n(z)$ satisfy an algebraic law of addition [30] (i.e. $f_j(x+y)$ belongs to $\overline{\mathbb{Q}}[f_1(x), \ldots, f_n(x), f_1(y), \ldots, f_n(y)]$: $j = 1, \ldots, n$) then for any point θ at which $f_i(z)$ are regular one of the numbers

$$f_i(\theta) : i = 1, \ldots, n$$

is transcendental. If, in addition, algebraic differential equations (1.1) are satisfied, either $f_1(\theta)$ or $f_2(\theta)$ is a transcendental number.

We want to find a measure of the transcendence of $f_2(\theta)$ if $f_1(\theta)$ is an algebraic number. It is possible to prove

$$|f_2(\theta) - \xi| > \exp(-c_1 \log H(\xi)^{c_2}) \qquad (1.3)$$

for algebraic ξ and $c_1 > 0$, $c_2 > 0$ depending only on $f_1(z)$, $f_2(z)$, θ and $d(\xi)$.

We want to prove a more strong transcendence measure:

$$|f_2(\theta) - \zeta| > H(\zeta)^{-c_3} : H(\zeta) \geq c_4 \qquad (1.4)$$

for an algebraic ζ and $c_3 > 0$, $c_4 > 0$ depending on $f_1(z)$,

$f_2(z)$ and $d(\zeta)$. We call the number $f_2(\theta)$ with the property (1.4) a 'normal' number. Our aim is to establish conditions on the differential equations satisfied by $f_1(z), f_2(z)$ that guaranteed 'normality' of $f_2(\theta)$. These conditions reduce to the possibility of converting nonlinear algebraic differential equations like (1.1) to linear ones by substitutions

$$f_1(z) = w\ ,\ f_2(z) = g(w) \tag{1.5}$$

We will work below only with Fuchsian linear differential equations. In this case the most important arithmetical conditions are G-function and (G,C)-function conditions [35],[17].

<u>Definition 1.6</u> (C. Siegel) Let

$$f(z) = \sum_{n=0}^{\infty} a_n z^n\ ,$$

then $f(z)$ is called a G-function if

a) a_n belong to a fixed algebraic number field $\mathbb{K}$ and $\overline{|a_n|} = \max_\sigma |a_n^{(\sigma)}| \le C_5^n$ for $C_5 > 0$;

b) $\text{denom}(a_0,\dots,a_n) \le C_6^n$ for $C_6 > 1$ and any $n = 1,2,3,\dots$.

This condition itself (contrary to an E-function condition) does not suffice for a transcendence proof. It should be improved to include denominators of Taylor series of $f(z)$ at other points [roughly speaking]. This way we come to a (G,C)-function condition introduced by A. Galochkin (1974) [25].

<u>Definition 1.7</u> Let $\mathbb{K}$ be an algebraic number field and $L \in \mathbb{K}(x)[\frac{d}{dx}]$ be a differential operator $L = d^r/dw^r - \Sigma_{j=0}^{r-1} u_j(w) d^j/dw^j$ for $u_j(w) \in \mathbb{K}(w)$: $j = 0,\dots,r-1$. Let for $m \ge 0$,

$$\frac{d^m}{dw^m} = \sum_{j=0}^{r-1} v_{m,j}(w) \frac{d^j}{dw^j} \pmod{L} \tag{1.8}$$

with $v_{m,j}(w) \in \mathbb{K}(w)$. Now L is called a (G,C)-operator if there is a polynomial $T(w) \in \mathbb{K}[w]$ such that

$$T^m(w).v_{m,j}(w) \in \mathbb{K}[w] \qquad (j = 0,\dots,r-1)$$

and the common denominator A_n of the coefficients of the polynomials

$$\frac{1}{m!} T^m(w).v_{m,j}(w) : j = 0,\ldots,r-1; m = 0,1,\ldots,n$$

is at most C_7^n (for $n = 1,2,\ldots$). See Corollary 12.8.

If $f(x)$ is a G-function satisfying a (G,C)-equation it is called a (G,C)-function.

The (G,C)-functions are interesting, since Siegel (1929) [35] showed that for algebraic ξ, $\xi \neq 0$ but $|\xi|$ very small in comparison with $H(\xi)$ and (G,C)-function $f(x)$, the number $f(\xi)$ cannot be rational. After Galochkin [25] (1974) and Flicker [24] (1977), Väáneen [40] (1979) in the p-adic case we obtain the following type of results (cf. [6]). Let $f_1(z),\ldots,f_n(z)$ be a system of (G,C)-functions satisfying a linear differential (G,C)-equation of order n for some fixed $\mathbb{K}$.

Let $\xi \neq 0$ be an algebraic number from $\mathbb{K}$ and let

$$|\xi| < \exp(-C_8.(\log H(\xi))^{\frac{n}{n+1} - \varepsilon}) \qquad (*)$$

for $C_8 = C_8(\mathbb{K},\bar{f},D,\varepsilon) > 0$. Then $f_1(\xi),\ldots,f_n(\xi)$ are not related by an algebraic relation with integer coefficients of degree $\leq D$ [provided $f_1(x),\ldots,f_n(x)$ are algebraically independent over $\mathbb{C}(x)$].

The existence of the condition (*) with huge constant C_8 (even for $\mathbb{K} = \mathbb{Q}$, $D = 1$ and simple $\bar{f}$ like binomials or logarithms C_8 is of the form e^{100}) does not allow us to get any interesting concrete results.

§2. We shall use the (G,C)-function condition not itself but as a necessary arithmetic addition to prove 'normality'.

Our main result is the following (notations of 1.2):

THEOREM 2.1. *Let* $f_1(z),f_2(z)$ *be algebraically independent meromorphic functions of orders* ρ_1,ρ_2 *and let for* $w = f_1(z)$, $g(w) = f_2(z)$, *the function* $g(w)$ *be the (G,C)-function from* $\mathbb{K}((w))$ *and satisfying a linear differential equation over* $\mathbb{K}(w)$ *and* 12.8. *Let* $\ell > [\mathbb{K}:\mathbb{Q}](\rho_1 + \rho_2) + C_{11}$ *and* $\theta_1,\ldots,\theta_\ell$ *be distinct complex numbers such that* $f_1(\theta_i) \in \mathbb{K} : i = 1,\ldots,\ell$. *Then for algebraic numbers* $\zeta_{1j},\ldots,\zeta_{\ell j}$ *from* $\mathbb{K}$ *we have* $(j = 2,\ldots,n)$:

$$\max_{\substack{i=1,\ldots,\ell \\ j=2,\ldots,n}} |f_j(\theta_i) - \zeta_{ij}| > (\max_{\substack{i=1,\ldots,\ell \\ j=2,\ldots,n}} H(\zeta_{ij}))^{-C_9} \qquad (2.2)$$

for $\max_{\substack{i=1,\ldots,\ell \\ j=2,\ldots,n}} H(\zeta_{ij}) > C_{10}$ *for* $C_9 > 0$, $C_{10} > 0$ *depending on*

$[\mathbb{K}:\mathbb{Q}]$, $f_1(z)$, $f_2(z)$, $\theta_1,\ldots,\theta_\ell$, and C_{11} on $g(w)$.

Here are some examples of the corollaries of the theorem. They solve old problems about the measure of transcendence of elliptic logarithms.

Let $\wp(x)$ be a Weierstrass elliptic function corresponding to an elliptic curve $y^2 = 4x^3 - g_2x - g_3$ $(y = \wp'(u)$, $x = \wp(u))$ over $\bar{\mathbb{Q}}$ $(g_2, g_3 \in \bar{\mathbb{Q}})$. Let u be an algebraic point on $\wp(x)$, i.e. $\wp(u) \in \bar{\mathbb{Q}}$. Then by Schneider's theorem [34] (1935) u is a transcendental number.

It has been known since 1949 [26] that

$$|u - \frac{p}{q}| > |q|^{-\log^4 \log|q|} : |q| \geq q_0$$

for rational integers p,q and the problem was to replace $\log^4 \log|q|$ by $C > 0$ (as for almost all numbers).

Using Pade approximants to elliptic integrals of the first and second kind we can present some corollaries:

Corollary 2.3 Let $\wp(x)$ have algebraic invariants and u be an algebraic point of $\wp(x)$. Then for algebraic α, β with $|\alpha| + |\beta| \neq 0$ the number $\alpha u + \beta\zeta(u)$ is a number with a 'normal' transcendence measure. Moreover for algebraic α , β , γ and $|\alpha| + |\beta| + |\gamma| \neq 0$:

$$|\alpha u + \beta\zeta(u) + \gamma| > H^{-C_{12}}$$

for $H = \max(H(\alpha), H(\beta), H(\gamma))$ and $C_{12} > 0$ depending only on $\wp(x)$, u and $[\mathbb{Q}(\alpha,\beta,\gamma) : \mathbb{Q}]$.

Remark Some numbers are not 'normal' according to our criteria, for example

$$\frac{\log \alpha}{u}$$

for the following reason; the number $(\log \alpha)/u$ comes from the pair of functions $f_1(z) = e^{z\gamma}$, $f_2(z) = \wp(z)$. However, Dwork [23] proved that substituting $\wp(z) = w$ you do not get $e^{\gamma z}$ as a G-function of w , whenever $\gamma \in \bar{\mathbb{Q}}$ is non-zero. Nevertheless $(\log \alpha)/u$ and the number π/ω are normal, since the latter comes from a pair of functions $(\wp(x)\ ,\ \zeta(x) - \frac{\eta}{\omega} x)$ that can be transformed into (G,C)-functions!

The main result we use is the upper bound on the number of zeros of the auxiliary functions

$$R(z) = P(z, f_1(z), \ldots, f_n(z))$$

where $f_1(z), \ldots, f_n(z)$ satisfy linear differential equations with rational function coefficients.

We present our general result that we formulate only in the case of Fuchsian equations (for proofs see [17],[19]):

THEOREM 2.4. *Let* $f_1(z), \ldots, f_n(z)$ *be a solution of a linear differential equations with rational function coefficients*

$$\frac{d}{dx} \vec{f} = \vec{f} . \hat{Q}(z) \quad ;$$

$\hat{Q}(z) \in M_n(\mathbf{C}(z))$ *having regular singularities only and* q *is a total degree of* $\hat{Q}(z)$ *(i.e. degree of* $g(z).\hat{Q}(z)$ *where* $g(z)$ *is the denominator of* $\hat{Q}(z)$ *). Let* $x_i : i \in I$ *be the system of points for which* $f_1(z), \ldots, f_n(z)$ *are analytic functions of* $z = x_i : i \in I$. *Let* $P(z_0, \ldots, z_n) \in \mathbf{C}[z_0, z_1, \ldots, z_n]$ *and*

$$R(z) = P(z, f_1(z), \ldots, f_n(z)) \not\equiv 0 \quad .$$

Then

$$\sum_{i \in I} \operatorname{ord} R(z)|_{z=x_i} \leq \prod_{j=0}^{n} d_{z_j}(P) + C_{14} . q . (\sum_{j=1}^{n} d_{z_j}(P))^n |I| + Q . (\sum_{j=1}^{n} d_{z_j}(P))^{n+1} \quad .$$

Here C_{14} depends only on n and $Q = 0$ if x_i are not apparent singularities of the equations; otherwise Q is computed in terms of local multiplicities of the equation at $z = x_i$.

For $n = 2$ this result was announced by Nesterenko; his proof and those of Brownawell-Masser [8] and Osgood [33] (nonsingular case) give bad dependence on $d_{z_j}(P) : j = 1, \ldots, n$ for $n \geq 3$.

§3. We can consider now transcendence results following from the multidimensional generalization of the Schneider-Lang theorem. In this case, instead of cardinalities of sets, we are speaking about 'degrees' of sets in $\mathbb{C}^n$. Here are the definitions.

Definition 3.1 Let $S \subset \mathbb{C}^n$, and for $K \geq 1$ we define

$$\Omega(S,K) = \min\{\deg P(\bar{x}) : P(\bar{x}) \in \mathbb{C}[\bar{x}]\}$$

and $\partial^{\vec{k}} P(\bar{w}) = 0$ for any $\bar{w} \in S$ and $|\vec{k}| \leq k-1\}$. Then $\Omega(S) = \Omega(S,1)$ is called a degree of S and

$$\Omega_0(S) = \lim_{K\to\infty} \frac{\Omega(S,K)}{K}$$

(the limit exists) is called a singular degree of S.

We note that for an arbitrary $S \subset \mathbb{C}^n$

$$\frac{1}{n}\,\Omega(S) \leq \Omega_0(S) \leq \Omega(S) \quad . \tag{3.2}$$

For a generic finite set S in $\mathbb{C}^n$ we have

$$\Omega(X) = \sqrt[n]{n!}\,.\,|S|^{1/n} \tag{3.3}$$

and it was conjectured that for $|S| \geq C(n)$ and generic S,

$$\Omega_0(S) = |S|^{1/n} \tag{3.4}$$

(which is true for $n = 2$ and $|S|^{1/n} \in \mathbb{Z}$).

The ordinary Bombieri-Lang [7] theorem has the form:

THEOREM 3.5 (Bombieri). *Let* $\mathbb{K}$ *be an algebraic number field and* $f_1(\bar{z}),\ldots,f_{n+1}(\bar{z})$ *be* $n+1$ *algebraically independent functions that are meromorphic functions in* $\mathbb{C}^n$ *of orders of growth* $\rho_1,\ldots,\rho_{n+1}$. *Let us assume that* $f_1(\bar{z}),\ldots,f_{n+1}(\bar{z})$ *satisfy algebraic differential equations over* $\mathbb{K}$ *; this means that there are functions* $f_{n+2}(\bar{z}),\ldots,f_m(\bar{z})$ *such that*

$$\frac{\partial}{\partial z_i} \text{ maps } \mathbb{K}[f_1,\ldots,f_{n+1},f_{n+2},\ldots,f_m]$$

into itself: $i = 1,\ldots,n+1$.

We define

$$S_{\mathbb{K}} = \{\bar{z} \in \mathbb{C}^n : f_j(\bar{z}) \in \mathbb{K} \text{ for all } j = 1,\ldots,n+1,\ldots,m\}. \tag{3.6}$$

Then the set $S_{\mathbb{K}}$ *has singular degree* $\Omega_0(S_{\mathbb{K}})$ *at most*

$$[\mathbb{K}:\mathbb{Q}] \sum_{i=1}^{n+1} \rho_i , \tag{3.7}$$

or degree $\Omega(S_{\mathbb{K}})$ *at most*

$$n[\mathbb{K}:\mathbb{Q}] \sum_{i=1}^{n+1} \rho_i . \tag{3.8}$$

This bound can be improved sometimes, and the following is applied as well (see [41a],[41b]):

Remark 3.9 If some of the functions $f_1(\bar{z}),\dots,f_{n+1}(\bar{z})$ satisfy an algebraic law of addition, then the definition of $S_{\mathbb{K}}$ can be improved. More precisely, if $f_1(\bar{z}),\dots,f_L(\bar{z})$ $(L \le n+1)$ satisfy an algebraic law of addition over $\mathbb{K}$, then $S_{\mathbb{K}}$ can be changed to

$$S_{L,\mathbb{K}} = \{\bar{z} \in \mathbb{C}^n: \ f_i(\bar{z}) \in \overline{\mathbb{Q}} \quad \text{for} \quad i = 1,\dots,L \quad \text{and}$$

$$f_j(\bar{z}) \in \mathbb{K} \quad \text{for} \quad j = L+1,\dots,m\}$$

(or $S_{L,\mathbb{K}} = \{\bar{z} \in \mathbb{C}^n: \ \partial^{\vec{k}} f_i(\bar{z}) \in \overline{\mathbb{Q}} \quad \text{for} \quad i = 1,\dots,L \quad \text{and}$

$$\partial^{\vec{k}} f_j(\bar{z}) \in \mathbb{K} \quad \text{for} \quad j = L+1,\dots,n+1 \quad \text{and} \quad \vec{k} \in \mathbb{N}^n\}) .$$

For example we can consider the case $f_i(\bar{z}) = z_i$: $i = 1,\dots,L$ $(L \le n)$. Then

$$\Omega_0(S_{L,\mathbb{K}}) \le [\mathbb{K}:\mathbb{Q}](\rho_1 + \dots + \rho_{n+1}) .$$

In applications we usually consider the case, when *all* $f_1(\bar{z}),\dots,f_m(\bar{z})$ $(m \ge n+1)$ satisfy an algebraic law of addition and the set $S_{\mathbb{K}}$ usually has a cube $\underbrace{S \times \dots \times S}_{n} = S^{*n}$ for $S \subset \mathbb{C}^n$ having degree and singular degree $|S|$. The corresponding result does not need potential theory for the proof, and was established in 1940 by Th. Schneider who used reduction to the one-dimensional case.

§4. We can prove now results on the measures of transcendence of numbers involved in the Schneider or Bombieri-Lang theorems.

For proving results of the 'normal' type we need special algebraic differential equations that are transformations of linear differential equations having the (G,C)-function property.

We consider the system of $n+1$ algebraically independent

functions $f_1(\bar{z}),\ldots,f_n(\bar{z})$, $f_{n+1}(\bar{z})$ that are meromorphic in $\mathbb{C}^n$. We make a change of variables

$$w_1 = f_1(\bar{z}),\ldots,w_n = f_n(\bar{z})$$

so that $f_{n+1}(\bar{z}) = g(\bar{w})$ is a transcendental function in

$$\bar{w} = (w_1,\ldots,w_n) .$$

We call a point $\bar{z}_0 = (z_1^0,\ldots,z_n^0)$ regular if the map $\bar{z} \to \bar{w} = \vec{f}(\bar{z})$ $(= (f_1(\bar{z}),\ldots,f_n(\bar{z}))$ is invertible at $\bar{z} = \bar{z}_0$, i.e.

$$J(\vec{f})\Big|_{\bar{z}=\bar{z}_0} = \det(\partial_{z_j} f_k(\bar{z}))^n_{j,k=1}\Big|_{\bar{z}=\bar{z}_0} \neq 0 .$$

If the Jacobian of $(f_1(\bar{z}),\ldots,f_n(\bar{z}))$ is non-zero at $\bar{z} = \bar{z}_0$, then $\bar{z}_0$ or $\bar{w}_0 = \vec{f}(\bar{z}_0)$ is called a regular point.

We consider the situation in which

$$g(\bar{w}) = f_{n+1}(\bar{z}) \ ; \ \bar{w} = (f_1(\bar{z}),\ldots,f_n(\bar{z}))$$

satisfies a Fuchsian linear differential equation in $\partial_{w_1},\ldots,\partial_{w_n}$. In other words, we consider the situation in which $g(\bar{w})$ satisfies n differential equations, with coefficients in $\mathbb{C}(w_1,\ldots,w_n)$ in partial derivatives $\partial_{w_1},\ldots,\partial_{w_n}$, and having regular singularities only. The order of the corresponding Fuchsian system of $g(\bar{w})$ is defined in a natural way.

<u>Definition 4.1</u> Let $g(\bar{w})$ satisfy a Fuchsian system in $\partial_{w_1},\ldots,\partial_{w_n}$. Then the order of the system is

$$\dim_{\mathbb{C}(w_1,\ldots,w_n)}\{\partial_w^{\vec{k}} g(\bar{w}) : \vec{k} \in \mathbf{N}^n ,\ \partial_w^{\vec{k}} = \partial_{w_1}^{k_1}\cdots\partial_{w_n}^{k_n} ,\ \vec{k} = (k_1,\ldots,k_n)\} .$$

The ideal in $\mathbb{C}(w_1,\ldots,w_n)[\frac{\partial}{\partial w_1},\ldots,\frac{\partial}{\partial w_n}]$ generated by all equations satisfied by $g(\bar{w})$ will be denoted by $\mathcal{D}$. In this case the order of $\mathcal{D}$ (as before) is nothing but

$$\dim_{\mathbb{C}(w_1,\ldots,w_n)}\mathbb{C}(w_1,\ldots,w_n)[\frac{\partial}{\partial w_1},\ldots,\frac{\partial}{\partial w_n}]/\mathcal{D} = \mathrm{ord}(\mathcal{D}) .$$

Also we are considering equations with coefficients from $\bar{\mathbb{Q}}(w_1,\ldots,w_n)[\frac{\partial}{\partial w_1},\ldots,\frac{\partial}{\partial w_n}]$; in this case the ideal is still denoted by $\mathcal{D}$ and is called a system of equations for $g(\vec{w})$.

If all solutions of $\mathcal{D}$ have polynomial (or at most logarithmic-polynomial) growth in $\mathbf{C}P^n$, then the system $\mathcal{D}$ is called a Fuchsian system of equations. Of course, the criteria for a Fuchsian system can be rewritten in an algebraic form. For example we can restrict $g(\vec{w})$ and $\mathcal{D}$ to a line $\vec{\ell} = (a_1t + b_1, \dots, a_nt + b_n)$ in $\mathbf{C}P^n$, then the corresponding function $g(t)$, $g(t) = g(a_1t + b_1, \dots, a_nt + b_n)$ satisfies a Fuchsian linear differential equation $\mathcal{D}_{\vec{\ell}} = \mathcal{D}(\vec{\ell})$ of the order $\le \mathrm{ord}(\mathcal{D})$. Here for a generic line ℓ, $\mathrm{ord}(\mathcal{D}_{\vec{\ell}}) = \mathrm{ord}(\mathcal{D})$.

One of the major problems that is connected with the qualitative version is the problem of a bound for the number of zeros of

$$P(w_1, \dots, w_n, g(\overline{w}))$$

at fixed points (counted with multiplicities) in terms of degrees of $P(\overline{w}, g(\overline{w}))$. These bounds have, in general, the form

$$\left(\prod_{i=1}^{n} d_{w_i}(P)\right) d_{w_{n+1}}(P)^{\chi}$$

for χ depending on $\mathrm{ord}(\mathcal{D})$. For Fuchsian linear differential equations the corresponding bounds are almost the best possible with $\chi = \mathrm{ord}(\mathcal{D})$. There are two methods that give us such bounds: 1) is to use the reduction of $\mathcal{D}$ to a one-dimensional Fuchsian equation on an algebraic curve in $\mathbf{C}P^n$ containing a fixed set of points; 2) to use the generalization of Fuchs' relations expressing the orders of singularities of solutions of Fuchsian systems in terms of $\mathrm{ord}(\mathcal{D})$ and the topology of singularities of $\mathcal{D}$ [42].

Later we shall formulate the conjecture expressing the best possible bound for the number of zeros. However for the present purposes the existing bounds are enough.

Now we generalize the definition of (G,C)-operators for the n-dimensional case.

<u>Definition 4.2</u> Let $g(\overline{w})$, $\overline{w} \in \mathbf{C}^n$ satisfies a system of linear differential equations over $\overline{\mathbb{Q}}$. Thus we fix an algebraic number field $\mathbb{K}$ and define an ideal $\mathcal{D}$ as

$$\mathcal{D} = \{L \in \mathbb{K}(w_1, \dots, w_n)[\frac{\partial}{\partial w_1}, \dots, \frac{\partial}{\partial w_n}] : Lg(\overline{w}) = 0\}$$

so that $\dim_{\mathbb{K}(w_1,\dots,w_n)} \mathbb{K}(w_1, \dots, w_n)[\frac{\partial}{\partial w_1}, \dots, \frac{\partial}{\partial w_n}]/\mathcal{D} = \mathrm{ord}(\mathcal{D})$ and we choose arbitrary $\mathrm{ord}(\mathcal{D}) = r$ elements $g_1(\overline{w}), \dots, g_r(\overline{w})$ from

$$\{\partial_{\overline{w}}^{\vec{k}} g(\overline{w}) : \vec{k} \in \mathbf{N}^n\}$$

that are linearly independent over $\mathbf{C}(w_1,\ldots,w_n)$. Then bu the definition of $r = \mathrm{ord}(\mathcal{D})$, we have

$$\partial_{\overline{w}}^{\vec{k}} g(\overline{w}) = \partial_{w_1}^{k_1} \ldots \partial_{w_n}^{k_n} g(\overline{w})$$
$$= \sum_{j=1}^{r} R_{\vec{k},j}(\vec{w}) g_j(\overline{w}) \tag{4.3}$$

for $R_{\vec{k},j}(\overline{w}) \in \mathbb{K}(\overline{w})$. By the construction of $R_{\vec{k},j}(\overline{w})$ we find that there is a $P(\overline{w}) \in \mathbb{I}_{\mathbb{K}}[\overline{w}]$ (i.e. polynomial in $\overline{w}$ with integer coefficients from $\mathbb{K}$) such that

$$R_{\vec{k},j}(\overline{w}).P(\overline{w})^{|\vec{k}|} \in \mathbb{K}[\overline{w}]: \quad j = 1,\ldots,r$$

and $|\vec{k}| = k_1 + \ldots + k_n$, $\vec{k} \in \mathbf{N}^n$. Then we consider a common denominator D_M of the coefficients of polynomials

$$\{\frac{1}{\vec{k}!} R_{\vec{k},j}(\overline{w}).P(\overline{w})^{|\vec{k}|}: \ |\vec{k}| \le M\}$$

(for $\vec{k}! = k_1! \ldots k_n!$) . We call an operator $\mathcal{D}$ to be a (G,C)-operator in $\mathbf{C}^n$ if

$$D_M \le C^M \quad \text{for} \quad M \ge M_0 ,$$

and assuming that sizes of polynomials $(\vec{k}!)^{-1} R_{\vec{k},j}(\overline{w}).P(\overline{w})^{|\vec{k}|}$ are growing not faster than a geometric progression, say $C^{|\vec{k}|}$

This, last, condition is not important and can be removed.

The simplest examples of (G,C)-functions $g(\overline{w})$ are functions $g(\overline{w}) = R(g_1(w_1),\ldots,g_n(w_n))$, where $g_i(w_i)$ are (G,C)-functions $(i = 1,\ldots,n)$ and $R(z_1,\ldots,z_n) \in \overline{\mathbf{Q}}[z_1,\ldots,z_n]$.

Here is one of the qualitative versions of the Bombieri-Lang theorem that gives us examples of 'normal' numbers. We can formulate the corresponding result in such a form:

THEOREM 4.4. *Let* $f_1(\overline{z}),\ldots,f_{n+1}(\overline{z})$ *be algebraically independent functions, meromorphic in* $\mathbf{C}^n$ *and of the order of growth* $\rho_1,\ldots,\rho_{n+1}$. *Let* $\mathbb{K}$ *be an algebraic number field and after the substitution*

$$\overline{w} = \vec{f}(\overline{z}) = (f_1(\overline{z}),\ldots,f_n(\overline{z}))$$

let the function

$$g(\overline{w}) \overset{\text{def}}{=} f_{n+1}(\overline{z})$$

satisfy a linear Fuchsian system (4.3) *over* $\mathbb{K}$ *and let* $g(\overline{w})$ *be a* (G,C)*-function satisfying an analogue of* 12.8.

Let $S \subset \mathbb{C}^n$ *be a finite set such that*

$$\Omega_0(S) > [\mathbb{K}:\mathbb{Q}](\rho_1+\ldots+\rho_{n+1}) + C_0 \quad or \quad \Omega(S) > n[\mathbb{K}:\mathbb{Q}](\rho_1+\ldots+\rho_{n+1}) + C_0$$

and such that

$$f_i(\overline{z}) \in \mathbb{K} \quad \text{for} \quad \overline{z} \in S \text{ , } i = 1,\ldots,n$$

(or $\vec{f}(S) \subseteq \mathbb{K}$) . *Then for an arbitrary system* $\xi_{\overline{x},k} : \overline{x} \in S$, $k = n+1,\ldots,m$, *of algebraic numbers from* $\mathbb{K}$ *we have*

$$\max_{\substack{\overline{x}\in S\\ k=n+1,\ldots,m}} |f_k(\overline{x}) - \xi_{\overline{x},k}| > c_1.\Big(\max_{\substack{\overline{x}\in S\\ k=n+1,\ldots,m}} H(\xi_{\overline{x},k})\Big)^{-c_2} \qquad (4.5)$$

for $c_1 > 0$, $c_2 > 0$ *depending on* $f_1(\overline{z}),\ldots,f_{n+1}(\overline{z})$, S *and* $[\mathbb{K}:\mathbb{Q}]$, $k = n+1,\ldots,m$, *and* c_0 *on* $g(\overline{w})$.

In other words, this bound (4.5) can be reformulated for the functions $g_j(\overline{w})$:

$$\max_{\substack{\overline{y}\in\vec{f}(S)\\ j=1,\ldots,r}} |g_j(\overline{y}) - \xi_{\overline{y},j}| > c_1\Big(\max_{\substack{\overline{y}\in\vec{f}(S)\\ j=1,\ldots,r}} H(\xi_{\overline{y},j})\Big)^{-c_2} .$$

The most interesting applications deal with functions satisfying algebraic laws of addition. In this case S can be taken as a discrete subgroup of $\mathbb{C}^n$, provided that $\dim_{\mathbb{C}} S \geq n$.

Here is one of the results one gets in this direction.

<u>Corollary 4.6</u> Let $f_1(\overline{z}),\ldots,f_{n+1}(\overline{z})$ be functions as in the Theorem. Let us assume that $f_1(\overline{z}),\ldots,f_m(\overline{z})$ satisfy the algebraic law of addition and let $\vec{v}_1,\ldots,\vec{v}_n$ be n linearly independent vectors in $\mathbb{C}^n$ such that

$$f_i(\vec{v}_j) \in \overline{\mathbb{Q}} : \quad i,j = 1,\ldots,n .$$

Then for algebraic numbers $\xi_{1k},\ldots,\xi_{nk}$ we have $(k = n+1,\ldots,m)$

$$\max_{\substack{i=1,\ldots,n\\ k=n+1,\ldots,m}} |f_k(\vec{v}_i) - \xi_{i,k}| > c_1'\Big(\max_{\substack{i=1,\ldots,n\\ k=n+1,\ldots,m}} H(\xi_{i,k})\Big)^{-c_2'}$$

where $c_1' > 0$, $c_2' > 0$ depends on $f_1(\overline{z}),\ldots,f_{n+1}(\overline{z})$; $\vec{v}_1,\ldots,\vec{v}_n$, and on the degree of

$$\mathbb{Q}(f_k(\vec{v}_j), d(\xi_{i,k}) : i,j = 1,\dots,n;\ k = n+1,\dots,m) .$$

The natural question which arises here is whether it is easy to verify the condition that

$$g(\overline{w}) = f_{n+1}.\vec{f}^{-1}(\overline{z})$$

satisfies a Fuchsian linear differential equation in $\partial_{\overline{w}}$ for $\overline{w} = \vec{f}(\overline{z})$. Hopefully, for the functions satisfying an algebraic law of addition, the functions $f_1(\overline{z}),\dots,f_{n+1}(\overline{z})$ arise from Abelian functions on an Abelian (or, in general, group) variety $\mathbb{A}$. Thus n functions $f_1(\overline{z}),\dots,f_n(\overline{z})$ can be considered as Abelian functions on $\mathbb{A}$. In this case the functions $f_1(\overline{z}),\dots,f_n(\overline{z})$ are inversions of Abelian (one-dimensional) integrals. These integrals lead to the usual Fuchsian (G,C)-operators. Depending on the choice of $f_{n+1}(\overline{z})$ we come to an n-dimensional Fuchsian (G,C)-operator or a Fuchsian operator which is not a (G,C)-operator. (The latter case may appear for $n = 1$ and $f_1(x) = \wp(x)$, $f_2(x) = \wp(\beta x)$; or $f_1(x) = e^x$, $f_2(x) = e^{\beta x}$ for $\beta \in \overline{\mathbb{Q}}\backslash\mathbb{Q}$).

The best applications of our Corollary 4.6 are connected with the cases of Abelian functions having complex or real multiplications. In this case starting from one nonzero vector $\vec{v} \in \mathbb{C}^n$ such that $f_i(\vec{v}) \in \overline{\mathbb{Q}}$ $(i = 1,\dots,n)$ we get n linearly independent vectors $\vec{v}_i \in \mathbb{C}^n$ $(\vec{v}_1 = \vec{v})$ $i = 1,\dots,n$ such that $f_i(\vec{v}) \in \overline{\mathbb{Q}}$ $(i = 1,\dots,n)$. In this case we can naturally take $f_1(\overline{z}),\dots,f_n(\overline{z})$ as n algebraically independent Abelian functions for an Abelian variety $\mathbb{A}$ of dimension n having n (real) multiplications. The function $f_{n+1}(\overline{z})$ can be taken as a quasi-periodic function of $\mathbb{A}$, algebraically independent with respect to $f_1(\overline{z}),\dots,f_n(\overline{z})$. Under natural assumption of the normalization of the action of endomorphisms of $\mathbb{A}$, the function $f_{n+1}(\overline{z})$ can be chosen as any of z_i $(i = 1,\dots,n)$.

Changing the variables $w_i = f_i(\overline{z})$ $(i = 1,\dots,n)$ we find that $f_{n+1}(\overline{z}) = z_i$ as a function of $w_1,\dots,w_n$ is a linear combination of the Abelian integrals of the first kind and the equations satisfied by $z_i(\vec{w})$ are Fuchsian linear differential equations, having solutions as (G,C)-functions. That is why the simplest application is connected with algebraic points of Abelian varieties with real multiplications:

Corollary 4.7 Let $\mathbb{A}$ be an Abelian variety defined over $\overline{\mathbb{Q}}$ having real multiplications in field $\mathbb{K}$, $\dim(\mathbb{A}) = [\mathbb{K}:\mathbb{Q}] = n$. Let $A_1(\bar{z}),\ldots,A_n(\bar{z})$ be corresponding (normalized) n Abelian functions of $\mathbb{A}$ and let $\vec{u} \in \mathbb{C}^n$ $(\vec{u} \neq \vec{0})$ be an algebraic point of $\mathbb{A}$, i.e. $A_i(\vec{u}) \in \overline{\mathbb{Q}}$: $i = 1,\ldots,n$. Then for any $j = 1,\ldots,n$ the number $(\vec{u})_j$ is a 'normal' transcendental number:

$$|(\vec{u})_j - \xi| > c_3 H(\xi)^{-c_4}$$

for $c_3 > 0$, $c_4 > 0$ depending on $\mathbb{A}$, $\vec{u}$ and $d(\xi)$ only.

The statement of the transcendence of $(\vec{u})_j$ belongs to D. Bertrand [5]. For example he represented in this form all periods of parabolic forms for $\Gamma_0(N)$. In particular, these numbers (periods of Eichler's integrals) are 'normal' numbers.

Moreover, our transcendence measures are applied to the measure of linear independence of Abelian logarithms for real multiplication case, following Bertrand-Masser methods (see [5]).

In this case we get a lower bound for the linear form in Abelian logarithms which is the best possible with respect to sizes of the algebraic coefficients. This bound has the famous 'normal' form H^{-c}. We achieve this bound without the introduction of the Δ-polynomials [1] (factorial polynomials), but simply by changing the variables. The sense of this operation is clear if one looks at the normal situations with linear forms in logarithms, where Δ-polynomials were introduced for the first time.

1) For monomials $\alpha_1^{(\lambda_1+b_1\lambda_{n+1})x_1} \cdots \alpha_n^{(\lambda_n+b_n\lambda_{n+1})x_n} = M$

we have

2) $\partial_{x_1}^{m_1}\cdots\partial_{x_n}^{m_n}.M = \log \alpha_1^{m_1}\cdots\log \alpha_n^{m_n}(\lambda_1+b_1\lambda_{n+1})^{m_1}\times\cdots\times(\lambda_n+b_n\lambda_{n+1})^{m_n}.M$,

while changing to new variables $\alpha_i^{x_i} = z_i$, $i = 1,\ldots,n$; we obtain

$$\partial_{z_1}^{m_1}\cdots\partial_{z_n}^{m_n}.M = m_1!\cdots m_n!\binom{\lambda_1+b_1\lambda_{n+1}}{m_1}\cdots\binom{\lambda_n+b_n\lambda_{n+1}}{m_n}Mz^{-m_1-\cdots-m_n}$$

which explains plainly the sense of the appearance of the factorial polynomials.

At the same time we observe a new phenomena; while for the usual Baker's method [1] for exponential polynomials we are concerned with

$$\partial_{x_1}^{m_1} \ldots \partial_{x_n}^{m_n} M|_{x_1=\ldots=x_n=k}, \quad k \in \mathbb{Z},$$

for binomial polynomials

$$M = z_1^{\lambda_1+b_1\lambda_{n+1}} \ldots z_n^{\lambda_n+b_n\lambda_{n+1}}$$

we consider

$$(m_1! \ldots m_n!)^{-1} \partial_{z_1}^{m_1} \ldots \partial_{z_n}^{m_n} M|_{z_1=\alpha_1^k,\ldots,z_n=\alpha_n^k}, \quad k \in \mathbb{Z}.$$

Now instead of considering auxiliary functions having zeros in $\mathbb{C}^n$ sitting on a line $(t,\ldots,t)$ at integer points $t \in \mathbb{Z}$, which is a degenerate situation, we are dealing with the auxiliary function having zeros (of high orders) on a transcendental curve $(\alpha_1^t,\ldots,\alpha_n^t)$, whose algebraic degree is ∞.

<u>Corollary 4.8</u> Periods (or quasi-periods) of Abelian varieties with real (complex) multiplications are 'normal' transcendental numbers. In particular, for rational (not integer a and b, $B(a,b)$ is 'normal' transcendental.

The transcendental part is due in the complex case to Th. Schneider [34] (1940) and in the real case to D. Bertrand [5] (1979). Thus

$$|B(a,b) - \frac{p}{q}| > |q|^{-\gamma(a,b)}$$

for $p,q \in \mathbb{Z}$, $q \neq 0$. Unfortunately, due to the methods used, $\gamma(a,b)$ is too big. For $B(1/2,b)$ we can obtain better results using explicit Padé approximants for ${}_2F_1(1;\beta;\gamma;x)$.

We can propose the most general conjecture about the possible number of zeros of auxiliary functions in n-dimensional situations. This conjecture seems to be reasonable (at least from the present point of view), and if confirmed will give us all results on linear independence of periods, quasi-periods, Abelian integrals - we hope to obtain with present analytic methods.

<u>Conjecture 4.9</u> Let $f(z_1,\ldots,z_n)$ be a solution of a Fuchsian linear differential equation in $\mathbb{C}^n$ of order ∂. Let $f(z_1,\ldots,z_n)$ be 'highly transcendental', i.e. the restriction of $f(z_1,\ldots,z_n)$ on any algebraic curve in $\mathbb{C}^n$, is a transcendental function. We consider an auxiliary function

$$R(\bar{z}) = P(z_1,\dots,z_n,f(\bar{z}))$$

where the degree of $P(\bar{z},x)$ in $\bar{z}$ is at most D and in x is at most D_0 (for $P(\bar{z},x) \in \mathbf{C}[\bar{z},x]$). Let $\vec{z}_i \in \mathbf{C}^n$: $i \in I$ be points different from the singularities of the equation satisfied by $f(\bar{z})$. Then (assuming, possibly, some conditions on the distributions of $\vec{z}_i$) we have

$$\sum_{i\in I} \{\text{ord } R(\bar{z})|_{\bar{z}=\bar{z}_i} - \partial\}^n \le c.D^n.D_0^{\partial}\ .$$

In fact, for applications we need to consider only the case when $f(z_1,\dots,z_n)$ is a linear combination of Abelian integrals (for example logarithms etc., i.e. $f(z_1,\dots,z_n) = \gamma_1 z_1 + \dots + \gamma_n z_n$ for $\gamma_1,\dots,\gamma_n$ linearly independent over $\mathbf{Q}$).

We do have a partial answer to this question, which seems to be too weak for applications but already nontrivial:

$$\sum_{i\in I} \{\text{ord } R(\bar{z})|_{\bar{z}=\bar{z}_i} - \partial\}^n \le c_1.D^n.D_0^{n\partial}\ ;$$

assuming that the set $\{\bar{z}_i: i \in I\}$ is generic with given $|I|$. The same bound holds for $R(\bar{z}) = P(z_1,\dots,z_n; f_1(\bar{z}),\dots,f_\partial(\bar{z}))$ and $f_1(\bar{z}),\dots,f_\partial(\bar{z})$ being algebraically independent (even 'highly algebraically independent') solutions of the same differential equation as $\vec{f}(\bar{z})$.

CHAPTER 2 Measures of the algebraic independence of several numbers. Normal measures of algebraic independence of two numbers connected with elliptic functions

The purpose of this part of the paper is to present a complete proof of the bound of the measure of the algebraic independence of two numbers. We purposely give a complete proof with all possible details and a detailed exposition of auxiliary results since our considerations are very general. The method of the proof presented can be applied to any algebraic independence proof where the Gelfond-Schneider method is used. We decided to present the proof for the case of the numbers $\frac{\pi}{\omega}$, $\frac{\eta}{\omega}$, where the type of transcendence is $\le 3+\varepsilon$ for any $\varepsilon > 0$ and the result has many important and interesting corollaries. The methods of proofs are elementary, using only intersection theory and are based on the first two papers [11a],[11b]. One can consider the proof below as a good introduction to more sophisticated studies of measures of n algebraically independent numbers in the Gelfond-Schneider method with $n \geq 3$.

§5. If we consider now the case of the measure of the algebraic independence of several numbers, the first problem that appears is the scarcity of examples of two numbers (of 'algebraic' or 'geometric' nature) or three algebraically independent numbers. However for those pairs for which a proof of algebraic independence exists, the measure of the algebraic independence can be obtained and results that we have now are already at the limit of analytical methods.

The first example of two algebraically independent numbers was constructed by Gelfond (1949) [26]:

α^{β} and α^{β^2} are algebraically independent for algebraic $\alpha \neq 0,1$ and a cubic irrational β.

In order to formulate measures of algebraic independence we introduce some notations. In what follows $P(x,y)$ denotes a nonzero polynomial from $\mathbb{Z}[x,y]$ of total degree $d(P) \leq d$ and height $H(P) \leq H$.

Essentially the first bound for $|P(\alpha^{\beta},\alpha^{\beta^2})|$ was given by Gelfond (1950) [26]. D. Brownawell (1977) [39] obtained

$$|P(\alpha^{\beta},\alpha^{\beta^2})| > \exp(-\exp(c_1(d + \log H)^3)$$

and we have now [16]:

$$|P(\alpha^{\beta},\alpha^{\beta^2})| > \exp(-c_2 H^{c_3 d^{1+\varepsilon}})$$

for $c_2 = c_2(\varepsilon) > 0$ and $c_3 = c_3(\varepsilon) > 0$ and any $\varepsilon > 0$.

A similar type of lower bound can be proved now for two new algebraically independent numbers $\wp(u\beta)$, $\wp(u\beta^2)$ where $\wp(x)$ is defined over $\overline{\mathbb{Q}}$ with complex multiplication and u is an algebraic point of $\wp(x)$ (numbers of Masser-Wüstholz, this Conference).

As we see, these numbers do not have a finite type of transcendence and nothing of 'normal' type. However there are several pairs of algebraically independent numbers for which we have finite type and 'normality' of the bounds [11],[12],[14],[15].

Measures of the algebraic independence presented below are new in the sense that the 'normality' of the measures is obtained. The type of transcendence is basically unchanged and is the same as known in 1978 [13].

We consider Weierstrass's elliptic function $\wp(x)$ with

algebraic invariants g_2, g_3 and corresponding Weierstrass ζ-functions $\zeta(x)$, $\zeta'(x) = \wp(x)$. Let (ω, η) be a pair consisting of a period and a quasi-period of $\wp(x)$, i.e. $\eta = 2\zeta(\omega/2)$ (here $\omega \in L\setminus 2L$).

The main example deals only with π/ω , η/ω :

THEOREM 5.1. *Let* $R(x,y) \in \mathbf{Z}[x,y]$; $R \not\equiv 0$. *Let* $\wp(x)$ *be an elliptic function with algebraic invariants. Let* ω *be a period of* $\wp(x)$ *and* $\eta = 2\zeta(\omega/2)$ *for a Weierstrass* ζ*-function* $\zeta(x)$. *Then*

$$|R(\pi/\omega, \eta/\omega)| > \exp(-C_0(\log H(R) + d(R) \log d(R))d(R)^2 \log^2 d(R))$$

where $d(R) \ge d_0$ *for some constant* $C_0 > 0$.

Proof Let $\wp(x)$ be a Weierstrass elliptic function with algebraic invariants g_2, g_3 and $\zeta(x)$ be the corresponding Weierstrass ζ-function, $\zeta'(x) = -\wp(x)$. Let (ω, η) be a pair consisting of a period and a quasi-period of $\wp(x)$ so that $\eta = 2\zeta(\omega/2)$ (and $\omega/2$ is not a period of $\wp(x)$). Let ω' be another period of $\wp(x)$, linearly independent with ω .

In order to get a bound for the measure of algebraic independence of π/ω , η/ω we work following [14], with the auxiliary function of the form

$$F(z) = P(\zeta(z) - \frac{\eta}{\omega} z, \wp(z))$$

for $P(x,y) \in \mathbf{Z}[\frac{\pi}{\omega}, \frac{\eta}{\omega}][x,y]$ and $d_x(P) \le L_1$, $d_y(P) \le L_2$

and $L_2 \ge L_1 > 1$.

The function $F(z)$ above has the form

$$F(z) = \sum_{\lambda_1=0}^{L_1} \sum_{\lambda_2=0}^{L_2} C_{\lambda_1,\lambda_2} (\zeta(z) - \frac{\eta}{\omega} z)^{\lambda_1} \wp(z)^{\lambda_2} .$$

Then one has the following representation for the derivatives of the function $F(z)$:

$$F^{(k)}(z) = \sum_{\lambda_1=0}^{L_1} \sum_{\lambda_2=0}^{L_2} C_{\lambda_1,\lambda_2} \sum_{s=0}^{k} \binom{k}{s} \{(\zeta(z) - \frac{\eta}{\omega} z)^{\lambda_1}\}^{(s)} \{\wp(z)^{\lambda_2}\}^{(k-s)}$$

Consequently, at points $z = \frac{\omega}{4} + m\omega'$ we have

$$F^{(k)}\left(\frac{\omega}{4}+m\omega'\right) = \sum_{\lambda_1=0}^{L_1}\sum_{\lambda_2=0}^{L_2} C_{\lambda_1,\lambda_2} F_{\lambda_1,\lambda_2,k,m}\left(\frac{\pi}{\omega},\frac{\eta}{\omega}\right) . \tag{5.2}$$

Here

$$F_{\lambda_1,\lambda_2,k,m}\left(\frac{\pi}{\omega},\frac{\eta}{\omega}\right) = \sum_{j_1=0}^{\lambda_1}\sum_{j_2=0}^{\lambda_2} f_{\lambda_1,\lambda_2,k,m,j_1,j_2}\left(\frac{\pi}{\omega}\right)^{j_1}\left(\frac{\eta}{\omega}\right)^{j_2} , \tag{5.3}$$

is a polynomial in π/ω , η/ω of degree at most λ_1 in π/ω and η/ω , with algebraic number coefficients $f_{\lambda_1,\lambda_2,k,m,j_1,j_2}$ in the field $\mathbb{K} = \mathbb{Q}(i,g_2,g_3,\wp(\omega/4),\wp'(\omega/4))$ having sizes at most

$$\max_{\substack{\lambda_1\le L_1,\lambda_2\le L_2\\ j_1,j_2\le\lambda_1}} \overline{|f_{\lambda_1,\lambda_2,k,m,j_1,j_2}|} \le$$
$$\le \exp\{C_1(L_2+k) + k\log(k+1) + L_1\log(m+1)\} \tag{5.4}$$

for $C_1 > 0$ depending on g_2,g_3 only.

From now on we will consider $F_{\lambda_1,\lambda_2,k,m}$ to be polynomials $F_{\lambda_1,\lambda_2,k,m}(x,y) \in \mathbb{K}[x,y]$ of sizes bounded as in (5.4).

The coefficients C_{λ_1,λ_2} , as polynomials $C_{\lambda_1,\lambda_2}(x,y)$, are constructed using the Thue-Siegel lemma (see [1],[38]). Here is one of the definitions of $C_{\lambda_1,\lambda_2}(x,y)$:

LEMMA 5.5. *Let* $L_2 \ge L_1$ *; then there exists a system of polynomials*

$$\{C_{\lambda_1,\lambda_2}(x,y):\ 0 \le \lambda_1 \le L_1,\ 0 \le \lambda_2 \le L_2\}$$

from $\mathbb{Z}[x,y]$ *, without a common factor, such that the following conditions are satisfied:*

$$d(C_{\lambda_1,\lambda_2}) \le L_1 ;\quad t(C_{\lambda_1,\lambda_2}) \le c_2L_2\log L_2 ; \tag{5.6}$$

$\lambda_1 = 0,1,\ldots,L_1$; $\lambda_2 = 0,1,\ldots,L_2$ *and the system of linear equations in* C_{λ_1,λ_2} *is satisfied:*

$$\sum_{\lambda_1=0}^{L_1}\sum_{\lambda_2=0}^{L_2} C_{\lambda_1,\lambda_2}(x,y)F_{\lambda_1,\lambda_2,k,m}(x,y) = 0 \tag{5.7}$$

for

$$k = 0,1,\ldots,[c_3L_2] ;\quad m = 0,1,\ldots,[c_4L_1] .$$

Here $c_3c_4 < (8.[\mathbb{K}:\mathbb{Q}])^{-1}$. In particular, for the function

$$F(z) = \sum_{\lambda_1=0}^{L_1} \sum_{\lambda_2=0}^{L_2} C_{\lambda_1,\lambda_2}(\frac{\pi}{\omega}, \frac{\eta}{\omega})(\zeta(z) - \frac{\eta}{\omega} z)^{\lambda_1}.\wp(z)^{\lambda_2}, \tag{5.8}$$

we have

$$F^{(k)}(\frac{\omega}{4} + m\omega') = 0 \tag{5.9}$$

for $k = 0,1,\ldots,[c_3L_2]$; $m = 0,1,\ldots,[c_4L_1]$.

Now for the function $F(z)$ (5.8), satisfying (5.9), we can use the usual methods of Transcendental Number Theory (see [11],[1]). However from the point of view of applications, we consider a more general situation. We shall not pursue generalizations too far, however, because for functions of the form $P(\zeta(z) - \frac{\eta}{\omega} z, \wp(z))$, one has a Small Value Lemma. If there is no Small Value Lemma, it is necessary to consider a more general situation, where we actually have a family of functions.

We stick to π/ω and η/ω , and have in this case the following analytic statement:

LEMMA 5.10. *Let* C_{λ_1,λ_2} $(0 \le \lambda_1 \le L_1,\ 0 \le \lambda_2 \le L_2)$ *be complex numbers such that*

$$\max_{\substack{0\le\lambda_1\le L_1 \\ 0\le\lambda_2\le L_2}} |C_{\lambda_1,\lambda_2}| \le \exp \overline{C}$$

and the function

$$G(z) = \sum_{\lambda_1=0}^{L_1} \sum_{\lambda_2=0}^{L_2} C_{\lambda_1,\lambda_2}(\zeta(z) - \frac{\eta}{\omega} z)^{\lambda_1}.\wp(z)^{\lambda_2},$$

satisfies

$$\frac{1}{k!} |G^{(k)}(\frac{\omega}{4} + m\omega')| < \exp(-E)$$

for $k = 0,1,\ldots,[c_5L_2]$; $m = 0,1,\ldots,[c_6L_1]$.

Then we have

$$\frac{1}{k'!} |G^{(k')}(\frac{\omega}{4} + m'\omega')| < \exp(-E + \overline{C} + c_7L_1^2L_2) + \exp(-c_9L_1^2L_2)$$

for $k' = 0,1,\ldots,[c_8L_2]$; $m' = 0,1,\ldots,[c_8L_1]$,

and we have at the same time a Small Value Lemma:

$$\max\{|C_{\lambda_1,\lambda_2}| : 0 \le \lambda_1 \le L_1,\ 0 \le \lambda_2 \le L_2\} \le$$

$$\le \max\{\frac{1}{k'!}\ |G^{(k')}(\frac{\omega}{4} + m'\omega')| : k' = 0,1,\dots,[c_{10}L_2]\ ,$$

$$m' = 0,1,\dots,[c_{10}L_1]\} \ .\ \exp(c_{11}L_1^2L_2)\ .$$

for some constant $c_{10} > 0$ *, with* $c_{11} > 0$ *depending on the choice of* c_{10} .

The proof of Lemma 5.10 is the usual one and depends mainly on the Schwarz lemma applied to a function

$$G_1(z) = \sigma(z)^{L_1+2L_2}\ G(z)\ ,$$

which is an entire function of z . Since $G(z)$ is periodic with a period ω , the function $G_1(z)$ has small values $< \exp(-E)$ of multiplicity K at points

$$z = \frac{\omega}{4} + m\omega' + n\omega$$

$m = 0,1,\dots,M$; $n = 0,1,\dots,M'$. We apply then, the usual Hermite interpolation formula to $G_1(z)$ in the region $|z| = 2M_1$.

There is one nontrivial point in the proof of Lemma 5.5. It concerns the improvement of the bound (5.6) to a better one; namely

$$t(C_{\lambda_1,\lambda_2}) \le c_2L_2 \log L_1 \qquad (5.11)$$

for $\lambda_1 = 0,\dots,L_1$; $\lambda_2 = 0,\dots,L_2$. This is the place where we use the new arguments of §2 concerning (G,C)-functions. Indeed, the bound

$$t(C_{\lambda_1,\lambda_2}) \le c_2L_2 \log L_1$$

is the main point in establishing the 'normal' measures of algebraic independence of π/ω , η/ω as announced above in 5.1:

$$\log|R(\frac{\pi}{\omega}, \frac{\eta}{\omega})| > C_0 \log H(R)d(R)^2\log^2(d(R) + 2)$$

for $R(x,y) \in \mathbb{Z}[x,y]$, $R \not\equiv 0$.

In order to get the bound for $t(C_\lambda)$, we make a change of variables:

$$w = \wp(z) \ , \quad f(w) = \zeta(z) - \frac{\eta}{\omega} z \ . \qquad (5.12)$$

Then $f(w)$ satisfies a differential equation

$$f'(w) = (-w - \frac{\eta}{\omega}) \sqrt{4w^3 - g_2 w - g_3} \ . \qquad (5.13)$$

The function $f(w)$ is a (G,C)-function. This means that the following condition is satisfied:

if $k \geq 1$, then

$$(\sqrt{4w^3 - g_2 w - g_3})^k f^{(k)}(w) = A_k(w)\sqrt{4w^3 - g_2 w - g_3} + B_k(w) \ , \qquad (5.14)$$

where

$$A_k(w) = A_{0,k}(w) + \frac{\eta}{\omega} A_{1,k}(w) \ , \ B_k(w) = B_{0,k}(w) + \frac{\eta}{\omega} B_{1,k}(w)$$

and $A_{0,k}(w)$, $A_{1,k}(w)$, $B_{0,k}(w)$, $B_{1,k}(w)$ are polynomials in w with coefficients from $\mathbb{Z}[g_2,g_3]$ of degrees (in g_2,g_3) at most $\gamma_1 k$.

Moreover, and this is most essential for applications, for $K \geq 1$ there exists an integer

$$D_K \in \mathbb{Z} \ ,$$

such that

$$|D_K| < \exp(\gamma_2 K) \qquad (5.15)$$

with γ_2 depending only on g_2,g_3, and for which the polynomials

$$D_K \cdot \frac{1}{k!} A_{0,k}(w), \ D_K \cdot \frac{1}{k!} A_{1,k}(w), \ D_K \cdot \frac{1}{k!} B_{0,k}(w),$$

$$D_K \cdot \frac{1}{k!} B_{1,k}(w) \quad : \quad k = 1,2,\dots,K \ ;$$

all have *integer* coefficients from the field $\mathbb{Q}(g_2,g_3)$.

We can now rewrite

$$F(z) = \sum_{\lambda_1=0}^{L_1} \sum_{\lambda_2=0}^{L_2} C_{\lambda_1,\lambda_2} (\zeta(z) - \frac{\eta}{\omega} z)^{\lambda_1} \wp(z)^{\lambda_2}$$

in terms of a w-variable:

$$F(z) = F_1(w) \ ,$$

where

$$F_1(w) = \sum_{\lambda_1=0}^{L_1} \sum_{\lambda_2=0}^{L_2} C_{\lambda_1,\lambda_2} f(w)^{\lambda_1} w^{\lambda_2} . \tag{5.16}$$

Then we have

$$F_1^{(k)}(w) = \sum_{\lambda_1=0}^{L_1} \sum_{\lambda_2=0}^{L_2} C_{\lambda_1,\lambda_2} \sum_{s=0}^{k} \binom{k}{s} \times$$

$$\times \lambda_2 \ldots (\lambda_2 - s + 1) w^{\lambda_2 - s} \{f(w)^{\lambda_1}\}^{(k-s)} .$$

Hence

$$\frac{1}{k!} F_1^{(k)}(w) = \sum_{\lambda_1=0}^{L_1} \sum_{\lambda_2=0}^{L_2} C_{\lambda_1,\lambda_2} \sum_{s=0}^{k} \binom{\lambda_2}{s} \times$$

$$\times w^{\lambda_2 - s} \frac{1}{(k-s)!} \{f(w)^{\lambda_1}\}^{(k-s)} . \tag{5.17}$$

Because of the (G,C)-function property of $f(w)$, there exists $\mathcal{D}_{L_1,k} \in \mathbb{Z}$ such that

$$|\mathcal{D}_{L_1,k}| < \exp(\gamma_3 K \log(L_1 + 1)) ,$$

and we have

$$\frac{\mathcal{D}_{L_1,k}}{(k-s)!} \{f(w)^{\lambda_1}\}^{(k-s)} = \sum_{j=0}^{\lambda_1} \sum_{i=0}^{1} f(w)^j \times$$

$$\times (4w^3 - g_2 w - g_3)^{i/2} . \mathcal{R}_{j,i,\lambda_1,s,k}(w) ,$$

with

$$\mathcal{R}_{j,i,\lambda_1,s,k}(w) = \mathcal{R}_{j,i,\lambda_1,s,k,0}(w) + \frac{\eta}{\omega} \mathcal{R}_{j,i,\lambda_1,s,k,1}(w) ,$$

where $\mathcal{R}_{j,i,\lambda_1,s,k,\chi}(w)$ is a polynomial of w with algebraic integer coefficients from $\mathbb{K}$ of degree at most $\gamma_4 k$, and of size at most $\gamma_5 (k \log(L_1 + 1) + L_1)$.

Hence the equations

$$\frac{1}{k!} F_1^{(k)}(w_0) = 0 : \quad k = 0,1,\ldots,K \tag{5.18}$$

can be substituted by

$$\sum_{\lambda_1=0}^{L_1} \sum_{\lambda_2=0}^{L_2} C_{\lambda_1,\lambda_2} \sum_{s=0}^{k} \binom{\lambda_2}{s} w_0^{\lambda_2-s} \times$$

$$\times \sum_{j=0}^{\lambda_1} \sum_{i=0}^{1} f(w_0)^j (4w_0^3-g_2w_0-g_3)^{i/2}.R_{j,i,\lambda_1,s,k}(w_0) = 0 :$$

$$k = 0,1,\ldots,K . \qquad (5.19)$$

Now we take a particular solution $f(w)$ of the differential equation (5.13) which corresponds to the following initial conditions:

$$w_0 = \wp(\frac{\omega}{4}) , \quad f(w_0) = m(\frac{\omega\eta' - \omega'\eta}{\omega}) \qquad (5.20)$$

for $m = 0,1,\ldots,M$, with $M = [c_4L_1]$ as above. We note that the differential equation (5.13) has a nontrivial monodromy, so a particular $f_0(w)$ defined in (5.20) can be considered to be one of the branches of $f(w)$.

Hence, we can determine polynomials $C_{\lambda_1,\lambda_2}(\pi/\omega , \eta/\omega)$ from the system of equations (5.18):

$$\frac{1}{k!} F_1^{(k)}(w) = 0$$

for $k = 0,1,\ldots,[c_3L_2]$ and $w = \wp(\frac{\omega}{4})$, $f(w) = m \frac{\omega\eta' - \omega'\eta}{\omega}$: $m = 0,1,\ldots,[c_4L_1]$. According to the previous discussion, this system of equations can be written in the usual form (5.19) as:

$$\sum_{\lambda_1=0}^{L_1} \sum_{\lambda_2=0}^{L_2} C_{\lambda_1,\lambda_2} (\frac{\pi}{\omega} , \frac{\eta}{\omega}) K_{\lambda_1,\lambda_2,k,m} (\frac{\pi}{\omega} , \frac{\eta}{\omega}) = 0 :$$

$$k = 0,1,\ldots,[c_3L_2]; \quad m = 0,1,\ldots,[c_4L_1] . \qquad (5.21)$$

Here $K_{\lambda_1,\lambda_2,k,m}(\pi/\omega , \eta/\omega)$ is a polynomial in π/ω and η/ω, of degrees $\le \lambda_1$ (in each of the variables) with algebraic coefficients from the field $\mathbb{K} = \mathbb{Q}(g_2,g_3,\wp(\omega/4),\wp'(\omega/4),i)$ of sizes at most

$$\gamma_6 (k \log(L_1 + 1) + L_2 + L_1 \log(m + 1)) .$$

We construct $C_{\lambda_1,\lambda_2}(\pi/\omega , \eta/\omega) \in \mathbb{Z}[\pi/\omega , \eta/\omega]$ using the Thue-Siegel lemma. One sees, however, that these $C_{\lambda_1,\lambda_2}(\pi/\omega, \eta/\omega)$ give the solution of the initial problem for $F(z)$. Indeed, by the definition of our change of variables we have

$$F_1^{(k)}(w_0) = 0 : \quad k = 0,1,\dots,[c_3L_2]$$

for $\quad (w_0, f(w_0)) = (\wp(\frac{\omega}{4}), m_0(\frac{\omega\eta' - \omega'\eta}{\omega}))$.

Hence, making the change of variables locally at w_0 , we get:

$$F^{(k)}(z_0) = 0 : \quad k = 0,1,\dots,[c_3L_2] \tag{5.22}$$

for $z_0 = \omega/4 + m_0\omega'$. The polynomials $C_{\lambda_1,\lambda_2}(\pi/\omega, \eta/\omega)$ are now the polynomials from $\mathbb{Z}[\pi/\omega, \eta/\omega]$, without the common factor and having bounds for height and degree:

$$d(C_{\lambda_1,\lambda_2}) \le L_1 ; \quad t(C_{\lambda_1,\lambda_2}) \le c_2 L_2 \log L_1 .$$

LEMMA 5.5'. *For* $L_2 \ge L_1$ *there exists a system of polynomials*

$$\{C_{\lambda_1,\lambda_2}(x,y) : \quad 0 \le \lambda_1 \le L_1,\ 0 \le \lambda_2 \le L_2\}$$

from $\mathbb{Z}[x,y]$ *, without a common factor, which satisfies all the conditions of Lemma 5.5, but the second inequality in* (5.6) *is replaced by a sharp one:*

$$t(C_{\lambda_1,\lambda_2}) \le c_2L_2 \log L_1 : \quad 0 \le \lambda_1 \le L_1,\ 0 \le \lambda_2 \le L_2 . \tag{5.23}$$

We apply Lemma 5.10 to the function $F(z)$ from Lemma 5.5'. In this case we obtain immediately the usual statement:

LEMMA 5.24. *For every* $L_2 \ge L_1 > 1$ *we have one of the following two possibilities: either*

i) there exists a polynomial $P_{L_1,L_2}(x,y) \in \mathbb{Z}[x,y]$ *such that*

$$d(P_{L_1,L_2}) \le c_{12}L_1 ; \quad t(P_{L_1,L_2}) \le c_{12}L_2 \log L_1 ;$$

$$-c_{14}L_1^2L_2 < \log|P_{L_1,L_2}(\frac{\pi}{\omega}, \frac{\eta}{\omega})| < -c_{13}L_1^2L_2 ;$$

or ii) there exists a system of polynomials $C_\ell(x,y) \in \mathbb{Z}[x,y]$: $\ell \in L_{L_1,L_2}$ *without a common (non-constant) factor such that*

$$d(C_\ell) \le c_{15}L_1 ; \quad t(C_\ell) \le c_{16}L_2 \log L_1 ,$$

$$\log|C_\ell(\frac{\pi}{\omega}, \frac{\eta}{\omega})| < -c_{17}L_1^2L_2$$

for all $\ell \in L_{L_1,L_2}$.

We use these polynomials to bound the measure of the algebraic independence of π/ω , η/ω . For example, if we want to bound only the type of the transcendence of π/ω , η/ω , we can use polynomials only for $L_1 = L_2$.

Let us assume that the polynomial $R(x,y) \in \mathbb{Z}[x,y]$ is irreducible over $\mathbb{Z}$ (other cases follow looking at the factorization) and

$$\log|R(\frac{\pi}{\omega}, \frac{\eta}{\omega})| = -t\phi(d) , \tag{5.25}$$

where $t = \log H(R) + d(R) \log d(R)$, $d = d(R)$, where $d \geq d_0$ is sufficiently large and $\phi(d) > Cd^2\log^2 d$ for a sufficiently large $C > 0$.

First of all, we need a zero-dimensional set, approximating $\overline{\theta} = (\pi/\omega , \eta/\omega)$:

LEMMA 5.26. *There exists a polynomial* $P(x,y) \in \mathbb{Z}[x,y]$ *relatively prime to* $R(x,y)$ *such that*

$$d(P) \leq c_{18}d^{1/3} \phi(d)^{1/3} \log^{1/3} d ;$$

$$t(P) \leq c_{19}td^{-2/3} \phi(d)^{1/3} \log^{1/3} d ; \tag{5.27}$$

$$\log|P(\frac{\pi}{\omega}, \frac{\eta}{\omega})| < -c_{20}t\phi(d) .$$

<u>Proof</u> We take a pair (L_1^0,L_2^0) with $L_2^0 \geq L_1^0$ such that

$$L_1^0 = [c_{21}(d\phi(d)\log d)^{1/3}] ,$$
$$L_2^0 = [c_{22}td^{-2/3}\phi(d)^{1/3}\log^{-2/3}d] \quad (\geq L_1^0) \tag{5.28}$$

and

$$c_{14}(L_1^0)^2L_2^0 < \frac{1}{2}\, t\phi(d) . \tag{5.29}$$

If the first case i) of Lemma 5.24 is true, then we have a polynomial $P_{L_1^0,L_2^0}(x,y) \in \mathbb{Z}[x,y]$ such that

$$d(P_{L_1^0,L_2^0}) \le c_{14}L_1^0 \ ; \quad t(P_{L_1^0,L_2^0}) \le c_{14}L_2^0 \log L_1^0$$

and

$$-c_{14}(L_1^0)^2L_2^0 < \log|P_{L_1^0,L_2^0}(\bar{\theta})| < -c_{13}(L_1^0)^2L_2^0 \ . \tag{5.30}$$

Then $P_{L_1^0,L_2^0}(x,y)$ is relatively prime to $R(x,y)$. Indeed, otherwise $R(x,y)|P_{L_1^0,L_2^0}(x,y)$. We see that in this case, the lower bound for $|P_{L_1^0,L_2^0}(\bar{\theta})|$ in (5.30) contradicts the bound for $|R(\bar{\theta})|$ and the definition of (L_1^0,L_2^0) in (5.29). Thus, in the case i) of Lemma 5.24 we put $P(x,y) = P_{L_1^0,L_2^0}(x,y)$.

In the case ii) of Lemma 5.24, corresponding to (L_1^0,L_2^0), we take a polynomial $C_{\bar{\lambda}_0}(x,y) : \bar{\lambda} \in L_{L_1^0,L_2^0}$ relatively prime to $R(x,y)$. Such a polynomial $C_{\bar{\lambda}_0}(x,y)$ exists since $R(x,y)$ is irreducible. Then according to the choice of (L_1^0,L_2^0) in (5,28),(5,29), the polynomial $P(x,y) \overset{\text{def}}{=} C_{\bar{\lambda}_0}(x,y)$ satisfies all the requirements of Lemma 5.26.

Using Lemma 5.26, we will later approximate $\bar{\theta}$ by elements of an irreducible component of the zero set S of $P(x,y),R(x,y)$:

$$S = \{(x_0,y_0) \in \mathbb{C}^2 : \ P(x_0,y_0) = R(x_0,y_0) = 0\} \ . \tag{5.31}$$

The main auxiliary result we use here is the statement that $\bar{\theta}$ is 'approximated' by some component S_0 of S. This component S_0 comes from the ideal decomposition

$$(P,R) = \mathcal{P}_1 \ldots \mathcal{P}_k$$

into primary components, and S_0 corresponds to the component $\mathcal{P}_{i_0}$ which is 'smallest' at the point $\bar{\theta}$ $(= (\pi/\omega \ , \eta/\omega))$.

However, it is much more convenient to use the geometric rather than algebraic language of ideals; thus we speak of irreducible varieties in place of primary ideals, and there is a notion of distance. For example one can use the Puiseux expansion to imagine how close a component can lie to a point $\bar{\theta}$.

Our arguments are completely general and are applied to a polynomial in n variables.

For this purpose we devote the next few pages to the

description of the type properties of zero-dimensional sets and to the generalization of the Liouville theorem.

§6. Let us formulate a version of the Liouville theorem in the case of an arbitrary set $S \subset \mathbb{C}^n$ of algebraic numbers of dimension zero, i.e. a set of common zeros of a zero-dimensional ideal in $\mathbb{Z}[x_1,\dots,x_n]$. Namely, we consider the following situation. We have an ideal I in $\mathbb{Z}[x_1,\dots,x_n]$, which is zero-dimensional in the sense that the set

$$S(I) = \{\vec{x}_0 \in \mathbb{C}^n : P(\vec{x}_0) = 0 \quad \text{for every} \quad P \in I\}$$

is a finite set. We are working now in the affine situation since projective considerations do not add anything. Every element of $S(I)$ has a prescribed multiplicity, defined for example in Schafarevitch's book [37].

Let $P_1,\dots,P_k$ be certain generators of I, whose degrees and types we know:

$$d(P_i) \le D_i , \quad t(P_i) \le T_i : \quad i = 1,\dots,k .$$

Naturally, $k \ge n$. By intersection theory (say, Bezout theorem) we have

$$|S(I)| \le D_1 \dots D_k .$$

This bound is far from optimal whenever $k > n$. In this case we can use even the following bound

$$|S(I)| \le (\max_{i=1,\dots,k} D_i)^n .$$

Instead of treating different cases, we assume already that

$$P_1(x),\dots,P_n(x)$$

have only finitely many common zeros. Then we consider an ideal $\mathrm{I} = (P_1,\dots,P_n)$ and $S(\mathrm{I})$ instead of $S(I)$. Let $\vec{x}_0 \in S(\mathrm{I})$ and $m(\vec{x}_0)$ be a multiplicity of $\vec{x}_0$ in $S(\mathrm{I})$. Then we have

$$\sum_{\vec{x}_0 \in S(\mathrm{I})} m(\vec{x}_0) \quad D_1 \dots D_n .$$

We can apply to I the theory of u-resultants in the form of Kronecker (cf. Van der Waerden [43] or Hodge-Pedoe [27]). One gets the following main statement:

LEMMA 6.1. *With the above notation, the coordinates of common zeros* $\vec{x}_0 \in S(I)$ *are bounded above in terms of* $D_1,\dots,D_n$ *and* $T_1,\dots,T_n$. *Namely, let*

$$\vec{x}_0 = (x_{10},\dots,x_{n0}) \quad \text{for} \quad \vec{x}_0 \in S(I) .$$

Then for every $i = 1,\dots,n$ *there is a rational integer* A_i, $A_i \neq 0$, *such that for any distinct elements* $\vec{x}^1,\dots,\vec{x}^\ell$ *of* $S(I)$ *and* $n_i \le m(\vec{x}^i)$: $i = 1,\dots,\ell$, *the number*

$$A_i \prod_{j=1}^{\ell} (\vec{x}^j)_i^{n_j}$$

is an algebraic integer. Moreover, for any $i = 1,\dots,n$ *we have*

$$|A_i| \prod_{\vec{x}\in S(I)} \max\{1,(\vec{x})_i\}^{m(\vec{x})} \le \exp(C_1 \sum_{r=1}^{n} T_r \prod_{s\neq r, s=1}^{n} D_s) , \qquad (6.2)$$

for a constant $C_1 > 0$ *depending only on* n.

The analogue of the Liouville theorem applied to the elements of the set $S(I)$ has the following form:

LEMMA 6.3. *Let, as before,* $I = (P_1,\dots,P_n)$ *where* $d(P_i) \le D_i$, $t(P_i) \le T_i$: $i = 1,\dots,n$ *and the set* $S(I)$ *is a finite one.*

Let $R(x_1,\dots,x_n) \in \mathbb{Z}[x_1,\dots,x_n]$, $R \not\equiv 0$. *Let us assume that for several distinct* $\vec{x}^1,\dots,\vec{x}^\ell$ *from* $S(I)$ *of multiplicities* $m_1,\dots,m_\ell$, *respectively, we have*

$$R(\vec{x}^j) \neq 0 : \quad j = 1,\dots,\ell .$$

Then we have the following lower bound:

$$\prod_{j=1}^{\ell} |R(\vec{x}^j)|^{m_j} \ge \exp\{-C_2\{\sum_{i=1}^{n} t(P_i)d(R) \prod_{s\neq i} d(P_i) + t(R)d(P_1) \dots d(P_n)\}\} , \qquad (6.4)$$

where $C_2 > 0$ *depends only on* n.

Proof of Lemma 6.3 (Compare also the proof of Lemma 4.1 (p. 17) of [11a] in the case $n = 2$.) We consider the following auxiliary object using the notation of Lemma 6.1:

$$A = \prod_{i=1}^{n} A_i^{d(R)} \prod_{\substack{\vec{x}\in S(I)\\ R(\vec{x})\neq 0}} R(\vec{x})^{m(\vec{x})} , \qquad (6.5)$$

in (6.5) the product is over only those elements $\vec{x}$ of $S(I)$ for which $R(\vec{x}) \neq 0$. This is a usual 'semi-norm' [14]. Then by the definition of the set $S(I)$ (invariant under the algebraic conjugation) and the choice of $A_i: i = 1,\dots,n$ in Lemma 6.1 we get: $A \in \mathbb{Z}$. From the form of A it follows that $A \neq 0$, so that

$$|A| \geq 1 . \qquad (6.6)$$

We represent A as a product of two factors: $A = \mathfrak{A}.\mathcal{B}$, where $\mathfrak{A}$ is the product in the left hand side of (6.4). The product $\mathcal{B}$ is bounded from above by Lemma 6.1:

$$|\mathcal{B}| \leq (\prod_{i=1}^{n} A_i \prod_{\substack{\vec{x}\in S(I)\\ \vec{x}\notin\{\vec{x}^j:j=1,\dots,\ell\}\\ R(x)\neq 0}} \max(1,|(\vec{x})_i|)^{m(\vec{x})})^{d(R)} \exp(2t(R)\sum_{i=1}^{n} d(P_i)) \leq$$

$$\leq \exp(C_4\{t(R)\prod_{i=1}^{n} d(P_i) + d(R)\sum_{i=1}^{n} t(P_i)\prod_{s\neq i} d(P_s)\}) . \qquad (6.7)$$

Combining (6.6) and (6.7) one gets (6.4).

In order to prove our results in a straightforward way, we make some agreements on the notations, that will simplify our symbolic mess.

If we start, in the general case, with the ideal $I = (P_1,\dots,P_n)$ in $\mathbb{Z}[x_1,\dots,x_n]$ of the dimension zero, then the set

$$S(I) = \{\vec{x}_0 \in \mathbb{C}^n : P_i(\vec{x}_0) = 0 : i = 1,\dots,n\}$$

is a set of vectors in $\mathbb{C}^n$ with algebraic coordinates. The set $S(I)$ is naturally divided into components S_α closed under the conjugation in $\mathbb{Q}$

$$S(I) = \bigcup_{\alpha\in A} S_\alpha . \qquad (6.8)$$

The partition of $S(I)$ into sets S_α can be made in such a way that all elements of S_α have the same multiplicity m_α of its occurring in $S(I)$. We call S_α an irreducible component of $S(I)$ and we have

$$\sum_{\alpha\in A} m_\alpha |S_\alpha| \le d(P_1) \ldots d(P_n) . \tag{6.9}$$

We can define a type and degree of the component S_α in (6.8). The degree is, naturally, $|S_\alpha|$ itself and the type is defined using the sizes of the coordinates of elements of $S(I)$.

For this we remind the reader of Lemma 6.1, where we had non-zero rational integers $A_i : i = 1,\ldots,n$, such that

$$A_i \prod_{\vec{x}\in S'} (\vec{x})_i^{n(\vec{x})}$$

is an algebraic integer for any $S' \subseteq S(I)$ and $n(\vec{x}) \le m(\vec{x}) : \vec{x} \in S'$; and we obtained a bound

$$|A_i| \prod_{\vec{x}\in S(I)} \max\{1, |(\vec{x})_i|^{m(\vec{x})}\} \le$$

$$\le \exp\{C_1 \sum_{j=1}^{n} t(P_j) \prod_{s\ne j} d(P_s)\} : i = 1,\ldots,n . \tag{6.10}$$

Naturally, the quantity

$$\log\{\prod_{i=1}^{n} |A_i| \prod_{\vec{x}\in S(I)} \max\{1, |(\vec{x})_i|\}^{m(\vec{x})}\} \tag{6.11}$$

can be called ***the size*** of $S(I)$. We can define in a similar way the size of the component S_α as

$$\log\{\prod_{i=1}^{n} |a_i^\alpha| \prod_{\vec{x}\in S_\alpha} \max\{1, |(\vec{x})_i|\}\} , \tag{6.12}$$

where a_i^α are smallest non-zero rational integers such that

$$a_i^\alpha \prod_{\vec{x}\in S_1} (\vec{x})_i$$

are algebraic integers for any $S_1 \subseteq S_\alpha : i = 1,\ldots,n$.

By the ***type*** of the set S_α we understand the sum of the degree $|S_\alpha|$ and its size. We denote the type of S_α by $t(S_\alpha)$. Similarly one can define a type of $S(I)$ as a sum of its degree and size. The type of $S(I)$ is also denoted by $t(S(I))$. We note that the degree of $S(I)$ is not $|S(I)|$ but rather $\sum_{\vec{x}\in S(I)} m(\vec{x})$, when elements of $S(I)$ are counted with multiplicities.

By the definition of types we have

$$\sum_{\alpha\in A} t(S_\alpha)\, m_\alpha \le t(S(I)) \le \exp\{C_2 \sum_{j=1}^{n} t(P_j) \prod_{s\ne j} d(P_s)\} . \tag{6.13}$$

We remark that our decomposition of $S(I)$ into the union of (disjoint) irreducible components is not unique. It is easier to work with our definition, one can define a canonical decomposition of $\mathcal{S}(I)$ into the maximal union of irreducible components over $\mathbb{Z}$.

The case $n = 2$ is very easy to understand using resultants. In this case we have two relatively prime polynomials $P(x,y)$, $Q(x,y) \in \mathbb{Z}[x,y]$ and the set S of their common zeros

$$S = \{(x,y) \in \mathbb{C}^2 : P(x,y) = Q(x,y) = 0\} .$$

Their coordinates can be determined using the resultants of $P(x,y)$, $Q(x,y)$. We make a change of the coordinates to a 'normal' [11a] form and get new polynomials $P'(x',y')$, $Q'(x',y')$. In 'normal' coordinates defined in [11] distinct elements of S have both their coordinates distinct.

Then resultants $R_1(x')$ and $R_2(y')$ of $P'(x',y'),Q'(x',y')$ (in 'normal' coordinates [11a]) with respect to y' and x', respectively, can be written as

$$\begin{aligned} R_1(x') &= a_1 \prod_{i=1}^{k} (x' - \zeta'_{1,i})^{m_i} ; \\ R_2(y') &= a_2 \prod_{i=1}^{k} (y' - \zeta'_{2,i})^{m_i} ; \end{aligned} \tag{6.14}$$

where $(\zeta'_{1,i},\zeta'_{2,i})$ is an element of S of the multiplicity m_i. In particular, one can represent $R_1(x')$, $R_2(y')$ in terms of the powers of irreducible polynomials:

$$R_1(x') = \prod_{\alpha\in A} P^1_\alpha(x')^{m_\alpha} ; \quad R_2(y') = \prod_{\alpha\in A} P^2_\alpha(y')^{m_\alpha} ,$$

where $P^1_\alpha(x') = a^\alpha_1 \prod_j (x' - \zeta^\alpha_{1j})$ and $P^2_\alpha(y') = a^\alpha_2 \prod_j (y' - \zeta^\alpha_{2j})$ and $(\zeta^\alpha_{1j} , \zeta^\alpha_{2j})$ are elements of S. We naturally define

$$S_\alpha = \{(\zeta^\alpha_{1j} , \zeta^\alpha_{2j})\} ,$$

so that $\bigcup_{\alpha\in A} S_\alpha = S$. We can now look on the type of S and S_α: $\alpha \in A$ from the point of view of resultants written in (6.14).

For example one can define a size of S as the sum sizes of R_1 and R_2. This will be equivalent to the previous definition of the size in (6.11)-(6.12).

Consequently we can say

$$t(S) \le t(R_1) + t(R_2) ,$$

while for $\alpha \in A$,

$$t(S_\alpha) \le t(P_\alpha^1) + t(P_\alpha^2) .$$

This definition of the type, expressed not in the coordinate form, is, as a matter of fact, more useful in the higher-dimensional case, when we are working with mixed ideals; non-normal intersections etc.

The notion of the component of a zero-dimensional set already introduced is used to remove the main difficulty in the proof. Namely, we are dealing with the situation when $S = S(I)$ is a zero-dimensional set for an ideal $I = (P,Q)$ in $\mathbb{Z}[x,y]$ and the algebraic manifold I is 'close' to the point $\bar{\theta} \in \mathbb{C}^2$ in the sense that $|P(\bar{\theta})| < \varepsilon$, $|Q(\bar{\theta})| < \varepsilon$. For the proof it is necessary to substitute this notion of 'closeness' by an estimate of the distance from $\bar{\theta}$ to elements of S (in the usual, say, L_1 metric) in terms of ε and $t(P)$, $t(Q)$. The best possibility would be to have a single element of S close to $\bar{\theta}$ (as is usual for one-dimensional proofs, see [29]). However, the rigorous proof of the existence of a single element of S, approximating $\bar{\theta}$ is non-elementary and requires the use of generalized Jacobians (for $\mathbb{Z}[x_1,\dots,x_n]$ at $n = 2$) or resolutions of singularities (for $n > 2$). Instead we use the elementary method and we are able to prove only the existence of a single component S_α of S approximating $\bar{\theta}$ in the sense that $\prod_{\bar{\zeta} \in S_\alpha} \|\bar{\theta} - \bar{\zeta}\| \sim \varepsilon$ (see the discussion in [11a]). Surprisingly, with this weak statement we are able to finish the proof with the same length of exposition as with the more sharp assertion (cf. [11b]). The main tool is the Liouville theorem in the form of Lemma 6.3 above.

Hence we devote special attention to a precise formulation of the result of an approximation of $\bar{\theta}$ by a component S_α of $S(I)$. We follow arguments from [11a],[11b] and give the corresponding result for zero-dimensional sets in $\mathbb{C}^2$. The same proof works without any changes for arbitrary zero-dimensional sets in $\mathbb{C}^n$. In the proof we use two elementary lemmas from [11] which we include in the proof to make it self-contained. The reader should excuse us for the lengthy formulation of Proposition 6.15 but this way we have a single auxiliary result for further references.

The main auxiliary result is the following:

<u>Proposition 6.15</u> Let $\bar{\theta} = (\theta_1,\theta_2) \in \mathbb{C}^2$ (with the ℓ^1-norm) and $P(x,y)$, $Q(x,y)$ be two relatively prime polynomials from $\mathbb{Z}[x,y]$.

Let $S = S(P,Q)$ be the set of the zeros of an ideal (P,Q) and

$$S = \bigcup_{\alpha\in A}' S_\alpha$$

its representation through irreducible components. If m_α is a multiplicity of S_α, then one has

$$\sum_{\alpha\in A} m_\alpha t(S_\alpha) \le t(S) \le 4(d(P)t(Q) + d(Q)t(P)) \le 8t(P)t(Q) .$$

Let us now assume that

$$\max\{|P(\overline{\theta})| , |Q(\overline{\theta})|\} \le \exp(-E) \tag{6.16}$$

for $E > 0$. One can define the logarithmic distance $E(\overline{\zeta})$ of $\overline{\zeta} \in S$ to $\overline{\theta}$ as

$$\|\overline{\theta} - \overline{\zeta}\| \le \exp(-E(\overline{\zeta})) . \tag{6.17}$$

Let us put

$$E(S,\overline{\theta}) = \sum_{\overline{\zeta}\in S, E(\overline{\zeta})>0} E(\overline{\zeta}) \quad \text{and} \quad E(S_\alpha,\overline{\theta}) = \sum_{\overline{\zeta}\in S_\alpha, E(\overline{\zeta})>0} E(\overline{\zeta}) \tag{6.18}$$

as the definition of (minus logarithm of) distance from $\overline{\theta}$ to S or S_α.

In order to express relations between E, $E(S,\overline{\theta})$ and $E(S_\alpha,\overline{\theta})$ we put

$$T = \gamma_0 (d(P)t(Q) + d(Q)t(P) + d(P)d(Q)\log(d(P)d(Q) + 2)) \tag{6.19}$$

for an absolute constant $\gamma_0 > 0$ $(\gamma_0 \le 4)$ such that

$$t(S) \le T .$$

Similarly, for every $\alpha \in A$ there is a bound T_α of $t(S_\alpha)$ of the form

$$t(S_\alpha) \le T_\alpha \le t(S_\alpha) + d(S_\alpha)\log(d(P)d(Q) + 2) . \tag{6.20}$$

In terms of T_α we can formulate results on $E(S,\overline{\theta})$, $E(S_\alpha,\overline{\theta})$. If we assume $E \ge 4T$, then

$$E(S,\overline{\theta}) = \sum_{\alpha\in A} m_\alpha E(S_\alpha,\overline{\theta}) \ge E - 2T . \tag{6.21}$$

We denote the element of S_α nearest to $\overline{\theta}$ by $\overline{\xi}_\alpha$,

$E(\overline{\xi}_\alpha) = \min_{\overline{\zeta}\in S_\alpha} E(\overline{\zeta})$. Then we have

$$E(\overline{\xi}_\alpha) \geq \min\{c_5 E(S_\alpha,\overline{\theta}) , c_5 \frac{E(S_\alpha,\overline{\theta})^2}{d(S_\alpha) T_\alpha}\} \tag{6.22}$$

and, for other $\overline{\zeta} \in S_\alpha$ close to $\overline{\theta}$ we have

$$\sum_{\overline{\xi}\in S_\alpha, E(\overline{\xi})\geq B_\alpha} E(\overline{\zeta}) \geq \frac{E(S_\alpha,\overline{\theta})}{4} \tag{6.23}$$

where

$$B_\alpha = c_6 E(S_\alpha,\overline{\theta})^{3/2} (d(S_\alpha)T_\alpha^{1/2})^{-1} \tag{6.24}$$

for $\alpha \in A$.

As an application of these bounds we have the following result, where $d(S_\alpha)$ is replaced by its upper bound T_α . We remark that, in the addition to (6.20),

$$\sum_{\alpha\in A} m_\alpha T_\alpha \leq T . \tag{6.25}$$

In particular, there exists $\alpha_0 \in A$ such that

$$\frac{E(S_{\alpha_0},\overline{\theta})}{T_{\alpha_0}} \geq \frac{E(S,\overline{\theta})}{T} \geq \frac{E}{T} - 2 . \tag{6.26}$$

Under the conditions (6.26) one has as a corollary of (6.22)-(6.24):

$$E(\overline{\xi}_{\alpha_0}) \geq \min\{c_5 E(S_{\alpha_0},\overline{\theta}), c_5 \frac{E(S_{\alpha_0},\overline{\theta})^2}{T_{\alpha_0}^2}\} ;$$

$$\sum\{E(\overline{\xi}): \overline{\xi} \in S_{\alpha_0}, E(\overline{\xi}) \geq c_6 (E(S_{\alpha_0},\overline{\theta})/T_{\alpha_0})^{3/2}\} \geq E(S_{\alpha_0},\overline{\theta})/4 . \tag{6.27}$$

Proof of Proposition 6.15 First of all we must change the system of the coordinates to a 'normal' one with respect to a system of polynomials $P(x,y)$, $Q(x,y)$. We use for this Lemma 2.1 [11b] and 3.2 [11a]. According to this lemma there is a *nonsingular* transformation

$$(\pi) \begin{cases} x = x'a + y'c \\ y = x'b + y'd \end{cases}$$

for rational integers a , b , c , d such that

$$\max(|a|,|b|,|c|,|d|) \leq M \leq \gamma_1 d(P)d(Q) , \tag{6.28}$$

and which are normal with respect to $P(x,y)$, $Q(x,y)$. From our point of view, it means, first of all that for $\bar{\theta}$ written in new coordinates (x',y') as $\bar{\theta}' = (\theta_1',\theta_2')$ and for *any* common zero $\bar{\zeta}$ of $P(x,y) = 0$, $Q(x,y) = 0$, i.e. element of S, we have in new coordinates $\bar{\zeta}' = (\zeta_1',\zeta_2')$

$$||\bar{\theta}' - \bar{\zeta}'||_{\ell_1} \leq 4M^2 \min\{|\theta_1' - \zeta_1'|, |\theta_2' - \zeta_2'|\} . \tag{6.29}$$

The property (6.29) is the central one that we use from all of the 'normality' properties.

In new, 'normal' variables, we consider the resultant $R(x')$ of $P'(x',y')$ ($\equiv P(x,y)$) and $Q'(x',y')$ ($\equiv Q(x,y)$) taken with respect to y'. The polynomial $R(x')$ is a polynomial of the degree $\leq d(P)d(Q)$, but the type of $R(x')$ might be slightly higher than that of $R(x)$. This explains why we change the type $t(S)$ to a slightly higher quantity T defined in (6.19). Namely, we take S written in new variables (x',y') as a set S'. Our initial definition of the type was certainly 'coordinate-dependent' and $t(S')$ may be different from $t(S)$. In order to get the bound of $t(S')$ one can take, together with $R(x')$, the polynomial $R_2(y')$, being the resultant of $P'(x',y')$, $Q'(x',y')$ with respect to x'. Then one can define $T = t(S')$ as

$$t(S') = t(R) + t(R_2) .$$

This quantity is bounded by the number T from (6.19):

$$t(R) + t(R_2) \leq \gamma_0\{d(P)t(Q) + d(Q)t(P) +$$

$$+ d(P)d(Q) \log(d(P)d(Q) + 2)\} = T .$$

The term $\log(d(P)d(Q) + 2)$ appears in (6.19) since we make the transformation (π) which changes the type of the polynomials as in (6.28).

This factor can be removed however from all the statements of 6.15, by a more tedious computation. For example it is rather easy to get the property (6.22) without studying the transformed resultants, working only with $\mathrm{Res}_y(P,Q)$ and $\mathrm{Res}_x(P,Q)$.

We prefer however to work with T and T_α because the proof is straightforward though some estimates suffer.

Irreducible components $S_\alpha' = \pi^{-1}(S_\alpha)$ (we are writing now in transformed coordinates) are connected with irreducible components of $R(x')$. Namely, we have

$$R(x') = \prod_{\alpha\in A} P_\alpha(x')^{m_\alpha}$$

where $P_\alpha(x')$ is an irreducible polynomial from $\mathbb{Z}[x']$. Here, zeros of $P_\alpha(x')$ are exactly the x'-projections of the set $S'_\alpha : \alpha \in A$.

Hence we can work with $P_\alpha(x')$ rather than with $S'_\alpha = \pi^{-1}(S_\alpha)$. According to (6.29) it is enough to bound above $|\zeta'_1 - \theta'_1|$ in order to bound $\|\bar{\zeta}' - \bar{\theta}'\|$ with $\bar{\zeta}' = (\zeta'_1, \zeta'_2) \in S'$. Here and everywhere in the proof of Proposition 6.15, $\bar{\zeta}' = \pi^{-1}(\bar{\zeta})$, $\bar{\theta}' = \pi^{-1}(\bar{\theta})$.

First of all we can evaluate $\log|R(\theta'_1)|$ in terms of $E \le -\min\{|\log|P(\bar{\theta})||, \log|Q(\bar{\theta})|\}$. For this we use the property of the resultants in the classical form. Namely, we can use the formula [26]:

$$|\mathrm{Res}_x(p,q)| \le \{d(p)H^+(q)|q(x_0)| + d(q)H^+(p)|p(x_0)|\} \times$$
$$\times H^+(q)^{d(p)-1} H^+(p)^{d(q)-1}$$

for arbitrary polynomials $p(x), q(x) \in \mathbb{C}[x]$ and $H^+(p) = \max\{1, H(p)\}$. We put $p(x') = P'(\theta'_1, x')$, $q(x') = Q'(\theta'_1, x')$ and $x_0 = \theta'_2$.

We obtain this way the inequality:

$$\log|R'(\theta'_1)| \le -E + T. \tag{6.30}$$

In particular, writing $R'(x)$ in the form

$$R'(x') = a \prod_{\alpha\in A} (x' - \zeta'^{\alpha}_1)^{m_\alpha} = a \prod_{\bar{\zeta}'\in S'} (x' - (\bar{\zeta}')_1)$$

for $a \in \mathbb{Z}$, and using (6.29) one immediately obtains (6.21).

In order to get the other statements (6.22)-(6.24) we simply use the irreducible polynomial $P_\alpha(x')$, whose zeros are x -projections of S' and apply to the polynomial $P_\alpha(x')$ the following lemma, taken from [11a],[11b]:

LEMMA 6.31 [11a],[11b]. *Let* $P(x) \in \mathbb{Z}[x]$ *be an irreducible polynomial over* $\mathbb{Q}$ *and*

$$R(x) = a \prod_{i=1}^{d(R)} (x - \eta_i).$$

Let us assume that $\theta \in \mathbb{C}$ *and*

$$|R(\theta)| \le \exp(-E). \tag{6.32}$$

We arrange zeros in an order such that

$$|\theta - \eta_i| < 1 \quad \text{for} \quad i = 1,\dots,r ,$$

and put

$$|\theta - \eta_i| = \exp(-e_i) \ : \ i = 1,\dots,r \tag{6.33}$$

and we assume

$$e_1 \geq \dots \geq e_r > 0 .$$

We have then:

$$\sum_{i=1}^{r} e_i \geq E \ ; \quad \sum_{i=1}^{r} (i-1)e_i \leq 8d(P)t(P) . \tag{6.34}$$

If now, $E \geq c_7 d(P)t(P)$ *, then*

$$e_1 \geq c_8 E \quad \textit{for some} \quad c_8 = c_8(c_7) > 0 . \tag{6.35}$$

If, however, $E < c_7 d(P)t(P)$ *for a sufficiently small* $c_7 > 0$ *, then we have*

$$E > c_9 E^2/d(P)t(P) \ : \ c_9 > 0 \tag{6.36}$$

and moreover,

$$\sum_{\substack{j=1 \\ e_j \geq B}}^{r} e_j \geq E/4 \quad \text{for} \quad B = c_{10}E^{3/2}(d(P)t(P)^{1/2})^{-1} . \tag{6.37}$$

Remark Here, in order to exclude trivial cases, we assume $E \geq c_{11}t(P)$.

Proof We define a resultant $\Delta = \Delta(P)$ of $P(x)$:

$$\Delta = a^{2d(P)} \prod_{\substack{i,j=1 \\ i \neq j}}^{d(P)} (\eta_i - \eta_j) . \tag{6.38}$$

Then $\Delta \in \mathbb{Z}$ and $\Delta \neq 0$, so that $|\Delta| \geq 1$.

From (6.32) we immediately obtain

$$\sum_{i=1}^{r} e_i \geq E . \tag{6.39}$$

Now keeping in mind the bound

$$|a| \prod_{i=1}^{d(P)} \max\{1, |\eta_i|\} \le (d(P) + 1)H(P) ,$$

from the definition of the resultant $\Delta(P)$ we obtain

$$- \frac{\log|\Delta|}{2} + \sum_{i=1}^{r} e_i(i - 1) \le T_1 = 8d(P)t(P) . \qquad (6.40)$$

We deduce (6.34) from (6.39),(6.40). The bound (6.36) follows from (6.39)-(6.40) as in Lemma 2.3 [11a].

In order to prove (6.37) we put

$$J_C = \{j = 1,\dots,r: e_j < C\} .$$

Let

$$A = E/8d(P) .$$

If

$$\sum_{\substack{j=1 \\ j \notin J_A}}^{r} e_j \ge 3/4\ E \qquad (6.41)$$

is not true, then

$$\sum_{\substack{j=1 \\ j \in J_A}}^{r} e_j \ge 1/4\ E .$$

Then

$$|J_A|\ A \ge \sum_{\substack{j=1 \\ j \in J_A}}^{r} e_j \ge 1/4\ E .$$

But

$$|J_A| \le r \le d(P) ,$$

i.e.

$$A \ge E_1/4d(P) ,$$

which is impossible. We define B as in (6.37):

$$B = c_8 E^{3/2}(d(P)t(P)^{1/2})^{-1}$$

for some small constant $c_8 > 0$. Then $B \geq A$ according to the previous remark.

Now we want to show (6.37). If (6.37) is not true, then by (6.41),

$$\sum_{\substack{i=1 \\ j \in J_B \setminus J_A}}^{r} e_j \geq E/2 . \tag{6.42}$$

But according to (6.34) we have

$$|J_B \setminus J_A|^2 A \leq c_{12} d(P) t(P) .$$

On the other hand, by (6.42),

$$|J_B \setminus J_A| B \geq E/2 .$$

Thus

$$B \geq \frac{E^{3/2}}{16 d(P) t(P)} .$$

which contradicts the choice of B with $c_{10} < 1/16$. Q.E.D.

Now at last we derive the statements (6.26)-(6.27). Indeed, if (6.26) are false for every $\alpha_0 \in A$, we get

$$\sum_{\alpha \in A} E(S_\alpha, \overline{\theta}) m_\alpha T < \sum_{\alpha \in A} T_\alpha m_\alpha E(S, \overline{\theta}) ,$$

which contradicts (6.25). Results (6.27) are a consequence of (6.22)-(6.24). The Proposition 6.15 is proved.

<u>Remark 6.43</u> In (6.27), the quantity $E(\overline{\xi}_{\alpha_0}) \leq -\log \|\overline{\theta} - \overline{\xi}_{\alpha_0}\|$ is bounded below by $c_5 \min\{E(S_{\alpha_0}, \overline{\theta}), (E(S_{\alpha_0}, \overline{\theta})/T_{\alpha_0})^2\}$. Of these two terms, 'usually' (in practice), $(E(S_{\alpha_0}, \overline{\theta})/T_{\alpha_0})^2$ is the smaller one. Indeed, if $E(S_{\alpha_0}, \overline{\theta})$ is large then (6.27) shows simply that 'most' of the measure $E(S_{\alpha_0}, \overline{\theta})$ is concentrated in a single term of it: $E(\overline{\xi}_{\alpha_0})$: $E(\overline{\xi}_{\alpha_0}) \geq c_5 E(S_{\alpha_0}, \overline{\theta})$.

However, one must take even this possibility into consideration.

§7. We return now to the set S of common zeros of $P(x,y)$, $R(x,y)$ constructed in §5, in Lemma 5.26. First of all we state the bound for the degree and size of the set S as defined in (5.31) and then we use Proposition 6.15 to obtain an irreducible component S_0 of S approximating $\overline{\theta}$.

The set S has the degree $d(S)$ (the number of elements counted with their multiplicities) bounded by

$$d(S) \le c_1' \, d^{4/3}\phi(d)^{1/3}\log^{1/3} d \, , \tag{7.1}$$

and its type bounded by

$$t(S) \le c_2' \, t d^{1/3}\phi(d)^{1/3}\log^{1/3} d \, . \tag{7.2}$$

Now we can get an irreducible component S_0 of S approximating $\overline{\theta}$:

LEMMA 7.3. *Let us denote*

$$T(S) = \gamma_0 (t(S) + d(S)\log(d(S) + 2)) \tag{7.4}$$

for $\gamma_0 \le 2$ *and denote for* $\overline{\xi} \in S$,

$$\|\overline{\theta} - \overline{\xi}\| = \exp(-E(\overline{\xi}))$$

if $\overline{\xi} \in S$. *There exists an irreducible component* S_0 *of* S *such that the following properties are satisfied. If*

$$T(S_0) = \gamma_0(t(S_0) + d(S_0)\log(d(S) + 2)) \, ,$$

then for

$$\sum \{E(\overline{\xi}) : \overline{\xi} \in S_0 \, , \, E(\overline{\xi}) > 0\} \overset{\text{def}}{=} E_0 \tag{7.5}$$

and for $\overline{\xi}_0 \in S_0$ *closest to* $\overline{\theta}$: $\|\overline{\xi}_0 - \overline{\theta}\| \le \|\overline{\xi} - \overline{\theta}\| : \overline{\xi} \in S_0$, *we have*

$$\sum \{E(\overline{\xi}) : \overline{\xi} \in S_0 \, , \, E(\overline{\xi}) \ge c_3' \frac{E_0^{3/2}}{d(S_0)T(S_0)^{1/2}}\} \ge E_0/4 \, ;$$

$$E(\overline{\xi}_0) \ge c_4' \frac{E_0^2}{d(S_0)T(S_0)} \, . \tag{7.6}$$

Moreover, the component S_0 *is chosen in a way such that*

$$\frac{E_0}{T(S_0)} \ge c_5' \frac{t\,\phi(d)}{T(S)} \, . \tag{7.7}$$

After the component S_0 has been chosen, our strategy is very simple. We take some (L_1,L_2) and use the auxiliary function $F_{L_1,L_2}(z)$ in order to show that the approximation of $\overline{\theta}$ by elements of S_0, given by Lemma 7.3, is too good and contradicts a suitably chosen small value in Lemma 5.10.

The choice of (L_1,L_2) is straightforward. We demand

$$L_1^2L_2 = \lambda E_0 \overset{\text{def}}{=} \lambda E(\overline{\xi}_0) \tag{7.8}$$

for a sufficiently small $\lambda > 0$.

The definition of L_1,L_2 can be made as follows:

Definition 7.9 We put

$$L_1 = \min\{d(S_0)\ ,\ [(\frac{\lambda E_0 d(S_0)\log d(S_0)}{T(S_0)})^{1/3}]\}$$

and

$$L_2 = [\frac{\lambda E_0}{L_1^2}] + 1\ .$$

LEMMA 7.10. *We have* $L_2 \geq L_1$ *and the following important inequalities hold. For some* $C > 0$ *we have:*

$$E_0\lambda > C\ T(S_0)L_1 \geq C\ t(S_0)L_1\ ; \tag{7.11}$$

and

$$E_0\lambda > C\ d(S_0)L_2 \log L_1\ . \tag{7.12}$$

Proof of Lemma 7.10 First of all we must show that $L_2 \geq L_1$, i.e. that the definition of the pair (L_1,L_2) is legitimate. Indeed, if $L_1 \leq d(S_0)$, then this is the consequence of definition 7.9 and $T(S_0) > d(S_0)\log d(S_0)$.

Let now $L_1 \geq d(S_0)$, i.e. $L_1 = d(S_0)$. We have in this case $\lambda E_0 d(S_0)\log d(S_0) \geq T(S_0)d(S_0)^3$, which implies $L_2 \geq L_1$.

Let us show other statements (7.11)-(7.12) of Lemma 7.10. First of all, if $L_1 \geq d(S_0)$, then (7.11)-(7.12) are obvious; the first follows from the definition of L_1; the second is obvious if $d(S_0)$ (or d) is sufficiently large.

Thus we can assume $L_1 \leq d(S_0)$. By the definition of E_0 we have

$$E_0/T(S_0) \geq c_6' \; E/T(S) ,$$

where $E = t\phi(d)$. We have also $T(S_0) \leq T(S)$, $d(S_0) \leq d(S)$. Consequently, by the bounds of $T(S), d(S)$ in terms of t, d , we get:

$$E_0^2 \geq c_7' \; T(S_0)^2 d(S_0) \log d(S_0) \; \frac{\phi(d)}{d^2 \log^2 d} \; . \tag{7.13}$$

Since $\phi(d) > C\, d^2 \log^2 d$ for some $C > 0$, we get (7.11) from (7.13), if we bound L_1 by a larger quantity, where E_0 is replaced by E_0 .

At the same time, (7.13) implies (7.12) as the definition of L_2 shows.

§8. Let us consider now the auxiliary function $F(z)$ that corresponds to the choice (L_1, L_2) from Definition 7.9 above. There are several possible cases to consider, depending on whether or not $C_{\lambda_1,\lambda_2}(\bar{\xi}_0) = 0$ for all λ_1, λ_2: $0 \leq \lambda_1 \leq L_1$, $0 \leq \lambda_2 \leq L_2$.

Instead of considering several possibilities, we consider at once the general situation. We work now with the polynomials $C_{\lambda_1,\lambda_2}(x,y)$ $(0 \leq \lambda_1 \leq L_1,\ 0 \leq \lambda_2 \leq L_2)$, meaning that they correspond to the pair (L_1, L_2) from Definition 7.9 only.

Since not all polynomials $C_{\lambda_1,\lambda_2}(x,y)$ are zeros (and this is the *only* property of $C_{\lambda_1,\lambda_2}(x,y)$ we use so far), there is always a partial derivative $\partial C_{\lambda_1,\lambda_2}(x,y)$ of some $C_{\lambda_1,\lambda_2}(x,y)$, which is non-zero at $\bar{\xi}_0$. Since the degree of $C_{\lambda_1,\lambda_2}(x,y)$ is bounded by L_1 , let us denote by the k_0 , $0 \leq k_0 < L_1$ the largest integer such that

$$\left. \frac{\partial^{k_1+k_2}}{\partial x^{k_1} \partial y^{k_2}} C_{\lambda_1,\lambda_2}(x,y) \right|_{(x,y) = \bar{\xi}_0} = 0 \tag{8.1}$$

for all rational integers k_1, k_2: $0 \leq k_1$, $0 \leq k_2$ such that $k_1 + k_2 \leq k_0$ and for all λ_1, λ_2: $0 \leq \lambda_1 \leq L_1$, $0 \leq \lambda_2 \leq L_2$.

In other words, there exists $k^0 \in \mathbb{N}^2$ such that $\|\vec{k}^0\|_{L_1} = k_0 + 1$ and

$$\left. \partial^{\vec{k}^0}_{(x,y)} C_{\lambda_1^0,\lambda_2^0}(x,y) \right|_{(x,y) = \bar{\xi}_0} \neq 0 \tag{8.2}$$

for some λ_1^0, λ_2^0: $0 \leq \lambda_1^0 \leq L_1$, $0 \leq \lambda_2^0 \leq L_2$; while (8.1) is satisfied.

Now, instead of the coefficients $C_{\lambda_1,\lambda_2}(\pi/\omega,\eta/\omega)$ of $F(z)$ we use different coefficients changing consequently, the auxiliary function $F(z)$. Namely, we define new polynomials $\tilde{C}_{\lambda_1,\lambda_2}(x,y)$ instead of $C_{\lambda_1,\lambda_2}(x,y)$:

$$\tilde{C}_{\lambda_1,\lambda_2}(x,y) = \frac{1}{\vec{k}^0!} \frac{\partial^{\|k^0\|}}{\partial x^{k_1^0} \partial y^{k_2^0}} C_{\lambda_1,\lambda_2}(x,y) , \tag{8.3}$$

where $\vec{k}^0 = (k_1^0,k_2^0)$ and, as usual, $\vec{k}^0! = k_1^0!k_2^0!$. Consequently we define new coefficients as $\tilde{C}_{\lambda_1,\lambda_2}(\overline{\xi}_0)$ and a new function $\tilde{F}(z)$ as:

$$\tilde{F}(z) = \sum_{\lambda_1=0}^{L_1} \sum_{\lambda_2=0}^{L_2} \tilde{C}_{\lambda_1,\lambda_2}(\overline{\xi}_0) \left(\zeta(z) - \frac{\eta}{\omega} z\right)^{\lambda_1} \wp(z)^{\lambda_2} . \tag{8.4}$$

This function becomes our new auxiliary function corresponding to a pair $(L_1,L_2) \in L$.

Let us list some of the properties of $\tilde{F}(z)$ and $\tilde{C}_{\lambda_1,\lambda_2}(x,y)$:

<u>Proposition 8.5</u> The polynomials $\tilde{C}_{\lambda_1,\lambda_2}(x,y)$ are polynomials from $\mathbb{Z}[x,y]$, not all zeros, of degrees $\le L_1$ and of type $\le c_2 L_2 \log L_1 + L_1 \log 4$. In particular,

$$\log|\tilde{C}_{\lambda_1,\lambda_2}(\overline{\xi}_0)| \le c_8' L_2 \log L_1 \tag{8.6}$$

for all $0 \le \lambda_1 \le L_1$, $0 \le \lambda_2 \le L_2$. In addition to (8.6), $\tilde{C}_{\lambda_1^0,\lambda_2^0}(\overline{\xi}_0) \ne 0$ according to (8.2).

The statement of Proposition 8.5 follows from definition 8.3 and Lemma 8.5, defining polynomials $C_{\lambda_1,\lambda_2}(x,y)$.

The function $\tilde{F}(z)$ has many interesting features as well. First of all we look at the equations defining $C_{\lambda_1,\lambda_2}(x,y)$ in (8.7):

$$\sum_{\lambda_1=0}^{L_1} \sum_{\lambda_2=0}^{L_2} C_{\lambda_1,\lambda_2}(x,y)F_{\lambda_1,\lambda_2,k,m}(x,y) = 0 : \tag{8.7}$$

for $k = 0,1,\ldots,[c_3L_2]$; $m = 0,1,\ldots,[c_4L_1]$.

The system of equations (8.7) defines $C_{\lambda_1,\lambda_2}(x,y)$ for a 'generic' (x,y) ; $\tilde{C}_{\lambda_1,\lambda_2}$ is the specialization of C_{λ_1,λ_2} , adjusted to the 'non-generic' point $\overline{\xi}_0$.

Indeed, we have:

<u>Proposition 8.8</u> For the system $\tilde{C}_{\lambda_1,\lambda_2}(\overline{\xi}_0)$ of complex

numbers, not all of which are zeros, the system of linear equations (8.7) but at $(x,y) = \bar{\xi}_0$ is satisfied:

$$\sum_{\lambda_1=0}^{L_1} \sum_{\lambda_2=0}^{L_2} C_{\lambda_1,\lambda_2}(\bar{\xi}_0)\ F_{\lambda_1,\lambda_2,k,m}(\bar{\xi}_0) = 0 : \tag{8.9}$$

for $k = 0,1,\dots,[c_3L_2]$; $m = 0,1,\dots,[c_4L_1]$.

To prove (8.9) we apply the derivative $\partial^{\|k^0\|}/\partial x^{k_1^0}\partial y^{k_2^0}$ to the system of equations (8.7) and then specialize (x,y) by $(x,y) = \bar{\xi}_0$. Using the definition (8.2) of $\vec{k}^0$ we get

$$\sum_{\lambda_1=0}^{L_1} \sum_{\lambda_2=0}^{L_2} \frac{\partial^{\|\vec{k}^0\|}}{\partial x^{k_1^0}\partial y^{k_2^0}} C_{\lambda_1,\lambda_2}(x,y)\, F_{\lambda_1,\lambda_2,k,m}(x,y)\Big|_{(x,y)=\bar{\xi}_0} =$$

$$= 0 : \quad k = 0,1,\dots,[c_3L_2];\ m = 0,1,\dots,[c_4L_1] .$$

This means, according to (8.3), the statement (8.9) exactly.

The statement (8.9) does not mean however that the function $\tilde{F}(z)$ has zeros at $z = \omega/4 + m\omega'$, because by (8.2):

$$\tilde{F}^{(k)}(\tfrac{\omega}{4} + m\omega') = \sum_{\lambda_1=0}^{L_1} \sum_{\lambda_2=0}^{L_2} \tilde{C}_{\lambda_1,\lambda_2}(\bar{\xi}_0)\ F_{\lambda_1,\lambda_2,k,m}(\tfrac{\pi}{\omega}, \tfrac{\eta}{\omega}) \tag{8.10}$$

and $\bar{\xi}_0 \neq (\pi/\omega , \eta/\omega) = \bar{\theta}$. However, by the choice of $\bar{\xi}_0$, $\bar{\xi}_0$ is rather close to $\bar{\theta} = (\pi/\omega , \eta/\omega)$, which shows that $\tilde{F}(z)$ has *small* values at $\omega/4 + m\omega'$. Namely, we have

<u>Proposition 8.11</u> For the function $\tilde{F}(z)$ defined in (8.4) we have: $\tilde{F}(z) \not\equiv 0$ and for $k = 0,1,\dots,[c_3L_2]$; $m = 0,1,\dots,[c_4L_1]$:

$$\log\left|\frac{1}{k!}\tilde{F}^{(k)}(\tfrac{\omega}{4} + m\omega')\right| < -E_0 + c_9'L_2 \log L_1 . \tag{8.12}$$

For the proof of Proposition 8.11 it is enough to remark that for every polynomial $Q(x,y) \in \mathbf{C}[x,y]$ of the type $\leq T$ we have

$$|Q(\bar{\theta})| \leq |Q(\bar{\xi}_0)| + \|\bar{\theta} - \bar{\xi}_0\| \exp(\gamma_3 T) , \tag{8.13}$$

where $\gamma_3 > 0$.

Now we can apply this remark (8.13) to (8.9) and (8.10). However because of our care over the degree in the estimates this may not be quite satisfactory; because the type of polynomials $F_{\lambda_1,\lambda_2,k,m}(x,y)$ is bounded, by (8.4), by a quantity $c_{10}'L_2 + c_3L_2 \log L_2$ rather than by $c_{11}' L_2 \log L_1$. We must use the change of variable and (G,C)-functions. According to this change of variables, $C_{\lambda_1,\lambda_2}(x,y)$ satisfy another system of equations, equivalent to (8.7):

$$\sum_{\lambda_1=0}^{L_1}\sum_{\lambda_2=0}^{L_2} C_{\lambda_1,\lambda_2}(x,y)\, K_{\lambda_1,\lambda_2,k,m}(x,y) = 0 \qquad (8.14)$$

for all $k = 0,1,\dots,[c_3L_2]$; $m = 0,1,\dots,[c_4\, L_1]$.

By the definition of $\vec{k}^0$, applying $\partial^{\|\vec{k}^0\|}/\partial x^{k_1^0}\partial y^{k_2^0}$ to (8.14) and specializing it at $(x,y) = \bar{\xi}_0$, we obtain the system of equations:

$$\sum_{\lambda_1=0}^{L_1}\sum_{\lambda_2=0}^{L_2} \tilde{C}_{\lambda_1,\lambda_2}(\bar{\xi}_0)\, K_{\lambda_1,\lambda_2,k,m}(\bar{\xi}_0) = 0 \qquad (8.15)$$

for all $k = 0,1,\dots,[c_3L_2]$; $m = 0,1,\dots,[c_4L_1]$.

Then, again in the new variables $w = \wp(z)$; $f(w) = \zeta(z) - \eta/\omega\, z$; we get the expression for $\tilde{\tilde{F}}^{(k)}(w)$ (for a given $m = 0,1,\dots,[c_4L_1]$):

$$\frac{1}{k!}\tilde{\tilde{F}}^{(k)}(w_0) = \sum_{\lambda_1=0}^{L_1}\sum_{\lambda_2=0}^{L_2} \tilde{C}_{\lambda_1,\lambda_2}(\bar{\xi}_0)\, K_{\lambda_1,\lambda_2,k,m}\left(\frac{\pi}{\omega},\frac{\eta}{\omega}\right) \qquad (8.16)$$

if $w_0 = \wp(\frac{\omega}{4})$; $f(w_0) = m\,\frac{\omega\eta' - \omega'\eta}{\omega} + \zeta(\frac{\omega}{4}) - \frac{\eta}{4}$.

However now the polynomials $K_{\lambda_1,\lambda_2,k,m}(x,y)$ have the better bound of the type: the type of $K_{\lambda_1,\lambda_2,k,m}(x,y)$ is bounded by $\gamma_6(k\,\log(L_1+1) + L_2 + L_1\,\log(m+1))$. Hence, returning to an old variable z , we get, using remark (8.13):

$$\log\left|\frac{1}{k!}\tilde{\tilde{F}}^{(k)}\left(\frac{\omega}{4} + m\omega'\right)\right| < -E_0 + c_{12}L_2\log L_1 \;;$$

$$k = 0,1,\dots,[c_3L_2] \;, \qquad (8.17)$$

where $\log\|\bar{\theta} - \bar{\xi}_0\| < -E_0$. In other words, (8.12) is proved.

The main algebraic statement that enables us to handle the presence of the 'gas' of zeros $\bar{\xi} \in S_0$ around $\bar{\theta}$ is the following very convenient,

LEMMA 8.18. *Let* $Q(x,y) \in \mathbb{K}[x,y]$. *We assume that*

$$\log|Q(\bar{\xi}_0)| < -\lambda E_0 \;, \qquad (8.19)$$

$E_0 = E(\bar{\xi}_0)$ *for some* λ , $0 < \lambda \le 1$. *Then for all* $\bar{\xi} \in S_0$ *with* $\|\bar{\theta}-\bar{\xi}\| < 1$, *we have*

$$\log|Q(\overline{\xi})| < -\lambda E(\overline{\xi}) + \gamma_4 t(Q) . \tag{8.20}$$

If now in addition to (8.19), $Q(\overline{\xi}_0) \neq 0$, *then*

$$\textstyle\sum^{+}\{\lambda E(\overline{\xi}) - \gamma_4 t(Q) : \overline{\xi} \in S_0\} \le c'_{13}\{t(Q)\, d(S_0) + d(Q)\, t(S_0)\} , \tag{8.21}$$

where $\sum^{+}$ *means that in the sum only positive elements are counted.*

Proof of Lemma 8.18   We have

$$|Q(\overline{\xi})| \le |Q(\overline{\xi}_0)| + \|\overline{\xi} - \overline{\xi}_0\| \exp(\gamma_3 t(Q)) .$$

Now for $\overline{\xi} \in S_0$, we have using the definition of $\overline{\xi}_0$,

$$\|\overline{\xi} - \overline{\xi}_0\| \le \|\overline{\xi} - \overline{\theta}\| + \|\overline{\xi}_0 - \overline{\theta}\| \le 2\|\overline{\xi} - \overline{\theta}\| \le \exp(-E(\overline{\xi}) + \log 2) .$$

Since $E_0 \ge E(\overline{\xi})$ for $\overline{\xi} \in S_0$, we get (8.20).

Let $Q(\overline{\xi}_0) \neq 0$. Then, since S_0 is an irreducible component, $Q(\overline{\xi}) \neq 0$ for all $\overline{\xi} \in S$. We apply now the Liouville theorem to

$$\Big|\prod_{\overline{\xi} \in S'} Q(\overline{\xi})\Big| ,$$

where $S' = \{\overline{\xi} \in S_0 : \lambda E(\overline{\xi}) - \gamma_4 t(Q) > 0\}$. We have, naturally,

$$\Big|\prod_{\overline{\xi} \in S'} Q(\overline{\xi})\Big| > \exp(-c'_{14}\{t(Q)\, d(S_0) + d(Q)\, t(S_0)\}) . \tag{8.22}$$

Now we get an upper bound for the same product, using (8.20):

$$\Big|\prod_{\overline{\xi} \in S'} Q(\overline{\xi})\Big| < \exp\{-\textstyle\sum(\lambda E(\overline{\xi}) - \gamma_4 t(Q) : \overline{\xi} \in S')\} ,$$

which establishes (8.21) with the combination of (8.22). Lemma 8.18 is proved.

Let us apply this lemma to the polynomials arising from the function $\tilde{F}(z)$. As usual in the situation of the Gelfond-Schneider method, we have two alternatives: either some $\tilde{F}(\frac{\omega}{4} + m\omega')$ is bounded below and above by quantities of the same order $\exp(-O(L_1^2 L_2))$, or all $\tilde{C}_{\lambda_1,\lambda_2}(\pi/\omega , \eta/\omega) = \tilde{C}_{\lambda_1,\lambda_2}(\overline{\theta})$ are bounded above by the quantity like $\exp(-O(L_1^2 L_2))$. This is exactly the statement of Lemma 5.10, as applied to $\tilde{F}(z)$.

We will use now Lemma 8.18 for the pair (L_1, L_2) constructed in (7.8) and 7.9.

We can considerably simplify the proof using another auxiliary

LEMMA 8.23. *Let* $Q(x,y) \in \mathbb{K}[x,y]$ *and*

$$t(Q) \le c'_{15} L_2 \log L_1 ;$$

$$d(Q) \le c'_{16} L_1 ;$$

together with

$$\log|Q(\overline{\theta})| < -c'_{17} L_1^2 L_2$$

for some constants $c'_{15}, c'_{16}, c'_{17} > 0$. *If* λ *in* (7.8) *is sufficiently small, then we have*

$$Q(\overline{\xi}_0) = 0 .$$

<u>Proof of Lemma 8.23</u> Indeed, if $Q(\overline{\xi}_0) \ne 0$, then $Q(\overline{\xi}) \ne 0$ for all $\overline{\xi} \in S_0$. By the choice of (L_1,L_2) in 7.9 we can write

$$c'_{18} L_1^2 L_2 = \lambda E_0$$

for some sufficiently small parameter λ in (7.8), $0 < \lambda < 1$. Then we use Lemma 8.18 and obtain

$$\begin{aligned} \textstyle\sum^+ \{\lambda E(\overline{\xi}) - c'_{19}\, t(Q) : \overline{\xi} \in S\} &\le \\ &\le c'_{20}\{t(Q)d(S_0) + d(Q)t(S_0)\} \le \\ &\le c'_{21}\{L_2 \log L_1\, d(S_0) + L_1\, t(S_0)\} . \end{aligned} \qquad (8.24)$$

Now we use the property of S_0 . Namely, we have

$$\textstyle\sum\{E(\overline{\xi}) : \overline{\xi} \in S_0,\ E(\overline{\xi}) \ge c'_3\ E_0^{3/2}/d(S_0)T(S_0)^{1/2}\} \ge E_0/4 .$$

However, by the choice of E_0 , T_0 in (7.7) one gets

$$E_0/T(S_0) \ge (c'_{22})^2 .$$

Thus

$$E_0^{3/2}/d(S_0)T(S_0)^{1/2} \ge c'_{22}\ E_0/d(S_0) .$$

Thus by Lemma 7.10, we have

$$\lambda c'_{23}\ E_0^{3/2}/d(S_0)T(S_0)^{1/2} > C'\ t(Q)$$

for a sufficiently large $C' > 0$. In particular, we get

$$\lambda E_0 \le c'_{24}\{L_2 \log L_1\ d(S_0) + L_1\ t(S_0)\}\ ,$$

which is contrary to the choice of L_1 , L_2 and Lemma 7.10. Lemma 8.23 is proved.

We are going to apply Lemma 8.23 in the following way:

COROLLARY 8.25. *Let all the conditions of Lemma 8.23 are satisfied. Then in addition to* $Q(\overline{\xi}_0) = 0$ *we have*

$$\log|Q(\overline{\theta})| < -\tilde{C}\ L_1^2 L_2 \tag{8.26}$$

for a suitable constant $\tilde{C} > 0$ *(depending only on* C *).*

<u>Proof of Corallary 8.25</u>. Since $Q(\overline{\xi}_0) = 0$, one obtains

$$|Q(\overline{\theta})| \le |Q(\overline{\xi}_0)| + \|\overline{\theta} - \overline{\xi}_0\| \exp(\gamma_3 t(Q)) \le \|\overline{\theta} - \overline{\xi}_0\| \exp(\gamma_3 t(Q))\ .$$

Now by the choice of (L_1, L_2) in 7.9 we get

$$\|\overline{\theta} - \overline{\xi}_0\| \le \exp(-E_0) \le \exp(-2\tilde{C}\ L_1^2 L_2)\ .$$

This, together with a bound on $t(Q)$ establishes (8.26). Corollary 8.25 is proved.

Corollary 8.25 is applied now to $\tilde{F}^{(k)}(\frac{\omega}{4} + m\omega')$. First of all, by Lemma 5.10, the following is true:

$$\log\left|\frac{1}{k!}\ \tilde{F}^{(k)}(\frac{\omega}{4} + m\omega')\right| < -c'_{25}\ L_1^2 L_2 \tag{8.27}$$

for $m = 0,1,\ldots,[c_{10}L_1]$ and $k = 0,1,\ldots,[c_{10}L_2]$. In order to get better bound on type, we make a change of variables: $\wp(z) = w$, $\zeta(z) - \frac{\eta}{\omega} z = f(w)$. Let us denote by $Q_{k,m}$ a number

$$Q_{k,m} = \frac{1}{k!}\ \tilde{F}^{(k)}(w)\Big|_{w=\wp(\omega/4),\ f(w) = m\frac{\omega\eta' - \omega'\eta}{\omega} + \zeta(\omega/4) - \eta/4}$$

for $k = 0,1,\ldots,[c_{10}L_2]$, $m = 0,1,\ldots,[c_{10}L_1]$. Then we have

$$Q_{k,m} = \sum_{\lambda_1=0}^{L_1} \sum_{\lambda_2=0}^{L_2} \tilde{C}_{\lambda_1,\lambda_2}(\bar{\xi}_0)\, K_{\lambda_1,\lambda_2,k,m}(\bar{\theta}) \ . \tag{8.29}$$

If one replaces in (8.29), $\bar{\theta}$ by $\bar{\xi}_0$, one gets a polynomial $Q'_{k,m}(\bar{\xi}_0)$ in $\bar{\xi}_0$ of degree at most $2L_1$, with the coefficients from $\mathbb{K} = \mathbb{Q}(g_2, g_3, \wp(\omega/4), \wp'(\omega/4), i)$ of the type at most $c'_{26} L_2 \log L_1$. Also, by the choice of $L_1^2 L_2$ and (8.27)-(8.28) we have

$$\log|Q'_{k,m}| < -c'_{27} L_1^2 L_2 \ . \tag{8.30}$$

Hence, by Lemma 8.23,

$$Q'_{k,m}(\bar{\xi}_0) = 0 \ . \tag{8.31}$$

Now we replace in (8.29), in $K_{\lambda_1,\lambda_2,k,m}$, $\bar{\xi}_0$ back by $\bar{\theta}$. We have

$$\log\|\bar{\theta} - \bar{\xi}_0\| < \lambda^{-1} L_1^2 L_2 \ ,$$

and so (8.31) implies by (8.26):

$$\log|Q_{k,m}| < -\tilde{C}\, L_1^2 L_2 \tag{8.32}$$

for a sufficiently large constant $C > 0$.

We return now to the old z-coordinates and deduce from (8.32), (8.28):

$$\log\left|\frac{1}{k!}\tilde{F}^{(k)}\left(\frac{\omega}{4} + m\omega'\right)\right| < -\frac{\tilde{C}}{2} L_1^2 L_2 \tag{8.33}$$

if $k = 0,1,\dots,[c_{10}L_2]$; $m = 0,1,\dots,[c_{10}L_1]$. We use now the Small Value Lemma 5.10, since $\tilde{C}$ is sufficiently large (with respect to c_{10}, c_{11}).

Consequently, there is an upper bound on the values of the coefficients $\tilde{C}_{\lambda_1,\lambda_2}(\bar{\xi}_0)$:

$$\log|\tilde{C}_{\lambda_1,\lambda_2}(\bar{\xi}_0)| < -\frac{\tilde{C}}{4} L_1^2 L_2 \ . \tag{8.34}$$

Here $\tilde{C}$ is sufficiently large and the comparison with Lemma 8.23 gives us at once

$$\tilde{C}_{\lambda_1,\lambda_2}(\bar{\xi}_0) = 0$$

for all $\lambda_1 = 0,1,\ldots,L_1$; $\lambda_2 = 0,1,\ldots,L_2$. This contradicts the choice of $\tilde{C}_{\lambda_1,\lambda_2}(\bar{\xi}_0)$ in 8.8. Theorem 5.1 is proved.

§9. We present below bounds for the measure of algebraic independence of other numbers connected with elliptic functions.

First of all, Theorem 5.1 has very interesting corollaries when the elliptic curve $y^2 = 4x^3 - g_2x - g_3$ has complex multiplication. In this case we have the normal measures of transcendence of numbers connected with values of Γ-function. For example:

COROLLARY 9.1. *For* $P(x) \in \mathbb{Z}[x]$, $P \not\equiv 0$ *we have*

$$|P(\Gamma(1/3))| > H(P)^{-c_1(d)d^2\log^2(d+1)}$$

In particular, for an algebraic ξ *of degree* $\le d$ *and height* $\le H(\xi)$ *we have*

$$|\Gamma(1/3) - \xi| > H(\xi)^{-c_2(d)} .$$

The same is true for $\Gamma(1/4)$, $\Gamma(1/6)$. But what is $c_2(d)$ for $d = 1$? Remember that for $\Gamma(1/2)$ ($= \sqrt{\pi}$) the best value of $c_2(d)$ at $d = 1$ is only 39.7799... [9b]. It is far from the natural conjecture that all numbers $\Gamma(p/q)$, $0 < p < q$, have measures of the irrationality $2 + \varepsilon$ (for any $\varepsilon > 0$).

Another example of 'normal' measures of the algebraic independence.

THEOREM 9.2. *Let* u *be an algebraic point of* $\wp(x)$. *Then*

$$|P(\zeta(u) - \tfrac{\eta}{\omega}u, \tfrac{\eta}{\omega})| > H^{-c\, d^8\log^9(d+1)}$$

if $\log H \ge c_4 d$

We remark that [13],[14]:

$(\frac{\pi}{\omega}, \frac{\eta}{\omega})$ has type of transcendence $\le 3 + \varepsilon$;

$(\zeta(u) - \frac{\eta}{\omega}u, \frac{\eta}{\omega})$ has type of transcendence $\le 9 + \varepsilon$ for any $\varepsilon > 0$.

THEOREM 9.3. *If* $\wp(z)$ *has complex multiplication by a quadratic imaginary number* β , *then for* $P(x,y) \in \mathbb{Z}[x,y]$, $P \not\equiv 0$,

$H(P) \leq H$, $d(P) \leq d$, $t = \log H + d$,

$$|P(\frac{\pi}{\omega}, e^{\pi i \beta})| > \exp(-c_5(\log H + d^3)^{7/3} \log^7(t+1)) .$$

One more example of the 'normal' bound for two algebraically independent numbers.

THEOREM 9.4. *If* $\wp(x)$ *has complex multiplication and* u *is an algebraic point of* $\wp(x)$ *, then*

$$|P(u,\zeta(u))| > H^{-c_6(d)}$$

with $c_6(d) > 0$ *depending on* d *and* u , $\wp(x)$ *only.*

In particular, any number $P(u,\zeta(u))$ is normal (like $\zeta(u)\cdot u$). However, we do not know whether this pair of algebraically independent numbers has finite type of transcendence.

§10. We present in this section normal measures of algebraic independence of numbers arising from elliptic and Abelian generalizations of the Lindemann-Weierstrass theorem. These are numbers that are values of Abelian functions (with complex multiplication) at algebraic points [13],[14],[15].

In the complex multiplication case the situation is almost like that for the exponent [11],[15a].

THEOREM 10.1 [15a]. *Let* $\wp(x)$ *be a Weierstrass elliptic function with algebraic invariants* g_2 , g_3 *and complex multiplication in* $\mathbb{F}$. *Let* $\alpha \in \overline{\mathbb{Q}}$, $\alpha \neq 0$ *and*

$$\ell = [\mathbb{F}(g_2,g_3,\alpha) : \mathbb{F}] .$$

If $P(x) \in \mathbb{Z}[x]$, $P(x) \not\equiv 0$, $H(P) \leq H$, $d(P) \leq d$ *and*

$$\log\log H \geq c_7\, d^3 ,$$

then

$$|P(\wp(\alpha))| > H^{-c_8 \ell^2 d} .$$

Moreover the constant c_8 *can be chosen as* $c_8 \leq 16$.

Of course this is still far from our conjecture.

CONJECTURE 10.1. *For $\wp(x)$ defined over Q and with a complex multiplication, and any $\varepsilon > 0$,*

$$|\wp(r) - p/q| > |q|^{-2-\varepsilon}: \ |q| \geq q(r,\varepsilon)$$

if $r \neq 0$, $r \in Q$.

The reason for this, and stronger conjectures, lies in the existence of effective Padé approximations and nonregular continued fraction expansions for elliptic and similar functions (like Painlevé transcendences). Already non-effective methods like Siegel's lemma give us detailed information about these approximations.

We have also several results on the measure of the algebraic independence of two numbers like $\wp(\alpha)$, $\wp(\beta)$ or $A_1(\alpha,0)$, $A_2(\alpha,0)$ in a CM-case. For example (cf. [15]):

THEOREM 10.3. *Let $\wp(z)$ be a Weierstrass elliptic function with algebraic invariants g_2, g_3 and with complex multiplication in the field $\mathbb{F}$. Let α, β be algebraic numbers linearly independent over $\mathbb{F}$ and*

$$\ell = [\mathbb{F}(g_2,g_3,\alpha,\beta): \mathbb{F}] .$$

If $P(x,y) \in \mathbb{Z}[x,y]$, $P \not\equiv 0$, $H(P) \leq H$, $d(P) \leq d$ *and*

$$\log \log H \geq c_9 \ d^5$$

then

$$|P(\wp(\alpha),\wp(\beta))| > H^{-c_{10}\ell^3 d^2} .$$

Here $c_{10} \leq 128$. *Of course, by Dirichlet's box principle* $|P(\wp(\alpha),\wp(\beta))| < H^{-c_{11}d^2}$ *infinitely often.*

We will formulate now our most general theorem which is the elliptic generalization of the Lindemann-Weierstrass theorem. This result is proved using the author's method of algebraic independence of n numbers [11b],[16]. The detailed proof is a lengthy one; for the case $n = 3$ we have the exposition given in §§5-8.

THEOREM 10.4. *Let* $\wp(x)$ *be a Weierstrass elliptic function with algebraic invariants and complex multiplication in* $\mathbb{K}$. *Let* $\alpha_1,\ldots,\alpha_n$ *be algebraic numbers linearly independent over* $\mathbb{K}$. *Then* $\wp(\alpha_1),\ldots,\wp(\alpha_n)$ *are algebraically independent.*

Moreover, for $P(x_1,\ldots,x_n) \in \mathbb{Z}[x_1,\ldots,x_n]$, $P \not\equiv 0$, *we have*

$$|P(\wp(\alpha_1),\ldots,\wp(\alpha_n))| > H(P)^{-c_{12}(\bar{\alpha}).d(P)^n}$$

where $c_{12}(\bar{\alpha}) > 0$ *depends only on* $[\mathbb{K}(g_2,g_3,\alpha_1,\ldots,\alpha_n):\mathbb{Q}]$ *provided that* $\log\log H(P) > c_{13}d(P)^{2n+1}$.

Let us look now at Abelian varieties of a CM-type. We have results similar to those in a one-dimensional case [15].

Let $\mathbb{A}$ be an Abelian variety defined over an algebraic number field $\mathbb{K}$ and having the dimension d ; we have Abelian functions $A_1(\bar{Z}),\ldots,A_d(\bar{Z})$, $B(\bar{Z})$ such that d/dz_i maps $\mathbb{K}[A_1,\ldots,A_d,B]$ into itself: $i = 1,\ldots,d$ and $A_1(\bar{Z}),\ldots,A_d(\bar{Z})$ are algebraically independent over $\mathbb{C}[\bar{Z}]$, where $A_i(\bar{Z})$ are regular at $\bar{Z} = \bar{0}$.

Now let $\mathbb{A}$ be of CM-type and $\mathbb{E}$ be the field of complex multiplication of $\mathbb{A}$, $[\mathbb{E}:\mathbb{Q}] = 2d$. We proved (in 1978) results on the algebraic independence of $A_1(\bar{\alpha}_j),\ldots,A_d(\bar{\alpha}_j)$; $j = 1,\ldots,k$, where $\bar{\alpha}_j \in \mathbb{Q}^d$: $j = 1,\ldots,k$ are linearly independent over $\mathbb{E}$ and having all coordinates but the k_0-th one as zeros.

In the two-dimensional case we have the following quantitative result:

THEOREM 10.5. *Let* $\mathbb{A}$ *be the two-dimensional* <u>*simple*</u> *Abelian variety of CM-type defined over* $\mathbb{K}$ *with complex multiplications in the field* $\mathbb{E}$, $[\mathbb{E}:\mathbb{Q}] = 4$. *Let* α,β *be algebraic numbers linearly independent over* $\mathbb{E}$ *and*

$$t = [\mathbb{L}:\mathbb{Q}] \quad \textit{where} \quad \mathbb{L} \supset \mathbb{K} \cup \mathbb{E}$$

and $\alpha,\beta \in \mathbb{L}$. *If* $P(x,y) \in \mathbb{Z}[x,y]$, $P \not\equiv 0$, $H(P) \le H$, $d(P) \le d$, *then assuming*

$$\log\log H \ge c_{14}d^3$$

we have

$$|P(A_1(\alpha,0),A_2(\beta,0))| > H^{-c_{15}t^3d^2} .$$

Question 10.6 If $\mathbb{A}$ is an arbitrary simple variety over $\overline{\mathbb{Q}}$ and $\overline{\alpha} \in \overline{\mathbb{Q}}^d$, $\overline{\alpha} \neq \overline{0}$ is it true that any of the numbers

$$A_i(\overline{\alpha}): i = 1,\dots,d$$

is transcendental?

We have some partial answer to this question. In connection with period relations we can ask about 'non-period relations':

Question 10.7 If $\mathbb{A}$ is a simple variety defined over $\overline{\mathbb{Q}}$, is it possible to find $\overline{\alpha} \in \overline{\mathbb{Q}}^d$, $\overline{\alpha} \neq \overline{0}$, such that

$$\text{deg. tr. } \mathbb{Q}(A_i(\overline{\alpha}): i = 1,\dots,d) < d \text{ ?}$$

Similarly to Theorem 10.4 one can formulate general statements about algebraic independence and normal measures of the measure of algebraic independence for more than two numbers connected with values of Abelian functions at algebraic points (cf. [15b]). One should note, however, that the transcendence of $\zeta(\alpha)$ or $\sigma(\alpha)$ is not yet proved for $\alpha \in \overline{\mathbb{Q}}$, $\alpha \neq 0$.

§11. One can try to prove results of the Siegel G-function type, without additional hypotheses on the relation between the size $H(z)$ of z and $|z|$. There are a few particular results of this sort in §2.

Here is a reasonable conjecture that can maybe be proved though there are serious difficulties in both the analytic and arithmetic parts:

CONJECTURE 11.1. *Let* $f_1(z),\dots,f_n(z)$ *be a fundamental system of solutions of a* (G,C)*-operator* $\mathcal{D} \in \mathbb{K}(z)[d/dz]$. *Then the set*

$$S_{\mathbb{K}}(\vec{f}) = \{w \in \mathbb{K}: f_i(w) \in \mathbb{K}: i = 1,\dots,n\}$$

has cardinality $\leq [\mathbb{K}:\mathbb{Q}]\cdot C$, *with* C *independent of* $[\mathbb{K}:\mathbb{Q}]$. *Moreover*

$$|\cup_{\mathbb{K},[\mathbb{K}:\mathbb{Q}]=D} S_{\mathbb{K}}(\vec{f})| \leq C\cdot D .$$

For example for $w \in \mathbb{K}$, $H(w) \geq C_1(d)$ *one of the numbers* $f_1(w),\dots,f_n(w)$ *is of degree* $\geq d$ *over* $\mathbb{K}$.

This conjecture is true for the case discussed above when the change of the variables $z \to h(t)$, $f_i(z) \to g(t)$ gives us meromorphic functions $h(t)$, $g(t)$ of finite order of growth.

However we can prove the conjecture in (at least one) non-trivial and interesting situation:

THEOREM 11.2. *Let* $b, c \in \mathbb{Q}$ $(b \notin \mathbb{Z}^-, b \notin c)$ *and function* ${}_2F_1(1,b;c;z)$ *is transcendental (this is true when, for example,* $b \notin \mathbb{Z}$, $c \notin \mathbb{Z}$, $b-c \notin \mathbb{Z}$). *Then the conjecture is true for the function* ${}_2F_1(1,b;c;z)$.

In particular, for an algebraic z , $d(z) \leq d$ *and* $H(z) \geq c_2(d;b;c)$ *implies that* ${}_2F_1(1,b;c;z)$ *is not algebraic of the degree* $\leq d$.

Functions ${}_2F_1(1,b;c;z)$ are rather complicated and reduce to many classical constants [14],[17]:

$${}_2F_1(1,2b\,;\,1+b\,;\,\tfrac{1}{2}) = b\,B(\tfrac{1}{2},b) \qquad (b \notin \mathbb{Z})$$

$${}_2F_1(1,b\,;\,b+1\,;\,-1) = 2b\{\psi(\tfrac{b+1}{2}) - \psi(\tfrac{b}{2})\} \qquad (b \notin \mathbb{Z}) .$$

Here

$$F_1(1,b;c;x) = \sum_{n=1}^{\infty} \frac{(b)_n}{(c)_n} x^n$$

and

$$(b)_n = b,\ldots,(b+n-1) .$$

One of the most intriguing questions is the determination of those linear differential equations that have a fundamental system of solutions which are G-functions or (G,C)-functions.

The following question is still unsolved even though we can attribute it to C. Siegel's paper (1929) [35].

Problem 11.3 Let $w_1(z),\ldots,w_n(z)$ be the fundamental system of functions satisfying a linear differential equation $\mathcal{D}w_i = 0$ $(i = 1,\ldots,n)$ for $\mathcal{D} \in \overline{\mathbb{Q}}(z)[\frac{d}{dz}]$, such that $w_i(z)$ are G-functions $(i = 1,\ldots,n)$. Prove that $w_1(z),\ldots,w_n(z)$ are (G,C)-functions, or, equivalently that $\mathcal{D}$ is a (G,C)-operator.

§12. The necessary condition for $\mathcal{D} \in \overline{\mathbb{Q}}(z)[\frac{d}{dz}]$ to be a (G,C)-operator is that it be a Fuchsian linear differential operator with algebraic regular singularities and rational exponents. The natural conjecture of E. Bombieri assumed that the converse is true. This conjecture was disproved by Dwork [23]. Dwork's studies show that the situation is nontrivial indeed. However, Dwork's counter-example has nothing to do with Problem 11.3. Indeed, Dwork's result shows that the simplest Fuchsian equation with four regular singularities, Lamé's equation, does not have a fundamental system of G-function solutions, thus it is not a (G,C)-operator.

In any case, Dwork's series of counterexamples shows that it is equally difficult to show that some operator is a (G,C)-operator, or that its solutions are G-functions [6],[23].

There is little hope that for equations with four or more singularities there is a simple algorithm to determine whether or not it is a (G,C)-equation. However, in applications to transcendental number theory we are mainly concerned with two classes of equations: a) equations 'coming from geometry', i.e. Picard-Fuchs equations for period structure; b) equations with three regular singularities only (like polylogarithmic functions of generalized hypergeometric functions). As for equations a) the (G,C)-property for these equations is well-known (Dwork-Katz [21]). For the class b) this has never been proved:

Problem 12.1 Let us consider the matrix linear differential equation with three regular singularities, that we normalize to be at $0,1,\infty$:

$$\frac{dY}{dz} = Y\left(\frac{U_0}{z} + \frac{U_1}{z-1}\right), \tag{12.2}$$

for $n \times n$ matrices U_0, U_1. Let U_0, U_1 belong to $M_n(\mathbb{Q})$ and

$$U_0\ ,\ U_1 \text{ and } U_\infty = -U_0 - U_1$$

all have rational eigenvalues. Prove that equation (12.2) is a (G,C)-equation, or that there is a fundamental system of solutions that are G-functions for every regular point of (12.2).

In many interesting cases we do have a positive solution of the problem: i) for 2×2 matrices; ii) for nilpotent U_1; iii) for idempotent $U_1 = U_1^2$.

The answer to the following question is the most useful for the solution of the problem:

Question 12.3 Let A and B be matrices in $M_n(\mathbb{Q})$ such that A, B and $A+B$ have rational eigenvalues. Let $N = n+m$, and we define

$$\binom{A\ ,\ B}{n\ ,\ m} \text{ as } \frac{N!}{n!m!} \text{ products}$$

of n matrices

$$A, A-1, \ldots, A-n+1 \tag{12.4}$$

and m matrices

$$B, B-1, \ldots, B-m+1 \tag{12.5}$$

such that the order of the sequences (12.4) and (12.5) is kept in each case. For example,

$$\binom{A\ ,\ B}{2\ ,\ 1} = A(A-1)B + AB(A-1) + BA(A-1)\ , \quad \text{etc.}$$

We are interested in the common denominator Δ_N of all elements of matrices

$$\left\{\frac{\binom{A\ ,\ B}{n\ ,\ m}}{N_1!} : n+m = N_1 \le N\right\}. \tag{12.6}$$

Is it true that $\Delta_N \ll C^N$ for some $C = C(A,B) > 1$ for $N \ge N_0$?

Note that for scalars A and B, $\binom{A,B}{n,m}/(n+m)! = \binom{A}{n}.\binom{B}{m}$ and $\binom{A}{k}$ are the usual binomial coefficients. We remark that the answer to the question for $n = 2$ is positive but nontrivial (i.e. Δ_N is not the power of an integer, but has a component like l.c.m.$\{1,\ldots,2N\}$).

As for positive results for equations with three regular singularities, we do have a positive solution for our problem 11.3 for this class of equations.

Now it is quite natural to consider p-adic criteria of G-function or (G,C)-function properties for solutions of linear differential equations from $\overline{\mathbb{Q}}(z)[\frac{d}{dz}]$ [21]-[23]. We study p-adic properties of formal power series solutions

$$w(x) = \sum_{n=0}^{\infty} b_n x^n , \qquad b_n \in \overline{\mathbb{Q}} \qquad (n = 0,1,...)$$

of Fuchsian linear differential equations $\mathcal{D}w = 0$ for $\mathcal{D} \in \overline{\mathbb{Q}}(x)[\frac{d}{dx}]$ in p-adic discs $D(0,r^-)$, $r \le 1$ and $D(t,r^-)$, $r \le 1$. Here a generic point t is lying in some transcendental extension of $\mathbb{Q}_p$ such that $|t|_p = 1$ and such that the residue class of t is transcendental over $\mathbb{F}_p$ [22].

For almost all p (with a natural class of exceptions) the solutions of $\mathcal{D}$ at $x = 0$ converge for

$$\mathrm{ord}_p\ x > \frac{1}{p-1}$$

and those at t converge for

$$\mathrm{ord}_p\ (x - t) > \frac{1}{p-1}$$

('typical exponential' convergence). Any improvement in the convergence at t implies results of the (G,C)-type .

For example from the results of Katz-Dwork [22], it follows that, under the condition that all solutions of $\mathcal{D}$ at the generic point t converge in a disk

$$\mathrm{ord}_p\ (x - t) > \frac{1}{p-1} - \varepsilon , \tag{12.7}$$

we get the following congruence:

$$\frac{d^{p.n}}{dx^{p.n}} \equiv 0(\mathrm{mod}\ \mathbb{F}_p((x))[\tfrac{d}{dx}]\overline{\mathcal{D}})$$

for $n = \mathrm{ord}(\mathcal{D})$ and $\overline{\mathcal{D}}$ being a reduction of $\mathcal{D}$ mod p . Now, following Dwork-Bombieri [23] we can ask:

Does (12.7) for almost all p imply that the solutions of $\mathcal{D}$ at t converge in $D(t,1^-)$ for almost all p ?

Moreover, the simplest case of this conjecture is still open. Let

$$(\frac{d}{dx})^p \equiv 0(\mathrm{mod}\ \mathbb{F}_p((x))[\tfrac{d}{dx}]\overline{\mathcal{D}})$$

for almost all p . Do solutions of $\mathcal{D}$ at t converge in $D(t,1^-)$ for almost all p ? Grothendieck's conjecture tells us that the solutions of $\mathcal{D}$ should be algebraic functions; in which case the positive answer is provided by the Eisenstein theorem.

The criteria of the (G,C)-operator can be formulated following Dwork, Robba [22]. Let us assume that $\mathcal{D} \in \mathbb{Q}(x)[\frac{d}{dx}]$ and all solutions of $\mathcal{D}$ at t converge for

$$\mathrm{ord}_p(x-t) > \varepsilon_p .$$

COROLLARY 12.8 (Dwork). *If* $\Sigma_p \varepsilon_p \log p < \infty$ *, then* $\mathcal{D}$ *is a* (G,C)*-operator and so* $\mathcal{D}$ *has a fundamental system of* G*-function solutions at any regular point of* $\mathcal{D}$.

The 'typical' ε_p in many applications, if less than $1/p-1$, should be $1/p(p-1)$ for almost all p , in which case $\Sigma_p \varepsilon_p \log p < \infty$.

However, it is not easy to check this condition even having a formal series expansion of solutions of $\mathcal{D}$ at one given point.

The property that the solutions of $\mathcal{D}$ are convergent at $D(t,1^-)$ (and so at $D(0,1^-)$ by the transfer theorems of Dwork-Robba [22]) implies remarkable restrictions on $\mathcal{D}$.

§13. In order to demonstrate the nontriviality of the question, which equations have G-function and (G,C)-function solutions, we consider Fuchsian equations (which are the most natural candidates for this). First of all, Fuchsian linear differential equations contain notorious *accessory* parameters. This means that equations cannot be reconstructed by knowledge of the local monodromy (i.e. local multiplicities at regular singularities), but require the information on the global monodromy. These parameters, in addition to local multiplicities, on which the complete reconstruction of the equation depends, are called *accessory*. Their number for the second-order equation with $n + 1$ regular singularities is, in general, $n - 2$. However, while the relation between local multiplicities and monodromy data is obvious (local multiplicities are the logarithms of the eigenvalues of the monodromy matrices), the dependence of accessory parameters on monodromy data is unknown.

We give now the method of the effective construction of auxiliary functions in the case of functions satisfying linear differential equations [17]. The idea of the method is based on the 'isomonodromy deformation' [19], when the monodromy group of the equation is fixed but apparent singularities are added and local multiplicities of the real singularities are changed by integers, in order to adjust the definition of auxiliary functions [17]-[19].

When the problem does not contain accessory parameters (as with hypergeometric functions [10] and Jordan-Pochammer equations), we can present explicitly the corresponding auxiliary function; its differential equation; power series and integral representation of remainder function and polynomial coefficients. In these cases we can determine the asymptotics of the remainder function and polynomial coefficients [17]; these asymptotics can be determined in terms of the Riemann surface associated with the differential equation [19]. In many interesting cases the asymptotics can be determined by a Riemann-Roch theorem.

In some cases for example $f_i(x) = (1-x)^{\omega_i}$: $i = 1,\dots,m$; $N_1 = \dots = N_m = N$, the asymptotics are governed by the m-valued function $(1-(1-x)^{1/m})^{m.n}$ corresponding to a Riemann surface of m sheets with $(1-x)^{1/m}$ [17].

It is a much more difficult problem to determine the denominators of polynomial coefficients of the auxiliary function (remainder function). This requires the knowledge of the p-adic properties of the differential equation (p-curvature ...) for each p separately.

These problems have now been solved completely for hypergeometric $({}_pF_q)$-functions [10].

In the case when the auxiliary function is constructed explicitly, as a Padé approximation, our results on linear independence and measure of irrationality take a very good form with small constants. For example we have [10]:

COROLLARY 13.1. *Let*

$$L_2(x) = \sum_{n=1}^{\infty} \frac{x^n}{n^2}$$

and $q \geq 14$ *be an integer. Then*

$$1,\ \log(1+\tfrac{1}{q}),\ L_2(\tfrac{1}{q})$$

are linearly independent over $\mathbb{Q}$. *Moreover,*

$$|L_2(\tfrac{1}{q}) - \tfrac{P}{Q}| > |Q|^{-\lambda(q)}$$

for integers P,Q *with* $\lambda(q) > 0$ *if* $q \geq 14$ *and* $\lambda(14) \leq 300$ *while* $\lambda(50) = 3.0\dots$ *etc.*

In comparison with the G-function (ineffective method) we would like to point out that in the G-function approach $L_2(1/q)$ becomes irrational only for $q \geq e^{127}$. For logarithms of algebraic numbers better results can be obtained (cf. [3],[9],[31],[32]). Here are measures of irrationality of numbers connected with π.

COROLLARY 13.2 [9]. *We have for* $\varepsilon > 0$ *and rational integers* p,q *for* $|q| \geq q_1(\varepsilon)$ *the following bounds*

$$|q.\pi\sqrt{3} - p| > |q|^{-\chi_1-\varepsilon}$$

for

$$\chi_1 = - \frac{\log(e(2\cos\pi/12)^2)}{\log(e(2\sin\pi/12)^2)} = 7.30998634\ldots ;$$

$$|q.\pi - p| > |q|^{-\chi_2-\varepsilon}$$

for

$$\chi_2 = 4 - 5\,\frac{\log(e^5(2\cos\pi/24)^6)}{\log(e^5(2\sin\pi/24)^6)} = 18.88999444\ldots .$$

Moreover for any $|q| \geq 2$ *we have*

$$|q.\pi\sqrt{3} - p| > |q|^{-7.31}$$

and

$$|q.\pi - p| \geq |q|^{-18.9} .$$

Explicit Padé approximations to binomial functions (or, in general, to algebraic functions) enable us to improve considerably the Liouville estimates for algebraic numbers of a particular nature (at least our methods give explicit expressions for auxiliary functions in all proofs of Thue-Siegel-Roth, which were also non-explicit).

For example among cubic roots of integers the following get effective estimates that are considerable improvements over Liouville:

$$\sqrt[3]{2},\ \sqrt[3]{3},\ \sqrt[3]{4},\ \sqrt[3]{6},\ \sqrt[3]{17},\ldots .$$

For example for integers p,q $(q \neq 0)$:

$$\left|\sqrt[3]{2} - \frac{p}{q}\right| > |q|^{-2.429709513\ldots} ,$$

$$\left|\sqrt[3]{3} - \frac{p}{q}\right| > |q|^{-2.692661368\ldots} ,$$

$$|\sqrt[3]{17} - \frac{p}{q}| > |q|^{-2.198220241...} .$$

For $\sqrt[3]{2}$, $\sqrt[3]{17}$, A. Baker (1964) [2] obtained effective improvements of the Liouville theorem (2.955 and 2.4 respectively in the exponent).

REFERENCES

1. A. Baker, Transcendental Number Theory, Cambridge University Press, 1975.

2. A. Baker, Rational approximations to $\sqrt[3]{2}$ and other algebraic numbers, Quart.J.Math. Oxford 15 (1964), 375-383.

3. A. Baker, Approximations to the logarithms of certain rational numbers, Acta Arith. 10 (1964), 315-323.

4. H. Bateman and A. Erdélyi, Higher transcendental functions, 3 v., McGraw-Hill, 1953.

5. D. Bertrand, Varietes Abeliennes et formes lineaires d'integrales elliptiques, Séminaire Delange-Pisot-Poitou, Exposé du 8 Octobre, 1979, 12 pp.

6. E. Bombieri, On G-functions (to appear).

7. E. Bombieri and S. Lang, Analytic subgroups of group varieties, Invent.Math. 11 (1970), 1-14.

8. D. Brownawell and D. Masser, Multiplicity estimates for analytic functions; part I, J. reine und ang.Math. 314 (1980), 200-216.

9. G.V. Chudnovsky, a) Approximations rationnelles des logarithmes de nombres rationnels, C.R.Acad.Sci. Paris, Serie A, v.288 (1979), A-607-A-609; b) Formules d'Hermite pour les approximants de Padé de logarithmes et fonctions binome et measures d'irrationalite, C.R. Acad.Sci. Paris, Serie A, v.288 (1979), A-965-A-967.

10. G.V. Chudnovsky, a) Un systeme explicite d'approximants de Padé pour les fonctions hypergeometriques generalisees, avec applications a l'arithmetique, C.R.Acad.Sci. Paris, Serie A, v.288 (1979), A-1001 -A-1004; b) Padé approximations to the generalized hypergeometric function, I, Journal de Math. Pures et Appl. Paris 58 (1979), 445-476.

11. G.V. Chudnovsky, a) Algebraic grounds for the proof of algebraic independence, I, Elementary algebra, Comm. Pure and Appl.Math. 34 (1981), 1-28; b) Algebraic grounds for the proof of algebraic independence, II, Comm. Pure and Appl.Math. (to appear).

12. G.V. Chudnovsky, Algebraic independence of constants connected with exponential and elliptic functions, Doklady Ukranian Academy of Sciences No.8 (1976), 697-700.

13. G.V. Chudnovsky, Algebraic independence of values of exponential and elliptic functions, Invited address, Proceedings of the International Congress of Mathematicians, Helsinki, Finland (1978), v.1, 339-350.

14. G.V. Chunovsky, Transcendences arising from exponential and elliptic functions, in Collection of translations of papers on Transcendental Number Theory by G.V. Chudnovsky, Mathematical Surveys, published by the American Mathematical Society, USA (to appear); Algebraic independence of constants connected with exponential and elliptic functions, March 1978, pp.1-60, in Collection of translations of papers on Transcendental Number Theory by G.V.

Chudnovsky, Mathematical Surveys, published by the American Mathematical Society, USA (to appear).

15. G.V. Chudnovsky, a) Algebraic independence of the values of elliptic function at algebraic point, Elliptic analogue of Lindemann-Weierstrass theorem, Inventiones Math. 61 (1980); b) Indépendence algébrique des valeurs d'une fonction elliptique en des points algébriques, C.R.Acad.Sci. Paris, Serie A, v.288 (1979), A-439-A-440.

16. G.V. Chudnovsky, Indépendence algébrique dans la méthode de Gelfond-Schneider, C.R.Acad.Sci. Paris, Serie A, 291 (1980), A-419-A-422.

17. G.V. Chudnovsky, Padé approximation and the Riemann monodromy problem, Cargese Lectures, June, 1979, in Bifurcation phenomena in Mathematical Physics and Related Topics, D. Reidel Publishing Company, Boston, USA, 1980, 448-510.

18. G.V. Chudnovsky, The inverse scattering method and applications to arithmetic, approximation theory and transcendental numbers, Lecce lectures, May 1979, Proceedings of the Lecce Conference on Nonlinear Evolution Equations and Dynamical Systems, Lecture Notes in Physics, 120 (1980), Springer, New York, 1980, 103-150.

19. G.V. Chudnovsky, Rational and Padé approximations to solutions of linear differential equations and the monodromy theory, Les Houches Lectures, September 1979, Proceedings of the Les Houches International Colloquium on Complex Analysis and Relativistic Quantum Field Theory, Lecture Notes in Physics, Springer, 126 (1980), 136-169.

20. P. Cijsouw, Transcendence measures, Amsterdam, 1972.

21. B.U. Dwork, On p-adic analysis, Belfer Grad. School, Yeshiva Univ., Annual Science Conference Proceedings 2 (1969), 129-154.

22. B.U. Dwork and P. Robba, Effective p-adic bounds for solutions of homogeneous linear differential equations, Trans.Am.Math.Soc. 259 (1980), 559-577.

23. B.U. Dwork, Arithmetic theory of differential equations, Symp. Math., v.24, Academic Press, 1981, 225-243.

24. Y. Flicker, On p-adic G-functions, J. London Math.Soc. 15 (1977), 395-402.

25. A.I. Galochkin, Lower bounds for polynomials of the values of a class of analytic functions, (Russian) Math.Sbornik 137 (1974), No.3, 396-417.

26. A.O. Gelfond, Transcendental and algebraic numbers, Dover, N.Y., 1960 (Russian edition 1951).

27. W.V.D. Hodge and D. Pedoe, Methods of algebraic geometry, 3 volumes, Cambridge University Press, 1952.

28. S. Lang, Introduction to Diophantine approximations, Addison-Wesley, 1966.

29. S. Lang, Introduction to transcendental numbers, Addison-Wesley, 1966.

30. S. Lang, Elliptic curves, Diophantine analysis, Springer, 1978.

31. K. Mahler, Applications of some formulae by Hermite to the approximation of exponentials and logarithms, Math.An. 168 (1967), 200-227.

32. M. Mignotte, Approximations rationelles de π et quelques autres nombres, Bull.Soc.Math. France 37 (1974), 121-132.

33. C.F. Osgood (to appear).

34. Th. Schneider, a) Arithmetische untersuchungen elliptischer integrale, Math.Ann. 113 (1937), 1-13; b) Zur theorie der Abelschen funktionen und integrale, J. reine angew.math. 183 (1941), 110-128.

35. C.L. Siegel, Uber einige Anwendungen Diophantischer Approximationen, Abh.Preuss.Akad.Wiss.Phys.-Mat.Kl. Berlin (1929), No.1.

36. C.L. Siegel, Transcendental numbers, Princeton, New Jersey, 1949.

37. I.R. Shafarevich, Basic algebraic geometry, Springer, N.Y., 1974.

38. Transcendence Theory, Proceedings of the Cambridge Conference 1976, Academic Press, N.Y., 1977.

39. D. Brownawell, On the Gelfond-Feldman measure of algebraic independence, Comp.Math. 38 (1979), 355-368.

40. K. Väänänen, On linear forms of a certain class of G-functions and p-adic G-functions, Acta Arithmetica (1980).

41. G.V. Chudnovsky, a) A new method for the investigation of arithmetical properties of arithmetic functions, Ann. of Math. 109 (1979), 353-376; b) Singular points on complex hypersurfaces and multidimensional Schwarz lemma, Seminare D-P-P, January 1978, 40 pp. Birkhauser Verlag 1980.

42. Equations Differentielles et Systemes de Pfaff dans le Champ Complexe, Lect.Not.Math. v.712, Springer-Verlag, 1979.

43. B.L. Van der Waerden, Modern algebra, 2 volumes, Ungar, N.Y., 1951.

Added in Proof. Measures of irrationality of §13 for numbers connected with π can be improved. E.g. $|q\pi/\sqrt{3}-p| > |q|^{-4.8}$ and $|q\pi^2-p| > |q|^{-7.5}$ for rational integers $p, q, |q| > q_0$.

THE RIEMANN ZETA-FUNCTION

D.R. Heath-Brown
Magdalen College
Oxford OX1 4AU

This is a survey of some of the most significant recent developments in the theory of the Riemann Zeta-function. However, certain topics, notably Apéry's work on $\zeta(3)$, and also the application of transcendence methods to the distribution of values of the Zeta-function, will be outside our scope. We will be concerned principally with the distribution of zeros, applications to primes, and mean-value theorems. It should be stressed that much of the work can be generalized to Dirichlet L-functions, and questions of uniformity then become important.

The Riemann Hypothesis is still undecided, but Brent [3] has found that the first 8.1×10^7 zeros are on the critical line $\mathrm{Re}(s) = \frac{1}{2}$, and are simple. One of the best theoretical approximations is due to Levinson [21], and states that at least 34% of the zeros are on $\mathrm{Re}(s) = \frac{1}{2}$. His method is poorly understood, and there is plenty of room for improvement. One fact about Levinson's method not noticed until recently is that the zeros produced not only lie on $\mathrm{Re}(s) = \frac{1}{2}$, but are also simple. This was first observed by Selberg, see also [11]. It is conjectured that all the zeros are simple, but previously it was not known even that infinitely many of them were.

Approximating to the Riemann Hypothesis in a rather different way, Selberg [25] proved that

$$N(\sigma,T) << T^{1-\frac{1}{4}(\sigma-\frac{1}{2})} \log T \quad ,$$

where $N(\sigma,T)$ is the number of zeros in the region $\sigma \le \mathrm{Re}(s) \le 1$, $0 < \mathrm{Im}(s) \le T$, or, because the zeros lie symmetrically about the critical line, in the region $0 \le \mathrm{Re}(s) \le 1-\sigma$, $0 < \mathrm{Im}(s) \le T$. Since there are, asymptotically, $\frac{T}{2\pi} \log T$ zeros in the rectangle $0 \le \mathrm{Re}(s) \le 1$, $0 < \mathrm{Im}(s) \le T$, it follows from Selberg's bound that 'almost all' zeros (that is, in the density sense) lie in the region

$$|\sigma - \tfrac{1}{2}| \le \frac{f(T)}{\log T} \quad ,$$

if $f(T) \to \infty$. Jutila [19] has recently improved Selberg's bound, to give

$$N(\sigma,T) \ll_\delta T^{1-(1-\delta)(\sigma-\frac{1}{2})} \log T ,$$

for any $\delta > 0$. We would like to replace $1-\delta$ by $2-\delta$, giving a sharp form of the Density Hypothesis, but this seems hopeless at present.

Both Levinson's theorem and the estimates of Selberg and Jutila require asymptotic formulae for mean-values. In the latter case one examines the integral

$$I = \int_T^{2T} |\zeta(\sigma+it)M(\sigma+it)|^2 \, dt ,$$

where

$$M(s) = \sum_{n=1}^{X} \frac{\mu(n)w_n}{n^s}$$

and w_n is a suitably chosen weight. One wants to have X as large as possible, while having an asymptotic formula for I. Classical methods allow $X = T^{\frac{1}{2}-\delta}$, but it is possible that the techniques mentioned later will permit larger values of X to be used, thus improving very slightly the 34% in Levinson's result and the $1-\delta$ of Jutila's estimate.

Other bounds for $N(\sigma,T)$ have already benefitted from advances in mean value estimates. Consider bounds of the shape

$$N(\sigma,T) \ll_\varepsilon T^{k(\sigma)(1-\sigma)+\varepsilon} , \quad \tfrac{1}{2} \le \sigma \le 1 ,$$

for any $\varepsilon > 0$. Some of the most significant estimates for $k(\sigma)$ are displayed in figure 1.

In the range $\frac{1}{2} \le \sigma \le \frac{3}{4}$, Ingham's 1940 result still stands, and it is the range $\frac{3}{4} \le \sigma \le 1$ where recent work has been concentrated, using Halász's method as developed by Montgomery and Huxley, see [14]. For $\frac{25}{28} \le \sigma \le 1$ we have $k(\sigma) \le 1/(\sigma - \frac{1}{4})$, see [9]. This is a natural shape for the bound to assume, since it stems from the use of the estimate

$$\int_0^T |\zeta(\tfrac{1}{2}+it)|^{12} \, dt \ll T^{2+\varepsilon} ,$$

(see [6]), in place of the classical result

$$\int_0^T |\zeta(\tfrac{1}{2}+it)|^{4} \, dt \ll T^{1+\varepsilon} ,$$

due to Ingham [15]. The density results have been applied, in [10], to give estimates for the differences $d_n = p_{n+1} - p_n$ between

Figure 1.

Bounds for $k(\sigma)$

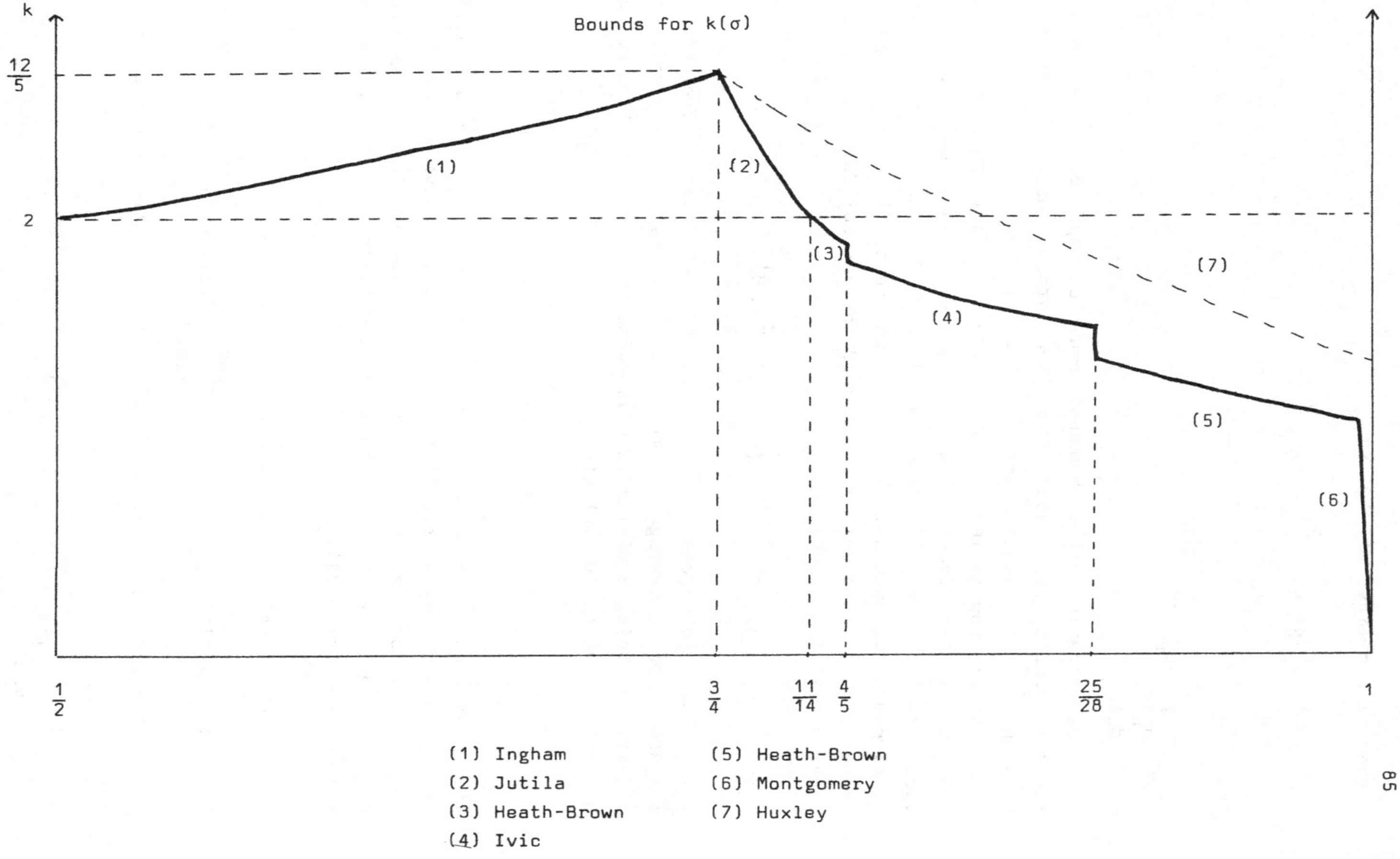

consecutive primes:

$$\sum_{p_n \le x} d_n^2 \ll x^{1+\frac{5}{18}+\varepsilon} ,$$

and

$$\sum_{\substack{p_n \le x \\ d_n \ge p_n^{\frac{1}{2}}}} d_n \ll x^{\frac{3}{4}+\varepsilon} .$$

It seems probable that such bounds could be improved by combining the zero density techniques with sieve methods, as in the proof that $d_n \ll p_n^{13/23}$, by Iwaniec and Jutila [16].

We turn now to mean value problems. This is the area in which the most exciting recent developments have taken place. All of the topics mentioned so far either could take advantage of the new ideas, or already have done so. To appreciate the significance of mean-value results one should bear in mind the general formula (see [22; Theorem 6.1] for example)

$$\int_0^T \left|\sum_1^N a_n n^{-\frac{1}{2}-it}\right|^2 dt = (T + O(N)) \sum_1^N \frac{|a_n|^2}{n} . \qquad (1)$$

Notice that the error term $O(N)$ is the length of the Dirichlet polynomial being averaged. Very roughly speaking one may think of $\zeta(\frac{1}{2}+it)$ as being a Dirichlet polynomial of length $|t|^{\frac{1}{2}}$. Now consider the mean square integral

$$\int_0^T |\zeta(\tfrac{1}{2}+it)|^2 dt = T(\log(\frac{T}{2\pi}) + 2\gamma - 1) + E(T) . \qquad (2)$$

The above remarks lead one to expect that $E(T) \ll_\varepsilon T^{\frac{1}{2}+\varepsilon}$, and this is indeed a classical result of Ingham [15]. However, Balasubramanian [2] has obtained $E(T) \ll_\varepsilon T^{\frac{1}{3}+\varepsilon}$. To appreciate the significance of this and later estimates consider the following (a simple extension of [6; Lemma 3]).

LEMMA. *Let* k *be a positive integer and let* $t \ge 2$. *Then*

$$|\zeta(\tfrac{1}{2}+it)|^k \ll_k (\log t)^2 (1 + \int_{-\log t}^{\log t} |\zeta(\tfrac{1}{2}+it+iu)|^k du) .$$

Differencing the formula (2), Balasubramanian's result implies that

$$\int_{-\log t}^{\log t} |\zeta(\tfrac{1}{2}+it+iu)|^2 du \ll t^{\frac{1}{3}+\varepsilon} ,$$

here we have discarded a vast amount of information from the mean value formula. None the less, an application of the lemma now yields

$$\zeta(\tfrac{1}{2}+it) \ll t^{\frac{1}{6}+\varepsilon} , \tag{3}$$

which is very close to being the best known bound for $\zeta(\frac{1}{2}+it)$. Moreover the analysis of $E(T)$ shows that precisely the same exponential sums that occur in the estimation of $\zeta(\frac{1}{2}+it)$ arise for $E(T)$ also. Thus the proof of Kolesnik's recent bound [20]

$$\zeta(\tfrac{1}{2}+it) \ll t^{35/216+\varepsilon} ,$$

may readily be adapted to yield

$$E(T) \ll T^{35/108+\varepsilon} .$$

One should not push this relationship too far however, since on the one hand we expect that

$$\zeta(\tfrac{1}{2}+it) \ll t^{\varepsilon} ,$$

(this is the Lindelöf Hypothesis), while on the other, Good [5] has shown that

$$E(T) = \Omega(T^{\frac{1}{4}}) . \tag{4}$$

It ought to be possible to improve this slightly, perhaps by a log factor, and to give $\Omega_{\pm}$ results. The result (4) may readily be deduced from the asymptotic formula

$$\int_0^T E(t)^2\,dt \sim cT^{\frac{3}{2}} ,$$

where c is a numerical constant (see [7]). One might conjecture that, in fact

$$E(T) \ll T^{\frac{1}{4}+\varepsilon} .$$

All this work has strong links with van der Corput's method for estimating exponential sums. It is now apparent that one may formulate a general estimate involving a parameter μ, of precisely the type one considers in practice, which leads immediately to the bounds

$$\zeta(\tfrac{1}{2}+it) \ll t^{\mu+\varepsilon}, \quad E(T) \ll T^{2\mu+\varepsilon},$$

$$\Delta(x) \ll x^{2\mu+\varepsilon}, \quad P(x) \ll x^{2\mu+\varepsilon}.$$

Here $\Delta(x)$ and $P(x)$ are the remainders in the classical lattice point problems for the divisor function and the circle respectively.

Another mean value theorem that is related to van der Corput's method is

$$\int_0^T |\zeta(\tfrac{1}{2}+it)|^{12}\, dt \ll T^{2+\varepsilon}, \tag{5}$$

for which see [6]. Here $\zeta(\frac{1}{2}+it)^6$ corresponds to a Dirichlet polynomial of length T^3, so that the trivial bound would be worse by a factor T. The lemma, discarding almost all of the interval $[0,T]$, again produces the estimate (3). From the proof of (3) one can extract a sum of the general form

$$\sum d(n)e^{i\sqrt{nt}}$$

which is estimated trivially in obtaining the exponent $\frac{1}{6}$. If instead we ask how often the sum can be large, and apply the Halász method, then we obtain (5). As mentioned earlier, (5) may be used in estimating $N(\sigma,T)$, and thus in problems involving the differences between primes.

In general we conjecture that

$$I_k = \int_0^T |\zeta(\tfrac{1}{2}+it)|^{2k}\, dt \sim C_k T(\log T)^{k^2}$$

for all non-negative real k. However this is only known to hold for $k = 0$, 1 and 2. Moreover no reasonable conjecture as to the value of C_k has been made. Ramachandra [23],[24] has given upper and lower bounds for I_k. This work is extended in [12]. We have

$$I_k \gg_k T(\log T)^{k^2}$$

for every rational $k \ge 0$, and

$$I_k \ll_k T(\log T)^{k^2}$$

for $k = n^{-1}$, if n is integral. On the Riemann Hypothesis

$$I_k \gg_k T(\log T)^{k^2}$$

for $k \geq 0$, and

$$I_k \ll_k T(\log T)^{k^2}$$

for $0 \leq k \leq 2$.

Perhaps the most important developments are those making use of Kloostermann sums, originally using Weil's estimate, and then, very recently, using the theory of modular functions to bound averages of such sums. Atkinson [1] treated a rather different kind of mean value estimate by a method which depended ultimately on the Kloostermann sum. More recently it was shown [8] that

$$\int_0^T |\zeta(\tfrac{1}{2}+it)|^4 \, dt = Tf(\log T) + O_\varepsilon(T^{7/8+\varepsilon}) \tag{6}$$

for any $\varepsilon > 0$, $f(\cdot)$ being a polynomial of degree 4 . Here $\zeta(\frac{1}{2}+it)^2$ corresponds to a Dirichlet polynomial of length T , so that classically one has an error term $O_\varepsilon(T^{1+\varepsilon})$. In this case the lemma yields

$$\zeta(\tfrac{1}{2}+it) \ll t^{7/32+\varepsilon} .$$

Thus we have a non-trivial bound, albeit weaker than (3), obtained by a method quite different from that of van der Corput.

Let us see, very briefly, how Kloostermann sums may arise. If the formula (1) is investigated more carefully we need not only $\sum |a_n|^2 n^{-1}$, but also non-diagonal sums, of the shape $\sum a_n \bar{a}_{n+1} n^{-1}$ for example. In our case, starting with

$$\zeta^2(s) = \sum_1^\infty d(n)n^{-s} ,$$

we are led to consider such sums as

$$\sum_{n\leq x} d(n)d(n+1) \tag{7}$$

(for which see Estermann [4]). Here we must examine

$$\sum_{\substack{n\leq x \\ n\equiv 0 (\mathrm{mod}\ u)}} d(n+1) \tag{8}$$

and we therefore investigate

$$\sum_{\substack{1 \\ m\equiv 1 (\mathrm{mod}\ u)}}^{\infty} d(m)m^{-s} .$$

This Dirichlet series may be extended to the entire complex plane, and has a functional equation relating it to

$$\sum_{a,b=1} S(a,\pm b;u)(ab)^{-s} \quad ,$$

here $S(a,\pm b;u)$ is a Kloostermann sum. Thus a bound for the size of $|S(a,\pm b;u)|$ leads to an asymptotic formula for (8), while for use in (7) we may consider averages of $S(a,\pm b;u)$ with respect to u. The Kloostermann sum can be brought in more directly, but we do not yet have as close a connection as we would like.

Iwaniec [17] has recently used Kloostermann sums in showing that

$$\int_0^T |\zeta(\tfrac{1}{2}+it)|^4 . |\sum_1^N a_n n^{-\frac{1}{2}-it}|^2 dt \ll_\varepsilon T^{1+\varepsilon} \sum_1^N |a_n|^2 n^{-1}$$

for $N \le T^{1/10}$. Classical methods do not allow one to prove any result of this type. We also have, [13],

$$\int_0^T |\zeta(\tfrac{1}{2}+it) \sum_1^N \mu(n) n^{-\frac{1}{2}-it}|^2 \, dt \ll_\varepsilon T^{1+\varepsilon}$$

for $N \le T^{8/15}$ - previously the range was $N \le T^{1/2}$. It seems likely that this last result could be adapted to give the type of relationship needed in the work of Levinson, or of Jutila, as described earlier. One might thus obtain some small improvements in the distribution of the zeros.

Finally, using deep results on modular functions, Iwaniec [18] has obtained

$$\int_T^{T+T^{2/3}} |\zeta(\tfrac{1}{2}+it)|^4 \, dt \ll T^{2/3+\varepsilon} \quad .$$

The lemma shows that this estimate too, contains the bound (3). It seems probable that the method of proof could be developed so as to obtain an error term $O_\varepsilon(T^{2/3+\varepsilon})$ in (6).

These ideas deserve full exploration and have potential applications in many diverse areas of analytic number theory.

REFERENCES

1. F.V. Atkinson, The mean value of the zeta-function on the critical line, Proc. London Math.Soc. (2) 47 (1941), 174-200.

2. R. Balasubramanian, An improvement on a theorem of Titchmarsh on the mean square of $|\zeta(\frac{1}{2}+it)|$, Proc. London Math.Soc. (3) 36 (1978), 540-576.

3. R.P. Brent, On the zeros of the Riemann zeta-function in the critical strip, Math.Comp. 33 (1979), 1361-1372.

4. T. Estermann, Über die Darstellungen einer Zahl als Differenz von zwei Produkten, J. Reine Angew.Math. 164 (1931), 173-182.

5. A. Good, Ein Ω-Resultat für das quadratische Mittel der Riemannschen Zetafunktion auf der kritische Linie, Invent.Math. 41 (1977), 233-251.

6. D.R. Heath-Brown, The twelfth power moment of the Riemann Zeta-function, Quart.J.Math. Oxford Ser. (2) 29 (1978), 443-462.

7. D.R. Heath-Brown, The mean value theorem for the Riemann Zeta-function, Mathematika 25 (1978), 177-184.

8. D.R. Heath-Brown, The fourth power moment of the Riemann Zeta-function, Proc. London Math.Soc. (3) 38 (1979), 385-422.

9. D.R. Heath-Brown, Zero density estimates for the Riemann Zeta-function and Dirichlet L-functions, J. London Math.Soc. (2) 19 (1979), 221-232.

10. D.R. Heath-Brown, The differences between consecutive primes, III, J. London Math.Soc. (2) 20 (1979), 177-178.

11. D.R. Heath-Brown, Simple zeros of the Riemann Zeta-function on the critical line, Bull. London Math.Soc. 11 (1979), 17-18.

12. D.R. Heath-Brown, Fractional moments of the Riemann Zeta-function, J. London Math.Soc. (2) 24 (1981), 65-78.

13. D.R. Heath-Brown, A mean value theorem for the Riemann Zeta-function, unpublished.

14. M.N. Huxley, On the difference between consecutive primes, Invent.Math. 15 (1972), 164-170.

15. A.E. Ingham, Mean-value theorems in the theory of the Riemann Zeta-function, Proc. London Math.Soc. (2) 27 (1926), 273-300.

16. H. Iwaniec and M. Jutila, Primes in short intervals, Arkiv för Matematik 17 (1979), 167-176.

17. H. Iwaniec, On the mean value for Dirichlet polynomials and the Riemann-Zeta-function, preprint.

18. H. Iwaniec, Mean values for Fourier coefficients of cusp forms and the Riemann Zeta-function, preprint.

19. M. Jutila, Zeros of the Zeta-function near the critical line, preprint.

20. G. Kolesnik, On the order of $\zeta(\frac{1}{2}+it)$ and $\Delta(R)$, preprint.

21. N. Levinson, More than one third of zeros of Riemann's Zeta-function are on $\sigma = \frac{1}{2}$, Advances in Math. 13 (1974), 383-436.

22. H.L. Montgomery, Topics in multiplicative number theory (Springer, Berlin, 1971).

23. K. Ramachandra, Some remarks on the mean value of the Riemann Zeta-function and other Dirichlet series II, and III, preprints.

24. K. Ramachandra, Progress towards a conjecture on the mean value of Titchmarsh series, preprint.

25. A. Selberg, Contributions to the theory of the Riemann Zeta-function, Arch.Math. Naturvid 48 (1946), 89-155.

ON EXPONENTIAL SUMS AND CERTAIN OF THEIR APPLICATIONS

C. Hooley
Department of Pure Mathematics
University College
Cardiff

1. INTRODUCTION

The analytical theory of numbers has profited enormously from the estimates given by Weil for exponential sums of the form

$$\sum_{\substack{g(X,Y)\equiv 0, \bmod p \\ 0\le X,Y<p}} e^{2\pi i f(X,Y)/p} ,$$

where $f(X,Y)$ and $g(X,Y)$ are polynomials with rational integral coefficients. To quote but one instance, we may mention the striking applications to the Riemann zeta function and divisor problems that have resulted from the estimate for the Kloosterman sum, which of course is the sum obtained by taking $f(X,Y) = AX + BY$ and $g(XY) = XY - 1$.

Weil's estimates often yield weakish non-trivial results for similar sums in more variables. However, it was widely anticipated that the answer to the problem of finding best possible estimates for the general sum

$$\sum_{\substack{g_1(X_1,\ldots,X_N),\ldots,g_s(X_1,\ldots,X_N)\equiv 0, \bmod p \\ 0\le X_1,\ldots,X_N<p}} e^{2\pi i f(X_1,\ldots,X_N)/p} \tag{1}$$

would have to await the resolution of the generalized Weil conjectures. Nevertheless, the promulgation of Deligne's fundamental work in 1974 [4] did not of itself bring an automatic or universal solution to the question of bounding these sums, progress in the first instance having been limited to special situations. Needing such estimates in general circumstances, the author therefore sought an appropriate method and in February 1979 succeeded in finding one that depended on Deligne's results in a very simple way. Announced in our paper [10] on two h-th powers, the method was applied on that occasion to meet the particular needs of the moment, an account of the method in its full generality having been reserved for a later publication that will now shortly appear. Subsequently, however, Katz has developed alternative but much deeper procedures dealing with similar questions; his results give much insight into the structure of L-functions over algebraic varieties but do not seem to improve substantially upon the estimates obtainable by our method.

The aim of the present communication is to supply an accessible introduction to our method by giving an almost self-contained account of the estimation in the important special case where $s = 1$ and $N = 3$. The only part of Deligne's work to which we currently appeal is his theorem on the Riemann hypothesis for the L-functions for algebraic varieties over finite fields. If this theorem and one of Weil's estimates be granted, then the treatment is remarkably straightforward and utilizes only that part of the theory of the L-functions which can be established without the intervention of deep algebraic geometry. All the same, the L-function theory required is not to be found in a single place and has not always been expressed in terms that are over familiar to those who might wish to apply it to the analytical theory of numbers. We therefore lead into our method gradually by developing the relevant theory in its generality as simply and succinctly as possible; at two points here we make use of references to Serre's elegant summary [15] of Dwork's work on the zeta functions over varieties. This introductory part, which it is hoped satisfies a demand the author has gathered exists, has little originality save in the expository sense, it being worth drawing particular attention to the proof of Theorem 4 and the absence of Newton polygons from the treatment.

After the estimations for the case where $s = 1$ and $N = 3$ have been completed, we continue by commenting briefly on other avenues of approach and on how the method can be extended to treat the general sum (1). Finally, we briefly summarize some of the results in the theory of numbers to which these estimates have led us.

The author wishes to express his gratitude to Professor Bombieri, with whom he has had instructive correspondence and conversations on the subject of exponential sums.

2. DEFINITIONS OF L-FUNCTIONS AND THEIR EQUIVALENCE

The exponential sums (1) are related to sums and associated zeta or L-functions that are all determined by affine varieties defined over the finite field $\mathbb{F}_p$ of p elements. In fact, however, the theory we develop can easily be extended to cover the situation where the ground field is $\mathbb{F}_{p^r}$, while some of what we do is also applicable to more general forms of variety and to schemes.

Our primary concern will be with varieties in an affine space $\mathbb{A}^N$ of N dimensions that are defined to be the common zeros of a finite number of polynomials

$$g_i(x) = g_i(x_1,\dots,x_N) , \qquad i = 1,\dots,s , \qquad (2)$$

in $\mathbb{F}_p[x] = \mathbb{F}_p[x_1,\dots,x_N]$, or what is equivalent, the common zeros of an ideal $\mathfrak{a}$ in $\mathbb{F}_p[x]$. In such circumstances, we often write the variety as $V = V(\mathfrak{a})$, there being no presumption that V or $\mathfrak{a}$ be irreducible. Although V is defined over $\mathbb{F}_p$, we do not restrict its points correspondingly, since, as we shall see, particular significance attaches to the points on V that appertain to given algebraic extensions of $\mathbb{F}_p$.

As an aid to our work, we shall incidentally be involved with some special quasi-affine varieties $V^* = V^*(\mathfrak{a})$ that are defined exactly as V above save that the points satisfying $x_1\dots x_N = 0$ are excluded.

Finally, there are the projective varieties $V^H = V^H(\mathfrak{a})$ in projective space $\mathbb{H}^N$ that are defined through the solution rays of an homogeneous ideal $\mathfrak{a}$ in $\mathbb{F}_p[x_0,\dots,x_N]$.

The formation of the L-functions over an (affine) variety V depends on simple well known properties of maximal ideals ([17], Chapter XIII). Every point $(\xi_1,\dots,\xi_N)$ defining an algebraic extension $\mathbb{F}_p(\xi_1,\dots,\xi_N)$ of $\mathbb{F}_p$ is the generic point of a (proper) maximal ideal $\mathfrak{m}$ in $\mathbb{F}_p[x]$, while conversely any maximal ideal $\mathfrak{m}$ belongs to precisely r such algebraic points $(\xi_1,\dots,\xi_N)$ that are all generic and that are conjugate. Moreover, we then have $\mathbb{F}_p(\xi_1,\dots,\xi_N) \cong \mathbb{F}_p[x_1,\dots,x_N]/\mathfrak{m}$ so that $\mathbb{F}_p(\xi_1,\dots,\xi_N) = \mathbb{F}_{p^r}$, where $r = \deg \mathfrak{m}$ is the degree of the residue class field, mod $\mathfrak{m}$.

Thus all algebraic points on V are obtained by considering all the maximal ideals $\mathfrak{m}$ that contain $\mathfrak{a}$, each ideal $\mathfrak{m}$ of degree r corresponding to r conjugate points that are exactly of degree r over $\mathbb{F}_p$. In consequence, if we emulate the definition of the classical zeta functions, we are led to the simplest L-function and obtain the zeta function

$$Z(V,T) = \prod_{\mathfrak{m} \supset \mathfrak{a}} (1 - T^{\deg \mathfrak{m}})^{-1}$$

where T in the first place is an indeterminate, the product being defined since there are at most p^{Nr} points on V having co-ordinates in $\mathbb{F}_{p^r}$. This simple example has been principally mentioned because it is easily adapted to meet the needs of an homogeneous variety, the main change being the replacement of $\mathfrak{m}$ by a maximal homogeneous ideal (i.e. an homogeneous ideal that is contained in no larger homogeneous ideal other than the unit ideal) which is different from the one generated by $x_0,\dots,x_N$.

We go on to the more general L-functions over V, which no longer can be easily interpreted in the projective situation. To define these, let $f = f(x)$ belong to $\mathbb{F}_p[x]$ and, for $\alpha \in \mathbb{F}_{p^r}$, let $\sigma(\alpha) = \sigma_r(\alpha)$ be the trace of α taken from $\mathbb{F}_{p^r}$ to $\mathbb{F}_p$, interpreting it when necessary to be any rational integer in the residue class, mod p, that it represents. Then, extending the previous definition, we let

$$L(V,f,T) = L(T) = \prod_{m \supset u} (1 - T^{\deg m} e^{2\pi i \sigma\{f(m)\}/p})^{-1} , \tag{3}$$

where for brevity $\sigma\{f(m)\}$ denotes the unique trace of $f(\xi)$ (taken from $\mathbb{F}_p(\xi)$ to $\mathbb{F}_p$) for any of the zeros ξ of m. Furthermore, substituting formally $T = p^{-s}$, we can write

$$L(s) = \prod_{m \supset u} \left(1 - \frac{e^{2\pi i \sigma\{f(m)\}}}{(Nm)^s}\right)^{-1}$$

as a closer analogue of the classical case, where Nm is the cardinality of $\mathbb{F}_p[x]/m$.

Little change in the definition is needed in order to define the L-functions over the varieties V^*. Letting V^* be derived from V by the restriction $x_1 \ldots x_N \neq 0$ (V is not necessarily unique but this is immaterial) and then letting m^* be a maximal ideal whose zeros ξ satisfy $\xi_1 \ldots \xi_N \neq 0$, we simply obtain

$$L(V^*,f,T) = L^*(V,f,T) \tag{4}$$

by restricting m to m^* in the product on the right of (3).

In either case, extending a previous remark about $Z(V,T)$, we have the following useful lemma by comparing the products for $L(T)$ with the majorant product

$$\prod_{r=1}^{\infty} (1 - T^r)^{-p^{Nr}} .$$

LEMMA 1. *The infinite products for* $L(T)$ *and* $L^*(T)$ *converge when* T *is specialized to complex values* z *with* $|z| < p^{-N}$, *the functions* $L(z)$ *and* $L^*(z)$ *thus obtained being regular in this region.*

The importance of the functions just introduced is that they have alternative formulations in terms of the exponential sums

$$S_r = S_r(V,f) = \sum_{\substack{x \in V \\ x \in \mathbb{F}_{p^r}}} e^{2\pi i \sigma_r\{f(x)\}/p} \tag{5}$$

and

$$S_r^* = S_r(V^*, f) = \sum_{\substack{x \in V^* \\ x \in \mathbb{F}_{p^r}}} e^{2\pi i \sigma_r\{f(x)\}/p}$$

in whose definitions the symbolism $x \in \mathbb{F}_{p^r}$ denotes the condition $x_1, \ldots, x_N \in \mathbb{F}_{p^r}$. We have in fact

THEOREM 1. *If* T *be an indeterminate, then*

$$L(T) = \exp\left(\sum_{r=1}^{\infty} \frac{S_r T^r}{r}\right)$$

and

$$L^*(T) = \exp\left(\sum_{r=1}^{\infty} \frac{S_r^* T^r}{r}\right) \quad .$$

Let the symbol $x \,|||\, d$ denote the condition that $\mathbb{F}_{p^d}$ be the smallest extension of $\mathbb{F}_p$ to which x belongs. Then the argument of the exponential in the right-side of the first proposed identity equals

$$\sum_{r=1}^{\infty} \frac{T^r}{r} \sum_{d|r} \sum_{\substack{x \in V \\ x|||d}} e^{2\pi i (r/d)\sigma_d\{f(x)\}/p}$$

$$= \sum_{d=1}^{\infty} \frac{1}{d} \sum_{\substack{x \in V \\ x|||d}} \sum_{m=1}^{\infty} \frac{T^{md}}{m} e^{2\pi i m \sigma_d\{f(x)\}/p}$$

$$= -\sum_{d=1}^{\infty} \frac{1}{d} \sum_{\substack{x \in V \\ x|||d}} \log(1 - T^d e^{2\pi i \sigma_d\{f(x)\}/p})$$

$$= \sum_{m \supset u} \log(1 - T^{\deg m} e^{2\pi i \sigma\{f(m)\}/p})^{-1} \quad ,$$

from which the result follows. Obvious modifications in the proof then give the second result.

The polynomials in the definition (1) give rise by reduction, mod p, to polynomials over $\mathbb{F}_p$, it not being customary to change the notation after the transition. Hence the sums (1) and $S(V,f) = S_1(V,f)$ are obviously the same because of (5) and the meaning attributed to the trace function in the exponentials.

3. PROPERTIES OF POWER SERIES

The theory rests heavily on the use of power series having coefficients in a field with several valuations. In particular, some important conclusions about such series are derived by comparing their convergence properties with respect to various completions of the field.

As usual $\mathbb{Q}_p$ is the field of p-adic numbers that is the completion of $\mathbb{Q}$ in its p-adic valuation. We then form the complete algebraically closed field Ω_p that is the completion of the algebraic closure of $\mathbb{Q}_p$, the valuation of an element $a \in \Omega_p$ being denoted by $|a|_p$. The valuation ring consisting of the elements in Ω_p with valuations not exceeding 1 is denoted by Λ_p; $\mathfrak{p}$ is the maximal ideal in Ω_p whose elements have valuations less than 1 (van der Waerden [17] is a useful background reference for the theory of fields with valuations).

A function $f(t)$ is said to be (p-adically) regular in the region $|t|_p \le r$ if it be given by a power series with coefficients in Ω_p.

$$\sum_{n=0}^{\infty} a_n t^n$$

that converges for all t in that region. If $r = \infty$, then $f(t)$ is entire. A function is meromorphic for $|t|_p \le r$ if it be the quotient of two functions that are regular in the region.

First, we need the introductory

LEMMA 2. *If the power series*

$$F(t) = 1 + \sum_{n=1}^{\infty} a_n t^n = \sum_{n=0}^{\infty} a_n t^n \qquad (a_n \in \Omega_p)$$

converge for $|t|_p \le 1$, *then we have the identity* $F(t) = G(t)H(t)$, *where* $G(t)$ *is a polynomial*

$$1 + b_1 t + \dots + b_\ell t^\ell$$

and where $H(t)$ *is a power series that is both convergent and invertible for* $|t|_p \le 1$. *Here* ℓ *is the largest* n *for which* $|a_n|_p$ *attains its maximum value, while* $|b_\ell|_p = |a_\ell|_p$. *Also, the zeros of* $G(t)$ *are just those zeros of* $F(t)$ *within* $|t|_p \le 1$.

Since $|a_n|_p \to 0$ as $n \to \infty$ in virtue of the convergence of the given series for $|t|_p = 1$, the number ℓ described in the lemma certainly exists. Dividing $F(t)$ by a_ℓ to form

$$f(t) = \sum_{n=0}^{\infty} c_n t^n \quad ,$$

we then have $f(t) \equiv g_0(t).1$, mod $\mathfrak{p}$, where

$$g_0(t) = c_0 + \dots + c_{\ell-1} t^{\ell-1} + t^\ell \in \Lambda_p[t] \; . \tag{6}$$

Thus we almost have the conditions given in the reducibility criterion in §76 of [17] with $h_o(t) = 1$. The change of $f(t)$ from a polynomial to a power series does not affect the method and we therefore deduce that

$$f(t) = (d_o + \dots + d_{\ell-1}t^{\ell-1} + t^{\ell})h(t) = g(t)h(t) \quad , \tag{7}$$

where

$$g(t) \equiv g_o(t)\ , \bmod p\ , \qquad h(t) \equiv 1\ , \bmod p\ , \tag{8}$$

and where $h(t)$ enjoys the properties attributed above to $H(t)$.

Here $h(t)$ has no zeros for $|t|_p \le 1$, while $g(t)$ has no zeros for $|t|_p > 1$ since in the latter circumstance it is clear that $|g(t)|_p > 1$ by (6) and (8). Hence the zeros of $g(t)$ are exactly those zeros of $f(t)$ that lie in $|t|_p \le 1$. Also, by (8) and by equating the constant terms in (7), we deduce that $|d_o|_p = |c_o|_p = |a_\ell|_p^{-1}$.

To complete the proof, it only remains to multiply $f(t), g(t), h(t)$ by $a_\ell, d_o^{-1}, a_\ell d_o$, respectively.

The lemma obviously implies that a regular function can only have a finite number of zeros in a given disc if it be given by a formally non-zero power series. We thus have

COROLLARY 1. *A p-adic regular function has a unique power series development i.e. the principle of equating coefficients is valid.*

We also have

COROLLARY 2. *A function p-adically meromorphic in a given disc is the quotient of a regular function and a polynomial.*

The next lemma, which is also a corollary of Lemma 2, is a slightly restricted version of a p-adic analogue of Jensen's theorem that is due to Bombieri [1]. Our proof, however, is described in rather different language from that used in [1].

LEMMA 3. *Let the zeros of the p-adic entire function*

$$f(t) = 1 + \sum_{n=1}^{\infty} d_n t^n = \sum_{n=0}^{\infty} d_n t^n$$

be $r_1, r_2, \ldots$. *Then, for any p-adic value*† $R > 0$,

$$\prod_{|r_i|_p \le R} \frac{R}{|r_i|_p} = \max_n |d_n|_p R^n \quad .$$

Writing $R = |\rho|_p$ for some $\rho \in \Omega_p$, we consider the function

$$f_\rho(t) = f(t\rho) = \sum_{n=0}^{\infty} d_n \rho^n t^n$$

that is regular for $|t|_p \le 1$. If ℓ be the maximal value of n for which the maximum of $|d_n|_p R^n$ is attained, then Theorem 1 shews that $f_\rho(t)$ has a factor $1 + e_1 t + \ldots + e_\ell t^\ell$ where $|e_\ell|_p = |d_\ell|_p R^\ell$. Since the zeros of the polynomial factor are the numbers r_i/ρ for which $|r_i|_p \le R$, we infer that

$$\prod_{|r_i|_p \le R} \frac{R}{|r_i|_p} = |e_\ell|_p = |d_\ell|_p R^\ell$$

in accordance with what is required.

The third lemma in the section is a classical result due essentially to Hadamard and É. Borel [3].

LEMMA 4. *Let*

$$f(t) = \sum_{n=0}^{\infty} a_n t^n$$

be a formal power series with coefficients in a given field K *and set*

$$D_{m,q} = \begin{vmatrix} a_m & \cdots & a_{m+q} \\ a_{m+q} & & a_{m+2q} \end{vmatrix} \quad .$$

Then a necessary and sufficient condition that $f(t)$ *be a quotient of two polynomials*†† *is that there be an integer* $q \ge 0$ *such that* $D_{m,q} = 0$ *for all sufficiently large* m .

---

†There is no difficulty in treating the case where R is any positive real number.

††In the first instance, possibly over an extension field of K . But see the remarks following the proof.

First, if $f(t)$ be rational so that $f(t) = g(t)/h(t)$ where $g(t)$ is a polynomial and $h(t) = 1 + h_1 t + \dots + h_r t^r$, then the series for

$$h(t)f(t) = (1 + h_1 t + \dots + h_r t^r)(a_0 + a_1 t + \dots)$$

terminates. Hence, for any $r+1$ consecutive values of $m > m_0$, we have

$$a_{m+r} + h_1 a_{m+r-1} + \dots + h_r a_m = 0 \quad , \qquad (9)$$

from which the condition is seen to be necessary by the elimination of $1, h_1, \dots, h_r$.

Conversely, if $D_{m,q} = 0$ for all sufficiently large m, let r be the minimal value of q having this property, where we may already assume that $r > 0$ since $f(t)$ is a polynomial when $r = 0$. Next, since

$$D_{m+2,r-1}D_{m,r-1} - D^2_{m+1,r-1} = D_{m+2,r-2}D_{m,r} \qquad (D_{n,s} = 1 \text{ for } s = -1)$$

by Jacobi's theorem on the adjugate, the vanishing of $D_{m,r-1}$ for some large m would give $D_{m+1,r-1} = 0$ and hence the impossible conclusion that $D_{n,r-1} = 0$ for all $n \geq m$. Thus $D_{m,r-1} \neq 0$ for all large m. Hence any set of r consecutive equations (9) for $m > m_0$ has a unique solution in $h_1, \dots, h_r$; moreover, the following set has the same unique solution, since any set of $r+1$ consecutive equations is soluble on account of the vanishing of $D_{m,r}$. The set of all (9) for $m > m_0$ therefore has a (unique) solution and we conclude that $f(t)$ is rational.

In preparation for the succeeding lemma, we should note that the sufficiency part of the above proof provides an explicit construction for $g(t)$ and $h(t)$ in terms of the coefficients a_n of the expansion of a rational function $f(t)$; the polynomials $g(t)$, $h(t)$ thus found are relatively prime because of the minimal choice of r, while their coefficients belong to the given field K because of the genesis of $h_1, \dots, h_r$ as quotients of determinants. We can now state the following generalization of a theorem first enunciated by Fatou.

LEMMA 5. *If* $f(t)$ *be a rational function with a power series expansion*

$$1 + \sum_{n=1}^{\infty} a_n t^n$$

whose coefficients are integers in a finite algebraic extension K *of* $\mathbb{Q}$ *, then the zeros and poles of* f(t) *are the reciprocals of algebraic integers.*

Heeding the introductory remarks and after multiplying the initially found numerator and denominator by a suitable (rational) integer, we have

$$f(t) = g^*(t)/h^*(t) \quad , \tag{10}$$

where $g^*(t)$ and $h^*(t) = h_0^* + \ldots + h_r^* t^r$ are relatively prime and have coefficients that are integers in K . Next, forming the resultant of $g^*(t)$ and $h^*(t)$ or otherwise, we find polynomials $\ell_1(t)$ and $\ell_2(t)$ with integral algebraic coefficients such that

$$g^*(t)\ell_1(t) + h^*(t)\ell_2(t) = \text{alg. integer } \alpha \quad ,$$

whence $\alpha/h^*(t)$ has a power series expansion

$$b(t) = \sum_{n=0}^{\infty} b_n t^n$$

whose coefficients are integers in K . Let now the ideals A and B be the contents of $h^*(t)$ and $b(t)$, respectively, so that $A = (h_0^*, \ldots, h_r^*)$ and $B = (b_0, \ldots, b_n, \ldots)$. Then, since

$$A.B = (\alpha),\ h_0^* b_0 = \alpha,\ A|(h_0^*),\ B|(b_0),$$

we infer that $A = (h_0^*)$ and hence that $h_j = h_j^*/h_0^*$ is an algebraic integer for each j . Thus $h^*(t) = h_0^*(1+h_1t+\ldots+h_rt^r) = h_0^*h^\dagger(t)$, say, whence $g^*(t) = h_0^*g^\dagger(t)$ by (10), the coefficients of $g^\dagger(t)$ being also algebraic integers in K . Hence $f(t) = g^\dagger(t)/h^\dagger(t)$ and the lemma is substantiated.

The culmination of this section is a lemma that embodies a simplified special case of an important general theorem proved by Dwork ([6], Theorem 3). Related to a classical result of Borel [3], our lemma is particularly easy to prove but nevertheless suffices for most applications involving exponential sums.

LEMMA 6. *Letting* ε *be a primitive root of unity in the field of complex numbers, form a power series*

$$f(u) = \sum_{n=0}^{\infty} a_n u^n \quad ,$$

where $a_0, \ldots, a_n, \ldots$ *are integers in* $\mathbb{Q}(\varepsilon)$. *Next, if the conjugates of* $a_0, \ldots, a_n, \ldots$ *be denoted by* $a_0^{(j)}, \ldots, a_n^{(j)}$ $(j = 1, \ldots, p-1)$, *we suppose that the complex functions*

$$f_j(z) = \sum_{n=0}^{\infty} a_n^{(j)} z^n$$

are all regular in some complex region $|z| < \delta$. *Finally, identifying* $\mathbb{Q}(\varepsilon)$ *with an isomorphic sub-field of* Ω_p , *we also suppose that*

$$f_0(t) = \sum_{n=0}^{\infty} a_n t^n$$

is a p-adically meromorphic function for all t *(it is necessarily regular for* $|t|_p < 1$ *because* $|a_n|_p \le 1$ *).*

Then $f(u)$ *is a rational function.*

Take $\delta < 1$ and set $r = 3\delta^{-1}$. Then, for $|t|_p \le r$, Corollary 2 to Lemma 2 tells us that $f_0(t) = g(t)/h(t)$, where

$$g(t) = \sum_{n=0}^{\infty} b_n t^n$$

is regular and where

$$h(t) = 1 + h_1 t + \ldots + h_e t^e \qquad (e = e(r))$$

(we may assume the constant term in $h(t)$ to be 1 , since $f_0(t)$ is regular for $|t|_p \le 1$). Hence, for $|t|_p < 1$, we have $g(t) = h(t) f_0(t)$ and so, since this equality is an identity by Corollary 1 to Lemma 2, we deduce that

$$b_{m+e} = a_{m+e} + h_1 a_{m+e-1} + \ldots + h_e a_m \ .$$

Next, if $D_{m,q}$ be defined in terms of a_n as in the statement of Lemma 4, then obviously

$$D_{m,q} = \begin{vmatrix} a_m \ldots a_{m+e-1} & b_{m+e} \ldots b_{m+q} \\ \\ a_{m+q} \quad a_{m+q+e-1} & b_{m+q+e} \quad b_{m+2q} \end{vmatrix}$$

for $q \ge e$. Hence, since $|a_s|_p \le 1$ always and since $|b_s|_p \le 1/r^s$ for $s > s_0$ by the regularity of $g(t)$ on $|t|_p = r$, we conclude that

$$|D_{m,q}|_p < r^{-m(q+1-e)} \qquad (m > m_0) \qquad (11)$$

in view of the non-Archimedean metric.

On the other hand, in the comparable Archimedean situations, the inequality $|a_s^{(j)}| < (2\delta^{-1})^s$ that is valid for $s > s_0$ implies that

$$|D_{m,q}^{(j)}| < (q+1)!(2\delta^{-1})^{(q+1)(m+q)} \qquad (m > m_0)$$

with an obvious meaning for $D_{m,q}^{(j)}$. Hence

$$|\mathrm{Nm}\, D_{m,q}|^{1/(p-1)} < (q+1)!(2\delta^{-1})^{(q+1)(m+q)} \tag{12}$$

for $m > m_0$.

In all, we infer from (11) and (12) that

$$\{|\mathrm{Nm}\, D_{m,q}|_p |\mathrm{Nm}\, D_{m,q}|\}^{1/(p-1)} < (q+1)!(2\delta^{-1})^{q(q+1)}\{(\tfrac{1}{2}r\delta)^{q+1}r^{-e}\}^{-m}$$

because ε is of degree $p-1$ over $\mathbb{Q}_p$. In this

$$(\tfrac{1}{2}r\delta)^{q+1}r^{-e} = (\tfrac{3}{2})^{q+1}r^{-e} > 1$$

for a suitably chosen value of $q \geq e$, whence

$$|\mathrm{Nm}\, D_{m,q}|_p |\mathrm{Nm}\, D_{m,q}| \to 0 \tag{13}$$

as $m \to \infty$. The left-side of (13) being a non-negative rational integer, we deduce that $D_{m,q} = 0$ for $m > m_0$ and thereby conclude the truth of the propostion from Lemma 4.

4. THE RATIONALITY PRINCIPLE FOR THE L-FUNCTIONS

First proved for zeta-functions by Dwork in his fundamental paper [6], the rationality principle was extended to the L-functions by Bombieri in 1966 [1]. The necessary prolegomena being complete, we propose to sketch in simple terms a proof of a slightly modified version of Bombieri's result.

The first step is to shew that every L-function is equal to an L-function whose underlying variety is an affine space $\mathbb{A}^m$ of suitable dimension. If $g_j(x_1,\ldots,x_n) = 0$ $(j = 1,\ldots,s)$ be the equations defining V, let

$$\phi(x_1,\ldots,x_{n+s}) = f(x_1,\ldots,x_n) + \sum_{1\leq j\leq s} x_{n+j}g_j(x_1,\ldots,x_n) \quad .$$

Then, since

$$\sum_{u \in \mathbb{F}_{p^r}} e^{2\pi i \sigma_r(hu)/p} = \begin{cases} p^r , & \text{if } h = 0 , \\ 0 , & \text{if } h \neq 0 , \end{cases} \tag{14}$$

for $h \in \mathbf{F}_{p^r}$, we easily infer that

$$p^{rs} S_r(V,f) = S_r(\mathbb{A}^{n+s},\phi)$$

with the required assertion to the effect that

$$L(V,f,p^s T) = L(\mathbf{A}^{n+s},\phi,T) \ . \tag{15}$$

It therefore being sufficient to establish the rationality theorem for the special category of L-functions, we revert to N-dimensional space and consider the L-function $L(\mathbf{A}^N,f,T) = L(f,T)$, where f is a given polynomial of degree d and where the corresponding exponential sum is

$$S_r(f) = \sum_{x_1,\ldots,x_N \in \mathbb{F}_{p^r}} e^{2\pi i \sigma_r\{f(x_1,\ldots,x_N)\}/p} \ .$$

Next, we express $L(f,T)$ in terms of functions $L^*(T)$ by letting A denote any sub-set of $1,\ldots,N$ and by then defining f_A to be the polynomial in $N - \nu(A)$ variables which is obtained by setting $x_i = 0$ when $i \in A$, $\nu(A)$ being the cardinality of A . If for any polynomial $g(u_1,\ldots,u_t)$ in t variables we write

$$S_r^*(g) = \sum_{\substack{u_1,\ldots,u_t \in \mathbb{F}_{p^r} \\ u_1,\ldots,u_t \neq 0}} e^{2\pi i \sigma_r\{g(u_1,\ldots,u_t)\}/p}$$

and let $L^*(g,T)$ be the corresponding L-function as in Theorem 1, then obviously

$$S_r(f) = \sum_A S_r^*(f_A)$$

and therefore

$$L(f,T) = \prod_A L^*(f_A,T) \ . \tag{16}$$

Moreover, by a combinatorial principle,

$$S_r^*(f) = \sum_A (-1)^{\nu(A)} S_r(f_A)$$

so that

$$L^*(f,T) = \prod_A \{L(f_A,T)\}^{(-1)^{\nu(A)}} \quad . \tag{17}$$

Thus the rationality principles for the functions L and L^* (defined over $\mathbb{A}^n$ and $\mathbb{A}^{*n}$, respectively) are equivalent.

Having disposed of the preliminaries, we sketch the remaining part of the proof with the aid of a reference to §§3, 4, and 5 of Serre's summary [15] of Dwork's work on $Z(T)$. First, since it will be needful to study $L^*(T) = L^*(f,T)$ quâ a function over Ω_p, we embed $\mathbb{Q}(e^{2\pi i/p})$ in Ω_p and therefore identify $e^{2\pi i/p}$ with a primitive p-th root of unity $\varepsilon = 1+\lambda$ in Ω_p. Next, for $y = (y_1,\ldots,y_N)$ and $u = (u_1,\ldots,u_N)$ with non-negative integers u_j, let y^u denote the monomial $y_1^{u_1} \ldots y_N^{u_N}$, where the substitution of a capital letter for y will indicate a specialization to $\bar{\mathbb{F}}_p$; also, to avoid possible confusion, we should point out the symbolism $y^{p^r-1} = 1$ is to mean that $y_j^{p^r-1} = 1$ $(j = 1,\ldots,N)$ in the apposite field.

Virtually following §5 in [15] in writing

$$f(x) = \sum_{1\le i\le R} a_i x^{w_i} \quad ,$$

we immediately transform our sum $S_r^*(f)$ into

$$\begin{aligned} &\sum_{x^{p^r-1}=1} \prod_{1\le i\le R} \theta_r(a_i x^{w_i}) \\ &= \sum_{X^{p^r-1}=1} \prod_{0\le j<r} \prod_{1\le i\le R} \theta(A_i X^{p^j w_i}) \\ &= \sum_{X^{p^r-1}=1} G(X)G(X^p) \ldots G(X^{p^{r-1}}) \text{ , say,} \end{aligned} \tag{18}$$

after comparison with Serre's equations (38), (42), (43), and (44). This in turn being equal to

$$(p^r-1)^N \mathrm{Tr}(\Psi^r) \tag{19}$$

by Proposition 4 in [15], we deduce that

$$\sum_{r=1}^{\infty} \frac{S_r^*(f)T^r}{r} = \sum_{0\le j\le N} (-1)^{N-j}\binom{N}{j} \sum_{r=1}^{\infty} \frac{(p^jT)^r}{r} \mathrm{Tr}(\Psi^r) \quad ,$$

which on exponentiation becomes the *identity*

$$\{L^*(T)\}^{(-1)^{N+1}} = \prod_{0\le j\le N} \{\Delta(p^jT)\}^{(-1)^j\binom{N}{j}} \quad , \tag{20}$$

where

$$\Delta(T) = \det(1 - T\Psi) \quad .$$

Hence $L^*(t)$ is p-adically meromorphic for all t in Ω_p because $\Delta(t)$ is p-adically entire by Serre's Propostion 5.

But Lemma 1 means that the power series expansion of $L^*(T)$ and its conjugates over $\mathbb{Q}$ all give rise to regular functions of z for $|z| < p^{-N}$, while the definition of $L^*(T)$ related to (3) and (4) ensures that the coefficients in this power series are integers in $\mathbb{Q}(e^{2\pi i/p})$. All the conditions in Lemma 6 therefore obtain, and we deduce that $L^*(f,T)$ is rational. Hence $L(f,T)$ is rational, the following extension of Dwork's theorem therefore following in the light of the initial discussion.

THEOREM 2. *The* L-*functions* $L(V,f,T)$ *defined in* (3) *are rational functions. In more detail, these functions are of the form* $g(T)/h(T)$ *, where* $g(T),h(T)$ *are (relatively prime) polynomials in* $\mathbb{Q}(e^{2\pi i/p})[T]$ *and where the zeros of* $g(T)$ *and* $h(T)$ *are the reciprocals of algebraic integers.*

The second part of the above statement is a corollary of Lemma 5 and the comments preceding it.

COROLLARY. *Theorem 2 is valid for the* L-*functions* $L^*(V,f,T)$.

This is easily inferred from Theorem 2 and a decomposition similar to (but not identical with) that used near the beginning of the section. But the only functions of L^* type that we shall need in the sequel are the special ones $L^*(f,T)$ over $\mathbf{A}^N$, and for these the result in the corollary is seen in fact to be anterior to that of Theorem 2 provided that an appropriate appeal be made to Lemma 5.

Before discussing the applications of Theorem 2, it is of some interest to dilate on an aspect of the procedure that led to it. We should note that the identity (20), which can now also be regarded as an equality in Ω_p, can be expressed as

$$\delta^N\Delta(T) = \{L^*(f,T)\}^{(-1)^{N+1}} = L^{**}(f,T) \text{ , say,} \tag{21}$$

where δ is the difference operator defined by $\delta\psi(T) = \psi(T)/\psi(pT)$. Hence, if

$$C(m) = C_N(m) = \binom{N+m-1}{N-1} \tag{22}$$

be the coefficient of x^m in the expansion of $(1-x)^{-N}$, then the usual formulae in the calculus of finite differences give Bombieri's formula [1]

$$\Delta(T) = \prod_{m=0}^{\infty} \{L^{**}(f,p^m T)\}^{C(m)} \tag{23}$$

as the only inversion of (21) that satisfies the necessary conditions in the p-adic metric. But the problem of inversion can be avoided by simply beginning with the equality

$$\mathrm{Tr}(\Psi^r) = (-1)^N (1-p^r)^{-N} S_r^*(f)$$

that is a restatement of (19). Then, by the binomial expansion of $(1-p^r)^{-N}$ that is convergent in the p-adic metric, we have in $\Omega_p[[T]]$ the identity

$$-\sum_{r=1}^{\infty} \frac{\mathrm{Tr}(\Psi^r)T^r}{r} = (-1)^{N+1} \sum_{r=1}^{\infty} \sum_{m=0}^{\infty} C(m) p^{rm} \frac{S_r^*(f)T^r}{r}$$

$$= (-1)^{N+1} \sum_{m=0}^{\infty} C(m) \sum_{r=1}^{\infty} \frac{S_r^*(f)(p^m T)^r}{r} ,$$

from which (23) is immediately recouped.

As a result of Theorem 2, the reciprocals $\omega_1, \ldots, \omega_l$ and $\omega_{l+1}, \ldots, \omega_\kappa$ of the zeros (counted according to multiplicities) of $h(T)$ and $g(T)$ intervene and are termed the characteristic values of $L(V,f,T)$, the characteristic values of $L^*(V,f,T)$ being defined in like manner. The provenance of the L-functions means that we normally locate the characteristic values in the complex field; yet, of course, they could equally well be assigned to Ω_p, thus providing an alternative interpretation that we shall exploit later.

On recalling Theorem 1 and taking logarithms, we obtain the important equation

$$S_r(V,f) = \omega_1^r + \ldots + \omega_l^r - \omega_{l+1}^r - \ldots - \omega_\kappa^r \tag{24}$$

that can be a valuable source of less recondite estimations for exponential sums. Its use, however, is usually contingent on information about the number and magnitudes of the characteristic values, and it is therefore to these two questions that we now address ourselves. Now admitting an utterly elementary answer, the first question is not actually wholly unconnected with the second question and its resolution by Deligne. It is therefore appropriate to give at once a brief summary of the relevant part of Deligne's fundamental researches, thereby not only dealing with the second matter but giving an opportunity of putting our previous work in a different context.

5. THE MAGNITUDES OF THE CHARACTERISTIC VALUES: DELIGNE'S THEORY

Deligne's profound work on the generalized Weil conjectures, of which the rationality principle formed a part, supplies inter alia an alternative proof of the Dwork theorem for the zeta functions. Owing, however, to their new origin in a Lefschetz trace formula, the theorem and its generalizations emerge in a slightly different form in which the polynomials forming the rational function are not necessarily relatively prime. The natural characteristic values now correspond to the polynomials in uncancelled form and have an important significance related to the Betti numbers. Being valid for the uncancelled characteristic values, Deligne's later researches (vid. [16] give a fortiori the following important theorem that, amongst other things, gives the answer to our second question.

THEOREM 3. (Deligne). *We have (Riemann hypothesis)*

$$|\omega_j| = p^{m_j/2} \qquad (j = 1,\ldots,n) \quad ,$$

where m_j *is a non-negative integer. Moreover,* ω_j *and its conjugates over* $\mathbf{Q}$ *have the same modulus, or in other words, all Archimedean valuations of* $\mathbf{Q}(\omega_j)$ *lead to the same value for* ω_j *. Also*

$$p^{\dim V}/\omega_j$$

is an algebraic integer.

Before quitting this short section, we should mention that the elementary (23) provides some information about the magnitudes of the ω_j because the p-adic holomorphy of $\Delta(t)$ obviously implies some simple connections between the poles and zeros of $L^{**}(f,T)$.

6. THE NUMBER OF CHARACTERISTIC VALUES

Surprising as it may now seem, the full resolution of the first question did not occur until after the harder one had been settled. Indeed, for some time the only explicit results in the required direction were some pertaining to various special situations to which Deligne's work had particular relevance. The matter, however, was resolved by Bombieri [2] in 1978 by an argument in which the only essentially deep component was the use of the final part of Theorem 3. However, as pointed out by Dwork and Bombieri, there are at least two ways in avoiding any appeal to Deligne's work. Therefore, following Bombieri's suggested procedure in spirit, we give an entirely elementary proof of his result, the main significance of which is that the number of characteristic values can be bounded by a number independent of p. Our method has much in common with [2] although there are some significant differences - especially in regard to the description, which has been couched in the simplest possible language.

The estimate, it should be added, does not give a bound for the number of uncancelled characteristic values in Deligne's theorem. Yet, it is understood that such an estimate is also now available as a consequence of some recent abstruse work by Katz.

The theorem is derived by first obtaining an analogous estimate for the number, $\nu^*(f)$, of characteristic values α of the function $L^*(\mathbb{A}^N,f,T) = L^*(T)$, where as before the degree of f is d. As a preliminary, we shall need

LEMMA 7. *All Archimedean valuations of each characteristic value* α *of* $L^*(T)$ *are less than or equal to* p^N.

Any conjugate of α over $\mathbb{Q}$ is a characteristic value of one of the conjugates $L^*(af,T)$ of $L^*(f,T)$ over $\mathbb{Q}$ $(a = 1,\dots,p-1)$. The result is then immediate from Lemma 1, since $L^*(af,z)$ is regular and non-zero for $|z| < p^{-N}$.

This elementary lemma is used to provide a surrogate for the p-adic estimate for the α that can be deduced from the last part of Theorem 3. To this end, embedding the field $\mathbb{Q}(e^{2\pi i/p},\alpha_1,\dots,\alpha_u)$ in Ω_p where $u = \nu^*(f)$, let a typical factor - irreducible over $\mathbb{Q}(e^{2\pi i/p})$ and with constant term 1 - in the numerator or denominator of the rational $L^*(T)$ have leading term c and zeros with reciprocals $\beta_1,\dots,\beta_v$, all of which are algebraic integers by the corollary to Theorem 2. We next consider the set of (induced) values $|\beta_i|_p$, using an argument that obviates the need to demonstrate that this (unordered) set is actually independent of the embedding. On the one hand, since $c = \pm\beta_1\cdots\beta_v$, we have

$$|\beta_1|_p \cdots |\beta_v|_p = |c|_p = (|\mathrm{Nm}\ c|_p)^{1/(p-1)} ,$$

where the norm is from $\mathbb{Q}(e^{2\pi i/p})$ to $\mathbb{Q}$. But, since any conjugate $c^{(j)}$ of c over $\mathbb{Q}$ is a product $\pm\beta_1'\ldots\beta_v'$ of conjugates of $\beta_1,\ldots,\beta_v$ over $\mathbb{Q}$, we also have

$$|\mathrm{Nm}\ c| = \prod_j |c^{(j)}| \le p^{Nv(p-1)}$$

by Lemma 7, whence

$$|\beta_1|_p \ldots |\beta_v|_p \ge |\mathrm{Nm}\ c|^{-1/(p-1)} \ge p^{-Nv} .$$

Thus, since $|\beta_i|_p \le 1$, we infer that at least a half of $\beta_1,\ldots,\beta_v$ have values exceeding p^{-2N} and hence that at least a half of all the characteristic values of $L^*(T)$ have values exceeding p^{-2N}. Actually, by the last part of Theorem 3, all the values exceed p^{-N}; this, however, lies much deeper.

It now being for the most part convenient to use the exponential valuation $\mathrm{ord}\ \alpha$ in place of $|\alpha|_p = p^{-\mathrm{ord}\ \alpha}$, let $\rho(M)$ denote the number of zeros δ, say, of the entire function

$$\Delta(t) = \det(1 - t\Psi) = \sum_{m=0}^{\infty} \gamma_m t^m \tag{25}$$

such that $\mathrm{ord}\ \delta \ge -M$, where M will be a positive integer in the sequel. Then it is obvious from (20) and our previous paragraph that

$$\tfrac{1}{2}\nu^*(f) \le \sum_{0\le j\le N} \binom{N}{j}\rho(2N-j) \tag{26}$$

on considering the genesis of the zeros and poles of $L^*(T)$ as zeros of $\Delta(p^j t)$.

To estimate $\rho(M)$ we need quantitative information on γ_m that is obtained by tracing back our steps to §4 and to Serre's article. Letting $c(u) = u_1 + \ldots + u_N$ for $u = (u_1,\ldots,u_N)$, we consider series

$$E(x) = \sum_u e_u x^u$$

satisfying a condition of the type $\mathrm{ord}\ e_u \ge Bc(u)$, where $B > 0$ and where the product of two such series is a similar series that appertains to the same value of B. Thus, because

$$\theta(t) = \sum_{m=0}^{\infty} \beta_m t^m$$

with $\mathrm{ord}\ \beta_m \ge m/(p-1)$ as in Serre's (21) and (22), the function

$$G(x) = \prod_{1\le i\le R} \theta(A_i x^{w_i}) = \sum_u g_u x^u \text{ , say,}$$

appearing in Serre's (43) and our (18) satisfies the condition $\text{ord } g_u \ge c(u)/d(p-1)$. Next, ordering the vectors u in a sequence $\upsilon_1,\dots,\upsilon_m,\dots$ so that $c(\upsilon_m)$ is monotonic and non-decreasing, we can write

$$\Psi = \Psi_{p,G} = (g_{p\upsilon_i-\upsilon_j}) \quad , \qquad i,j \ge 1 \quad ,$$

with the convention that $g_{pu-v} = 0$ when $pu - v$ contains negative components. Hence, if we emulate Serre's p.08 with his M equal to $1/d(p-1)$ and with a slightly different notation, we conclude that

$$\text{ord } \gamma_m \ge \frac{1}{d} \sum_{1\le i\le m} c(\upsilon_i) \tag{27}$$

for $m \ge 1$, the inequality being trivial for $m = 0$.

The estimate for $\rho(M)$ is drawn from (27) by means of Lemma 3 with $R = p^{M+1}$. We obtain

$$\rho(M) \le \sum_{\text{ord } \delta \ge -M-1} (M+1 + \text{ord } \delta) = \max_m (m(M+1) - \text{ord } \gamma_m)$$
$$\le \max_m (m(M+1) - \frac{1}{d} \sum_{1\le i\le m} c(\upsilon_i)) \quad ,$$

the latter maximum being clearly attained when m is the last value of i for which $c(\upsilon_i)$ is equal to $d(M+1)$. Hence, the number of solutions of $c(\upsilon) = m$ being the binomial coefficient $C_N(m)$ by (22), we get

$$\rho(M) \le (M+1) \sum_{s\le d(M+1)} C_N(s) - \frac{1}{d} \sum_{s\le d(M+1)} sC_N(s)$$
$$= (M+1)\binom{N + d(M+1)}{N} - \frac{N}{d}\binom{N + d(M+1)}{N + 1}$$
$$= \frac{1}{d}\binom{N + d(M+1)}{N + 1} \tag{28}$$

by an elementary calculation.

From (26) and (28) we infer that

$$\tfrac{1}{2} d\nu^*(f) \le \sum_{0\le j\le N} \binom{N}{j}\binom{N + d(2N+1-j)}{N + 1} \quad ,$$

the right-side of which inequality is seen to be the coefficient of $z^{(2N+1)d-1}$ in the Taylor expansion of $(1+z^d)^N(1-z)^{-N-2}$. Hence, by one of Cauchy's inequalities applied to the circle $z = (1+d^{-1})^{-1}$,

$$\tfrac{1}{2}d\nu^*(f) \le \{1 + (1+d^{-1})^{-d}\}^N (1+d)^{N+2}(1+d^{-1})^{(2N+1)d-1}$$

$$= d\{1 + (1+d)^{-1})^d\}^N (1+d)^{N+1}(1+d^{-1})^{(N+1)d}$$

$$\le d(1+e)^N (1+d)^{N+1} e^{N+1} \quad ,$$

whence

$$\nu^*(f) \le \{e(e+1)(d+1)\}^{N+1} < (11d+10)^{N+1} \quad .$$

We thus have the very first part of the next theorem, the other parts then flowing from (16) and (15).

THEOREM 4. *The numbers of characteristic values of* $L^*(\mathbf{A}^N,f,T)$ *and* $L(\mathbf{A}^N,f,T)$ *do not exceed* $(11d+10)^{N+1}$ *and* $(11d+11)^{N+1}$ *, respectively. The number of characteristic values of* $L(V,f,T)$ *has an upper bound that depends only on* N *,* d *, and the number and degrees of the equations defining* V *.*

The importance of the theorem is that the bounds it provides are independent of p . Yet, the reader will observe that some improvements are certainly possible since certain mutable parameters have been chosen for convenience rather than for optimal effect. He will also note how an improved result could be proved with less effort if recourse were had to the inequality $\operatorname{ord} \alpha \le N$ that stems from Theorem 3 and from (17), albeit the resulting demonstration would be inherently much deeper than the one we gave. It is therefore not without interest to mention that a trivial corollary of our theorem is the weak bound $\operatorname{ord} \alpha \le N(11d+10)^{N+1}$, which in fact can be substantially strengthened by elementary arguments.

7. THE ESTIMATION OF THE EXPONENTIAL SUM WHEN s = 1 and N = 3

The comment at the end of §2 means we can always estimate the sum (1) by considering $S(V,f)$ instead. Everything up to this point having been valid for general values of s and N , we are ready to apply our method to the assessment of

$$S(V,f) = \sum_{\substack{g(X_1,X_2,X_3)\equiv 0,\,\mathrm{mod}\,p \\ 0\le X_1,X_2,X_3<p}} e^{2\pi i f(X_1,X_2,X_3)/p} \qquad (29)$$

under very broad conditions. Here V is the variety $g(x) = g(x_1,x_2,x_3) = 0$ defined over $\mathbf{F}_p$, the notation for polynomials still being in accord with the convention imposed at the end of §2.

In what follows $A_1, A_2, \ldots,$ and the constants implied by the O-notation are positive numbers depending at most on the degrees of f and g.

We first remark that it is entirely natural from several points of view to restrict discussion to the case where $g(x)$ contains only one factor that is irreducible over $\mathbb{F}_p$ and where therefore V is irreducible over $\mathbb{F}_p$. Indeed, if $V = \bigcup_{1 \le i \le r} V_i$ be an irredundant representation of V corresponding to the factorization of $g(x)$ into a product $\lambda \prod_{1 \le i \le r} \{g_i(x)\}^{m_i}$ of distinct polynomials $g_i(x)$ that are irreducible over $\mathbb{F}_p$, then

$$|S(V,f) - \sum_{1 \le i \le r} S(V_i,f)| \le \sum_{1 \le i < j \le r} S(V_i \cap V_j, 0) = O(p) \quad ,$$

since there are at most $A_1 p$ points $x \in \mathbb{F}_p$ on the (possibly reducible) curve $V_i \cap V_j$. Actually, if we be interested in best possible constants, there is usually little difficulty in replacing the $O(p)$ here by $O(p^{\frac{1}{2}})$. Also, since the sum is obviously unaffected by the change of $g(x)$ into $\prod g_i(x)$, it is instinctive to begin the problem with a square-free polynomial $g(x)$ so that the ideal $(g(x))$ would belong to V. However, this simplification cannot always be easily arranged because in practice $g(x)$ may arise as the reduction, mod p, of an assigned square-free polynomial in $\mathbb{Z}[x]$; there may then be certain values of p for which $g(x)$ fails to be square-free.

Next, having reduced the problem to one where V is irreducible over $\mathbb{F}_p$, we may in fact suppose that V is absolutely irreducible. If, to the contrary, V be reducible over $\mathbb{F}_{p^r}$ for some $r > 1$, then the ideal belonging to V is generated by a polynomial $G(x)$ that is irreducible over $\mathbb{F}_p$ but that equals $h(x)k(x)$ for non-constant polynomials $h(x), k(x)$ in $\mathbb{F}_{p^r}[x]$. Hence, if $\Omega_1, \ldots, \Omega_r$ be a basis for $\mathbb{F}_{p^r}$ over $\mathbb{F}_p$, then $h(x) = \Omega_1 h_1(x) + \ldots + \Omega_r h_r(x)$, where $h_1(x), \ldots, h_r(x)$ are in $\mathbb{F}_p[x]$ and have no non-trivial common factor owing to the irreducibility of $G(x)$ over $\mathbb{F}_p$ (thus there are certainly two of the $h_j(x)$ that are non-zero and that are not both constant). The zeros of $h(x)$ in $\mathbb{F}_p$ being the common zeros of $h_1(x), \ldots, h_r(x)$ in $\mathbb{F}_p$, we deduce from Hilbert's Nullstellensatz that these zeros lie on at most A_2 curves defined over $\mathbb{F}_p$. Since the same is true for $k(x)$, we deduce that

$$S(V,f) = O(p) \quad ,$$

thus dismissing the case where V is only relatively irreducible over $\mathbb{F}_p$.

Matters have been arranged so that conditions may now be imposed without unnecessarily restricting the scope of the method. We shall limit $f(x)$ and $g(x)$ in the forthcoming estimation by supposing that the generic variety $\mathcal{W}(t)$ given by

$$f(x) - t = 0 \quad , \quad g(x) = 0 \tag{30}$$

is subject to the conditions

(i) ***generically*** $\mathcal{W}(t)$ ***is an absolutely irreducible curve*** (31)

(ii) $\mathcal{W}(t)$ ***is a curve (possibly reducible) or the zero variety for all specializations of*** t ***to*** $\bar{\mathbb{F}}_p$. (32)

The first condition, which means that the curve $\mathcal{W}(t)$ is irreducible over the algebraic closure of the function field $\mathbb{F}_p(t)$, implies that $\mathcal{W}(t)$ is an absolutely irreducible curve for all specializations of t to $\bar{\mathbb{F}}_p$ save for those in a set having cardinality less than A_3. The nature of $\mathcal{W}(t)$ on the exceptional set is given by the second condition, which by the theory of ideals is tantamount to the stipulation that $\mathcal{W}(t)$ be never of dimension 2 (i.e. never contain a surface).

Writing

$$S_r(\mu) = S_r(\mathcal{V}, \mu f)$$

for brevity when $\mu \in \mathbb{F}_{p^r}$, we form the moment

$$M_r = \sum_{\substack{\mu \in \mathbb{F}_{p^r} \\ \mu \neq 0}} |S_r(\mu)|^2 \tag{33}$$

and estimate it by letting $N_r(\tau)$ for $\tau \in \mathbb{F}_{p^r}$ be the number of solutions in $\mathbb{F}_{p^r}$ of (30) when $t = \tau$. By the definition of $S_r(\mu)$ in (5) and then by the second part of (14), we have

$$\begin{aligned} S_r(\mu) = \sum_{\tau \in \mathbb{F}_{p^r}} N_r(\tau) e^{2\pi i \sigma_r(\mu\tau)/p} &= \sum_{\tau \in \mathbb{F}_{p^r}} (N_r(\tau) - p^r) e^{2\pi i \sigma_r(\mu\tau)/p} \\ &= S_r'(\mu) \text{ , say ,} \end{aligned}$$

when $\mu \neq 0$. Hence we infer that

$$M_r \leq \sum_{\mu \in \mathbb{F}_{p^r}} |S_r'(\mu)|^2 = p^r \sum_{\tau \in \mathbb{F}_{p^r}} (N_r(\tau) - p^r)^2$$

from (33) and the Parseval principle that stems from (14). Invoking the work of Lang-Weil to complete the assessment, we use the inequality

$$N_r(\tau) - p^r = O(p^{r/2}) \tag{34}$$

that is valid when the curve $\mathcal{W}(\tau)$ is absolutely irreducible (Lang-Weil [13]; see, also, Schmidt [14], pp.260-264 for a more elementary approach). Since by (32) the inequality $N_r(\tau) = O(p^r)$ is valid in respect of the $O(1)$ values of τ in $\mathbb{F}_{p^r}$ for which (34) fails, we conclude that

$$M_r = O(p^{3r}) + O(p^{3r}) = O(p^{3r}) \quad . \tag{35}$$

To exploit this inequality, let now μ be any non-zero element of $\mathbb{F}_p$ and express the application of Theorem 2 to $L(\mathcal{V},\mu f,T)$ in the form

$$L(\mathcal{V},\mu f,T) = g_\mu(T)/h_\mu(T) \quad ,$$

where $g_\mu(T)$ and $h_\mu(T)$ are polynomials with coefficients in $\mathbb{Q}(\sqrt[p]{1})$. Then, since a change in μ merely takes $g_\mu(T)$ and $h_\mu(T)$ into polynomials conjugate to them under a common automorphism of $\mathbb{Q}(\sqrt[p]{1})$, we may write the apposite form of (24) as

$$S_r(\mu) = \omega_{1,\mu}^r + \dots + \omega_{\iota,\mu}^r - \omega_{\iota+1,\mu}^r - \dots - \omega_{\kappa,\mu}^r \quad ,$$

where $\omega_{1,\mu},\dots,\omega_{\kappa,\mu}$ are conjugate to $\omega_{1,\nu},\dots,\omega_{\kappa,\nu}$ and where $\omega_{i,\mu} = \omega_{j,\mu}$ if and only if $\omega_{i,\nu} = \omega_{j,\nu}$. Hence, combining terms corresponding to common values of $\omega_{i,\mu}$ and then suitably reordering the subscripts, we infer from Theorem 4 that

$$S_r(\mu) = e_1\omega_{1,\mu}^r + \dots + e_k\omega_{k,\mu}^r \quad (\omega_{i,\mu} \neq \omega_{j,\mu} \ \text{ for } \ i \neq j) \ , \tag{36}$$

in which $e_1,\dots,e_k$ are integers that are independent of μ and r and that are bounded in number (may be zero) and magnitude by A_4.

Next, by Theorem 3, each modulus $|\omega_{j,\mu}|$ is a power of $p^{\frac{1}{2}}$ and is independent of μ. Let us now suppose that at least one of the numbers $\omega_{j,\mu}$ in (36) had modulus greater than p, a hypothesis that is independent of the value of μ chosen. Then, isolating those values of $\omega_{j,\mu}$ having maximal modulus $p^{H/2}$ with $H \geq 3$ and attributing to them the subscripts $1,\ldots,\ell$, we would have

$$S_r(\mu) = e_1\omega_{1,\mu}^r + \ldots + e_\ell\omega_{\ell,\mu}^r + O(p^{r(H-1)/2})$$

$$= p^{rH/2}(e_1z_{1,\mu}^r + \ldots + e_\ell z_{\ell,\mu}^r) + O(p^{r(H-1)/2}) \tag{37}$$

in which $\ell > 0$ and $z_{1,\mu},\ldots,z_{\ell,\mu}$ were distinct roots of unity. Thus we would infer that

$$p^{-rH}|S_r(\mu)|^2 = |e_1z_{1,\mu}^r + \ldots + e_\ell z_{\ell,\mu}^r|^2 + O(p^{-r/2})$$

and hence that

$$p^{-rH}M_r \geq p^{-rH}\sum_{\substack{\mu\in\mathbb{F}_p\\ \mu\neq 0}} |S_r(\mu)|^2$$

$$= \sum_{\substack{\mu\in\mathbb{F}_p\\ \mu\neq 0}} |e_1z_{1,\mu}^r + \ldots + e_\ell z_{\ell,\mu}^r|^2 + O(p^{1-(r/2)})$$

by restricting the summation in (33) to values of μ in $\mathbb{F}_p$. By (35), this would give

$$\sum_{\substack{\mu\in\mathbb{F}_p\\ \mu\neq 0}} |e_1z_{1,\mu}^r + \ldots + e_\ell z_{\ell,\mu}^r|^2 = O(p^{(3-H)/r}) + O(p^{1-(r/2)})$$

$$= O(1) + O(p^{1-(r/2)}) \quad ,$$

from which it would be concluded that

$$\sum_{\substack{\mu\in\mathbb{F}_p\\ \mu\neq 0}} \frac{1}{u}\sum_{r\leq u} |e_1z_{1,\mu}^r + \ldots + e_\ell z_{\ell,\mu}^r|^2 = O(1) + O(\frac{p^{\frac{1}{2}}}{u}) \quad . \tag{38}$$

But it is easily ascertained that, for given distinct roots of unity $z_1,\ldots,z_\ell$,

$$\lim_{u\to\infty} \frac{1}{u}\sum_{r\leq u} |e_1z_1^r + \ldots + e_\ell z_\ell^r|^2 = e_1^2 + \ldots + e_\ell^2 \quad . \tag{39}$$

Hence, letting $u \to \infty$ in (38), we would get

$$(p-1)(e_1^2 + \ldots + e_\ell^2) = O(1) \quad , \tag{40}$$

which would result in a contradiction for $p > A_5$.

The supposition made at the beginning of the previous paragraph having been discounted save when $p \le A_5$, we deduce from (36) that, for $p > A_5$,

$$S_r(V,f) = O(p^r)$$

and hence, in particular, that

$$S(V,f) = O(p) \quad .$$

Since the final inequality is trivially true for $p \le A_5$, we therefore obtain

THEOREM 5. *Let the exponential sum* $S(V,f)$ *be defined as in* (29). *Then*

$$S(V,f) = O(p)$$

if the conditions (31) *and* (32) *on* f *and* g *hold, the constant implied by O-notation depending at most on the degrees of* f *and* g .

The conditions (31) and (32) cannot both hold if V be reducible. In that case we must first use the reduction given at the beginning of the section and should then apply the criterion to each absolutely irreducible factor of $g(x)$ that already belongs to $\mathbb{F}_p[x]$.

That there must be some restriction akin to (31) and (32) is clear from a consideration of special cases. For instance, if $f(x) = g(x)$, then $|S(V,f)| = |S(V,o)| > A_6 p^2$ in general; here (32) fails. Also, if $f(x) = x_1^2$ and $g(x) = x_2 + x_3$, then $|S(V,f)| = p^{3/2}$ for $p > 2$ by the estimate for the Gauss sum; here (31) obviously fails. Moreover, the criterion satisfactorily treats all the exponential sums of type (29) that we have so far encountered in practical applications.

As already hinted, there are problems in analytical number theory where polynomials $f(x),g(x) \in \mathbb{Z}[x]$ arise and where then an estimate for the right-hand sum in (29) is needed for all or large p . In such cases we must consider the applicability of our procedures to the reductions of f and g , mod p , for each p . However, in practice it often suffices that we be granted the analogues of (31) and (32) with the role of $\mathbb{F}_p$ being replaced by $\mathbb{Q}$. In such instances criteria (31) and (32) will then hold in respect of $\mathbb{F}_p$ for all but a finite number of p , the estimate $O(p)$ for the sums emerging for all p provided that the constant implicit in the O-notation may depend on f and g and not merely their degrees. If such a process be precluded by the reducibility of $g(x)$ over $\mathbb{Q}$ or $\bar{\mathbb{Q}}$, then we must first factor $g(x)$ in $\mathbb{Q}[x]$ or $\bar{\mathbb{Q}}[x]$ and initiate comparisons between $\bar{\mathbb{F}}_p[x]$ and $\bar{\mathbb{Q}}[x]$ that are too complicated to be dwelt upon here.

8. VARIANTS OF THE METHOD

Our dependence on Deligne's and Weil's results can sometimes be abated when we seek estimates for certain exponential sums that have arisen out of problems in additive number theory. To begin with, there are several instances in which the bound (35) follows from estimates for the sum

$$\sum_{\tau \in \mathbb{F}_{p^r}} N_r^2(\tau) \tag{41}$$

that have been derived in an entirely elementary manner, one approach, but by no means the only one, being through the fact that (41) is the number of solutions of the equations

$$f(x) = f(y) , \quad g(x) = g(y)$$

in $\mathbb{F}_{p^r}$. When successful, such a procedure not only eliminates an appeal to Weil's work but may sometimes also give an asymptotic formula for M_r that enables us to simplify the rest of the work.

Let us amplify the last point by referring to the formula $M_r \sim p^{3r}$ which can often be obtained in situations indirectly associated with the theory of elliptic curves. Here we can obviously use (37) for $\mu = 1$ and $\mu = -1$ merely because we can suppose that $\omega_{i,1}$ and $\omega_{i,-1}$ are complex conjugates having the same modulus. Since the consequent restriction of the summation in (33) to these two values gives $2(e_1^2 + \ldots + e_\ell^2)$ in place of the left-hand side of (40), we would obtain the impossible inequality $2(e_1^2 + \ldots + e_\ell^2) \leq 1$ after injecting the estimate $M_r \sim p^{3r}$ into the treatment. Thus we can occasionally avoid using Deligne's result about the uniqueness of the Archimedean valuations of the characteristic values. An example of this method with some additional twists is described in our paper on two cubes [9]; the procedure there is actually simpler than that suggested here because we were not tied entirely to the structure of the previous section. The generalized Kloosterman sum, which has by now been estimated successfully in countless ways, can also be handled by this method.

The above routine can sometimes be adapted to suit our needs in surroundings where an immediately applicable formula for M_r is either unavailable or is not obtainable by elementary methods. To take a case in point, if we choose the sum with $g(x) = x_1^3 + x_2^3 + x_3^3 - n$ and $f(x) = a_1x_1 + a_2x_2 + a_3x_3$ $(a_1a_2a_3 \neq 0)$, then the inequality $M_r \leq 3p^{3r} + O(p^{5r/2})$ is the best that we have obtained by elementary methods when $p \equiv 1$, mod 3, even though the formula $M_r \sim p^{3r}$ is usually a consequence of the deeper theory of elliptic curves. But here there are two non-trivial cube roots of unity ρ_1 and ρ_2 in $\mathbb{F}_p$ and it is easy to see the six sums $S_r(\pm 1)$, $S_r(\pm\rho_1)$, $S_r(\pm\rho_2)$ all have the same modulus. Thus the method of the previous paragraph can be readily restored and could have been applied to the estimation of this exponential sum in our paper on Waring's problem for two squares and three cubes [11]. In the event, however, we avoided this attack because the method of the previous section was much quicker to apply.

Alternatively, we could look for an upper bound for the fourth moment

$$\sum_{\substack{\mu \in \mathbb{F}_{p^r} \\ \mu \neq 0}} |S_r(\mu)|^4 ,$$

which of course would be $O(p^{5r})$ if Theorem 5 were already supposed to be known. The consequence of such an estimate found in another way would be the bound $S_r(1) = O(p^{5r/4})$, from which the superior estimate $S(V,f) = O(p)$ would readily flow from (39) and the Riemann hypothesis. But little is usually gained by attempting to activate the mechanism, since the estimation of the fourth moment is normally extraordinarily difficult unless cohomology theory be introduced. None the less, the generalized Kloosterman sum easily lends itself to this treatment.

9. THE EXPONENTIAL SUMS IN THE GENERAL CASE

The nature of the underlying variety V becomes very important when we seek an estimate for (1) for general values of s and N. We can no longer expect good estimates based on $\dim V$ alone and, in particular, find that the plausible bound $O(p^{(\dim V)/2})$ is certainly false unless pretty stringent geometrical conditions be enforced.

Our method can be extended to the general exponential sum in such a way that it takes account of the geometrical structure. Two main cases are then revealed. The first is where an irreducibility condition similar to (31) and (32) is satisfied, the estimate $O(p^{\dim V - 1})$ being then valid for $\dim V \geq 2$; this is an improvement on the previously best known estimate $O(p^{\dim V - \frac{1}{2}})$ that is deducible from Weil's estimate for the case $\dim V = 1$. At the other extreme, there is the case where we are given a strong condition involving the non-singularity of the relevant varieties and where the estimate $O(p^{(\dim V)/2})$ actually holds. Intermediate situations also occur but these must be examined in the light of the inherent geometry.

A full account of these methods will appear in a forthcoming publication.

10. APPLICATIONS OF THE ESTIMATES

We have used the estimates to shew that, if $\nu(x)$ be the number of integers not exceeding x that can be represented as the sum of two non-negative h-th powers in at least two essentially different ways, then for odd $h \geq 3$

$$\nu(x) = O(x^{(5/3h)+\varepsilon})$$

in contrast to the trivial estimate $O(x^{2/h})$ ([9], [10], in which generalizations and consequences of this result are also discussed). Amongst other things, this is a contribution to the study of whether the Diophantine equation

$$X^h + Y^h = Z^h + W^h$$

can have non-trivial solutions when $h \geq 5$. A similar method involving exponential sums shews that the number of representations of n as the sum of four non-negative cubes is $O(n^{(11/18)+\varepsilon})$, simplifying a technically more complicated treatment that was devised before Deligne's results were proved [8].

The estimates also played a very material part in our derivation of an asymptotic formula for the mixed Waring's problem for two squares and three cubes. The existence theorem for this problem having only recently been established by Linnik through his ergodic method, the importance of our formula is that it shews the situation here conforms to the traditional pattern of results in the theory of Waring's problem. There are also other applications to Waring's problem on which there is not time to expand here.

Estimates for the generalized Kloosterman sum give improvements in known asymptotic formulae for the divisor sums

$$\sum_{\substack{n \leq x \\ n \equiv a, \bmod k}} d_r(n)$$

with particular emphasis on the uniformity of the remainder terms with respect to k . Such formulae, however, seem to have little application in practice since better results on average are obtainable through the use of the large sieve.

Estimates for the number of solutions of a congruence $g(x_1,\ldots,x_n) \equiv 0$, mod p , lying in a restricted region will be described in a later publication.

These examples probably by no means exhaust the potential applications of such estimates to the analytical theory of numbers. We foresee that the theory of exponential sums will have a profound influence on the analytical theory after Deligne's methods have been further studied.

REFERENCES

We have included in the list below some useful relevant papers to which we have not referred in the text.

[1] E. Bombieri, On exponential sums in finite fields, Amer.J.Math. 88 (1966), 71-105.

[2] E. Bombieri, On exponential sums in finite fields: II, Inventiones Math. 47 (1978), 29-39.

[3] E. Borel, Sur une application d'un théorem de M. Hadamard, Bull.Sci.Math. (2) 18 (1894), 22-25.

[4] P. Deligne, La conjecture de Weil: I, Publ.Math. IHES 43 (1974), 273-307.

[5] P. Deligne, Séminaire de géométrie du Bois-Marie S.G.A. 4½, Lecture Notes in Mathematics 569 (Springer, Berlin, 1966).

[6] B. Dwork, On the rationality of the zeta function of an algebraic variety, Amer.J.Math. 82 (1960), 631-648.

[7] B. Dwork, On the zeta function of a hypersurface: III, Ann. of Math. 83 (1966), 457-519.

[8] C. Hooley, On the representations of a number as the sum of four cubes: I, Proc. London Math.Soc. (3) 36 (1978), 117-140.

[9] C. Hooley, On the numbers that are representable as the sum of two cubes, J. Reine Angew.Math. 314 (1980), 146-173.

[10] C. Hooley, On another sieve method and the numbers that are a sum of two h-th powers, Proc. London Math.Soc. (3) 43 (1981), 73-109.

[11] C. Hooley, On Waring's problem for two squares and three cubes, J. Reine Angew.Math. 328 (1981), 161-207.

[12] N.M. Katz, An overview of Deligne's proof of the Riemann hypothesis for varieties over finite fields, Amer.Math.Soc., Proceedings of Symposia in Pure Mathematics, Vol.XXVIII, 1976.

[13] S. Lang and A. Weil, Number of points on varieties in finite fields, Amer.J.Math. 76 (1954), 819-827.

[14] W.M. Schmidt, Equations over finite fields; an elementary approach, Lecture Notes in Mathematics 536 (Springer Verlag, 1976).

[15] J.-P. Serre, Rationalité des fonctions ζ des variétés algébriques (d'aprés Bernard Dwork), Séminaire Bourbaki 1959/60, Exposé 198 (W.A. Benjamin, New York, 1966).

[16] J.-P. Serre, Majorations de sommes exponentielles, Soc.Math. de France, Asterisque 41-42 (1972), 111-126.

[17] B.L. Van der Waerden, Modern Algebra (English edition) (Frederick Ungar, 1949).

[18] A. Weil, Sur les courbes algébriques et les variétés qui s'en déduisent (Hermann, Paris, 1948).

ON THE CALCULATION OF REGULATORS AND CLASS NUMBERS OF QUADRATIC FIELDS

H.W. Lenstra, Jr.

Introduction

In this lecture we present a mathematical model that can be used to analyze Shanks's algorithm to determine the regulator of a real quadratic field, see [24]. Let me briefly describe the situation.

In an earlier paper [23], Shanks indicated a method to calculate the class group of an imaginary quadratic field. For this method, it is convenient to view the class group as a group of equivalence classes of quadratic forms, the group multiplication being *composition* of forms. A basic fact underlying the algorithm is that every equivalence class contains exactly one *reduced* form. In the real quadratic case, this is not true any more; here every equivalence class contains a whole *cycle* of reduced forms. Shanks observed [24], that the principal cycle, corresponding to the neutral element of the class group, displays a certain group-like behaviour with respect to composition. In this lecture, we introduce a group F whose properties can be used to give precise formulations and proofs of Shanks's observations. The group is defined as the set of orbits of quadratic forms under $\begin{pmatrix} 1 & \mathbb{Z} \\ 0 & 1 \end{pmatrix}$ rather than $SL_2(\mathbb{Z})$. It has a close relationship to a certain group of idele classes. For a different approach to the analysis of Shanks's methods we refer to Lagarias [7; 8; 9].

In the first few sections below we present the standard dictionary between ideal classes and classes of quadratic forms in the way we need it, cf. [1]. Each of the languages has its merits: the ideals can be used for smooth and conceptual definitions and proofs, and the forms are a convenient vehicle for computations. In section 7 we describe Shanks's algorithm for imaginary quadratic fields, the main ideas of which also play a role in the more complicated real quadratic case. Sections 8 to 12 are devoted to the group $\mathcal{F}$ mentioned above, and section 13 gives an informal description of how its properties can be used to calculate regulators and class numbers of real quadratic fields. The final section touches upon some applications of the material in this lecture.

The correctness and efficiency of most of the algorithms that we describe depend on the generalized Riemann hypothesis. It would be of interest to obtain explicit versions of all inequalities used, assuming the Riemann hypotheses. It would also be of interest to see what remains if no unproved hypotheses are assumed.

The quadratic field K that we consider is supposed to be given by its discriminant. Checking that a given integer is the discriminant of a quadratic field involves testing squarefree-ness. For this I know no essentially faster method than factoring the number, and there is a good reason not to do this: namely, one of the most efficient factoring algorithms is based on the connection between the factorizations of the discriminant and the elements of order two in the class group, and makes use of the ideas set forth in this lecture; see sec. 15 for references. The only way out is that we develop the entire theory for arbitrary orders in quadratic fields rather than just the maximal order.

Throughout this paper the terms "class group" and "regulator" are used in the *strict* (narrow) sense: see the definitions in sections 2 and 6, respectively, and the end of section 13.

We denote by $\mathbb{Z}$, $\mathbb{Q}$, $\mathbb{R}$ and $\mathbb{C}$ the ring of integers, the field of rational numbers, the field of real numbers, and the field of complex numbers, respectively. For a ring B with 1, we denote by B* the group of units of B. The reader should note the distinction between $\mathcal{N}$, $\mathcal{R}$, $\mathcal{P}$, $\mathcal{F}$, $\mathcal{G}$ and N, R, P, F, G.

1. Orders in quadratic fields

Let K be a quadratic field extension of $\mathbb{Q}$. Denote by σ the non-trivial field automorphism of K, and define the *norm* $N: K \to \mathbb{Q}$ by

$$N(\alpha) = \alpha \cdot \sigma(\alpha), \qquad \text{for } \alpha \in K.$$

Let A_0 be the ring of algebraic integers in K. An *order* in K is a subring A of A_0 with $1 \in A$ and with field of fractions K. Every order A in K satisfies $\mathbb{Z} \subset A \subset A_0$, and since $A_0/\mathbb{Z}$ is cyclic as an additive group, every order is determined by its index in A_0. This index is finite and called the *conductor* of A. Every positive integer f occurs as the conductor of an order A in K, namely $A = \mathbb{Z} + fA_0$. If $A = \mathbb{Z}e_1 + \mathbb{Z}e_2$, then the *discriminant* Δ of A is defined by $\Delta = (e_1\sigma(e_2) - e_2\sigma(e_1))^2$; this is an integer which does not depend on the choice of the basis e_1, e_2. We have $\Delta = f^2 \cdot \Delta_0$, where f is the conductor of A and Δ_0 is the discriminant of A_0; we call Δ_0 also the discriminant of K. The integer Δ is not a square, and $\Delta \equiv 0$ or $1 \bmod 4$. Conversely, any non-square integer Δ that is 0 or 1 mod 4 is the discriminant of a uniquely determined order in a quadratic field, namely $A = \mathbb{Z}[(\Delta + \sqrt{\Delta})/2] \subset K = \mathbb{Q}(\sqrt{\Delta})$. It will be convenient, in the sequel, to assume that K is embedded in $\mathbb{C}$; square-roots of real numbers will be assumed to lie on the non-negative part of the real or imaginary axis.

2. Invertible ideals

Let K be a quadratic field, and $A \subset K$ an order of discriminant Δ. The *product* $M \cdot M'$ of two subsets $M, M' \subset K$ is the additive subgroup of K generated by $\{x \cdot y: x \in M, y \in M'\}$. An *invertible A-ideal* is a subset $M \subset K$ with $A \cdot M = M$ for which there exists M' such that $M \cdot M' = A$. Its *inverse* $A \cdot M'$ is then also an invertible A-ideal, and the set of invertible A-ideals is a commutative group with respect to multiplication. We denote this group by I.

Let M be an invertible A-ideal, and $M \cdot M' = A$. We claim that

$$A = \{\alpha \in K: \alpha M \subset M\}. \tag{2.1}$$

The inclusion $\subset$ is obvious. Conversely, if $\alpha M \subset M$ then $\alpha = \alpha \cdot 1 \in \alpha A = \alpha M \cdot M' \subset M \cdot M' = A$, as required.

From $M \cdot M' = A$ we see that there exist $x_i \in M$, $y_i \in M'$ $(1 \le i \le t)$ such that $\Sigma_{i=1}^{t} x_i y_i = 1$. Then $Ax_1 + Ax_2 + \ldots + Ax_t$ coincides with M, since it has the same inverse. Hence M is finitely generated over A. It follows that M is a finitely generated subgroup of K, and that we can write $M = \mathbb{Z}\alpha + \mathbb{Z}\beta$, where $\alpha, \beta \in K$ are linearly independent over $\mathbb{Q}$.

Let conversely $M = \mathbb{Z}\alpha + \mathbb{Z}\beta$, with $\alpha, \beta \in K$ linearly independent over $\mathbb{Q}$. Put $\gamma = \beta/\alpha$ ($\notin \mathbb{Q}$), and choose $a, b, c \in \mathbb{Z}$ such that $a\gamma^2 - b\gamma + c = 0$, $\gcd(a, b, c) = 1$. From $M = (\mathbb{Z} + \mathbb{Z}\gamma)\cdot\alpha$ and $a\gamma\cdot\gamma = b\gamma - c \in \mathbb{Z} + \mathbb{Z}\gamma$ we see that $\mathbb{Z}[a\gamma]\cdot M = M$. Using that

$$\sigma(\gamma) = -\gamma + (b/a), \qquad \gamma\cdot\sigma(\gamma) = c/a,$$
$$\gcd(a, b, c) = 1$$

one calculates easily

$$M\cdot\sigma[M] = \mathbb{Z}[a\gamma]\cdot N(\alpha)/a, \tag{2.2}$$

so M is an invertible $\mathbb{Z}[a\gamma]$-ideal, with inverse $\sigma[M]\cdot a/N(\alpha)$. But by (2.1), the group M is an invertible ideal for at most one ring. We conclude that M is an invertible A-ideal if and only if $A = \mathbb{Z}[a\gamma]$, and, upon comparing the discriminants, if and only if $\Delta = b^2 - 4ac$.

In the sequel we shall always assume that $N(\alpha)/a > 0$. This can be achieved by changing the signs of a, b, c, if necessary. Further, we assume that in $\gamma = (b \pm \sqrt{\Delta})/(2a)$ the +-sign holds. This can be achieved by multiplying b and β by ± 1. We see that the invertible A-ideals are precisely the subgroups of K of the form

$$M = (\mathbb{Z} + \mathbb{Z}\frac{b + \sqrt{\Delta}}{2a})\cdot\alpha$$

where $\alpha \in K^*$, $a, b \in \mathbb{Z}$ are such that

$$\left.\begin{array}{l} c = (b^2 - \Delta)/(4a) \in \mathbb{Z}, \qquad \gcd(a, b, c) = 1, \\ N(\alpha)/a > 0. \end{array}\right\} \tag{2.3}$$

Given M, the numbers α, a, b are not unique. For α we can take any element of M that is part of a $\mathbb{Z}$-basis of M or, equivalently, that is *primitive*, i.e. does not belong to nM for any $n \in \mathbb{Z}$, $n > 1$. Given M and α, it is easy to check that a is uniquely determined, and that b is only uniquely determined modulo $2a$. Notice that $b \equiv \Delta \bmod 2$.

The *norm* $N(M)$ of $M \in I$ is defined by $N(M) = |\det(\phi)|$, where ϕ is any $\mathbb{Q}$-linear endomorphism of K for which $\phi[A] = M$. We have $N(A\alpha) = |N(\alpha)|$ for $\alpha \in K^*$, and if M is specified by α, a, b as above, then $N(M) = N(\alpha)/a$. From (2.2) we see that

$$M\cdot\sigma[M] = A\cdot N(M).$$

It follows that $N\colon I \to \mathbb{Q}^*_{>0}$ is a group homomorphism.

Let $M_1, M_2 \in I$, and let M_i be given by α_i, a_i, b_i as above, for $i = 1, 2$. We show how to calculate $M_3 = M_1\cdot M_2$. We choose

$$\alpha_3 = \alpha_1\alpha_2/d, \tag{2.4}$$

where d is the unique positive integer for which $\alpha_1\alpha_2/d$ is a primitive element of M_3. Since $M_3 \in I$, we have

$$M_3 = (\mathbb{Z} + \mathbb{Z}\frac{b_3 + \sqrt{\Delta}}{2a_3})\cdot\alpha_3$$

for certain $a_3, b_3 \in \mathbb{Z}$ satisfying the analogue of (2.3). From $N(M_1)N(M_2) = N(M_3)$ and $N(M_i) = N(\alpha_i)/a_i$ we see that

$$a_3 = a_1a_2/d^2. \tag{2.5}$$

The equality $M_1M_2 = M_3$ now becomes

$$\mathbb{Z} + \mathbb{Z}\frac{b_1 + \sqrt{\Delta}}{2a_1} + \mathbb{Z}\frac{b_2 + \sqrt{\Delta}}{2a_2} + \mathbb{Z}\frac{\frac{1}{2}(b_1b_2 + \Delta) + \frac{1}{2}(b_1 + b_2)\sqrt{\Delta}}{2a_1a_2}$$
$$= \mathbb{Z}\frac{1}{d} + \mathbb{Z}\frac{b_3 + \sqrt{\Delta}}{2a_1a_2}d. \tag{2.6}$$

Comparing the $\sqrt{\Delta}/2$-coordinate we see that $\mathbb{Z}a_1^{-1} + \mathbb{Z}a_2^{-1} + \frac{1}{2}(b_1 + b_2)\mathbb{Z}(a_1a_2)^{-1} = \mathbb{Z}(a_1a_2)^{-1}\cdot d$, i.e.

$$d = \gcd(a_2, a_1, \tfrac{1}{2}(b_1 + b_2)). \tag{2.7}$$

The integer b_3 is determined, modulo $2a_3$, by the property that $(b_3 + \sqrt{\Delta})d/(2a_1a_2)$ belongs to (2.6). Hence, if λ, μ, ν are integers such that

$$\lambda a_2 + \mu a_1 + \nu\tfrac{1}{2}(b_1 + b_2) = d \tag{2.8}$$

then

$$b_3 \equiv \frac{1}{d}(\lambda a_2b_1 + \mu a_1b_2 + \nu\tfrac{1}{2}(b_1b_2 + \Delta)) \bmod 2a_3. \tag{2.9}$$

From (2.7), (2.5), (2.8), (2.9) we see that a_3, b_3 can be calculated if a_1, b_1, a_2, b_2 are given. The gcd in (2.7) and integers λ, μ, ν such that (2.8) holds can be determined using the Euclidean algorithm. If in addition α_1, α_2 are given, α_3 can be calculated using (2.4). For computational purposes it is useful to note Shanks's formula [23]

$$b_3 \equiv b_2 + 2\frac{a_2}{d}(\lambda\frac{b_1 - b_2}{2} - \nu c_2) \bmod 2a_3$$

where $c_2 = (b_2^2 - \Delta)/(4a_2)$, and where $\lambda\frac{b_1 - b_2}{2} - \nu c_2$ may be taken modulo a_1/d. It is proved by eliminating μa_1 from (2.8) and (2.9).

If M is given by α, a, b, then it follows easily from (2.2) that M^{-1} is given by a/α (or $|a|/\alpha$), a, -b.

A *principal* A-ideal (in the strict sense) is an additive subgroup of K of the form $A\alpha$, with $\alpha \in K^*$, $N(\alpha) > 0$. The principal ideals are exactly the invertible ideals that have $a = 1$ for a suitable choice of α. They form a subgroup P of I. The *class group* C (in the strict sense) of A is defined by $C = I/P$. It is well known that this is a finite group, cf. sec. 4. Its order is called the *class number* (in the strict sense) of A, and denoted by h.

3. Quadratic forms

Let Δ be an integer. A *primitive integral binary quadratic form of discriminant* Δ is a polynomial $aX^2 + bXY + cY^2 \in \mathbb{Z}[X, Y]$ for which $\gcd(a, b, c) = 1$ and $b^2 - 4ac = \Delta$. For brevity, we shall simply speak of *forms*, or *forms of discriminant* Δ, and we impose the extra condition that $a > 0$ if $\Delta < 0$. Forms of discriminant Δ exist if and only if $\Delta \equiv 0$ or $1 \bmod 4$. From now on, we fix such an integer, and we assume for simplicity that Δ is not a square; see [4; 7] for the case that Δ is a square. We let $K = \mathbb{Q}(\sqrt{\Delta})$ and $A = \mathbb{Z}[(\Delta + \sqrt{\Delta})/2]$ be as in the preceding sections. We shall denote the form $aX^2 + bXY + cY^2$ by (a, b, c), or simply by (a, b), since c is determined by a, b and Δ.

The group $SL_2(\mathbb{Z})$ = {2×2-matrices over $\mathbb{Z}$ with determinant 1} acts on the right on $\mathbb{Z}[X, Y]$ as a group of ring automorphisms by $XT = tX + uY$, $YT = vX + wY$, for $T = \begin{pmatrix} t & u \\ v & w \end{pmatrix} \in SL_2(\mathbb{Z})$. This action transforms the set of forms of discriminant Δ into itself. Two forms are called *equivalent* if they are in the same orbit under $SL_2(\mathbb{Z})$. It is well known that there is a natural bijection

$$C = I/P \to \{\text{forms of discriminant } \Delta\}/SL_2(\mathbb{Z}).$$

This bijection maps the class of $M \in I$ to the $SL_2(\mathbb{Z})$-orbit of the form $N(X\alpha + Y\beta)/N(M)$, where α, β satisfy

$$M = \mathbb{Z}\alpha + \mathbb{Z}\beta, \qquad (\beta\cdot\sigma(\alpha) - \alpha\cdot\sigma(\beta))/\sqrt{\Delta} > 0. \tag{3.1}$$

If $M = (\mathbb{Z} + \mathbb{Z}(b + \sqrt{\Delta})/(2a))\alpha$ as in sec. 2, then a short calculation shows that the above form equals $aX^2 + bXY + cY^2$, where $c = (\Delta - b^2)/(4a)$. For further details, see [1].

The above bijection can be used to transport the group structure of C to the set of $SL_2(\mathbb{Z})$-orbits of forms of discriminant Δ. The product of the orbits of (a_1, b_1) and (a_2, b_2) is the orbit of (a_3, b_3), where a_3, b_3 are given by (2.7), (2.5), (2.9). The inverse of the orbit of (a, b) is the orbit of $(a, -b)$. For a different algorithm to multiply classes of quadratic forms, depending on "united" or "concordant" forms, we refer to [14, fifth supplement; 3; 7]. It will not suit our needs in sec. 8, cf. [6].

4. Reduction

A form (a, b, c) is called *reduced* if

$$|\sqrt{\Delta} - 2|a|| < b < \sqrt{\Delta} \qquad \text{if } \Delta > 0,$$

$$\left.\begin{array}{l} |b| \le a \le c \\ b \ge 0 \text{ if } |b| = a \text{ or } a = c \end{array}\right\} \text{ if } \Delta < 0.$$

We denote the set of reduced forms by R. For $(a, b) \in R$, we have

$$|a| < \sqrt{\Delta} \qquad \text{if } \Delta > 0,$$

$$0 < a < \sqrt{|\Delta|/3} \qquad \text{if } \Delta < 0.$$

It follows that the set R is *finite*.

We describe an efficient *reduction algorithm*, which for any form (a, b) of discriminant Δ produces a reduced form equivalent to it. The algorithm consists of successive applications of the following two types of elements of $SL_2(\mathbb{Z})$:

(i) $T = \begin{pmatrix} 1 & m \\ 0 & 1 \end{pmatrix}$, with $m \in \mathbb{Z}$. We have

$$(a, b)T = (a, b + 2am).$$

(ii) $T = \begin{pmatrix} 0 & -1 \\ 1 & 0 \end{pmatrix}$. We have

$$(a, b, c)T = (c, -b, a).$$

#Using (i), we can bring b in any interval J_a of length $2|a|$. For this interval we choose

$$J_a = \{x \in \mathbb{R}\colon -|a| < x \le |a|\}$$
if either $\Delta < 0$, or $\Delta > 0$ and $|a| \ge \sqrt{\Delta}$,

$$J_a = \{x \in \mathbb{R}\colon \sqrt{\Delta} - 2|a| < x \le \sqrt{\Delta}\}$$
if $\Delta > 0$ and $|a| < \sqrt{\Delta}$.

Taking the second choice for all a, when $\Delta > 0$, as Gauss does [4; 14], leads to a worse algorithm, as was noted by Lagarias [7].

If, after this application of (i), the form (a, b) is reduced, stop. Otherwise, replace (a, b, c) by $(c, -b, a)$, using (ii), and go to #.

It can be shown that no more than $O(\max\{1, \log(|a|/\sqrt{|\Delta|})\})$ applications of (i), (ii) are needed to reduce a form (a, b) by this algorithm, cf. [7].

It follows that any form is equivalent to a reduced form. Since R is finite, this implies that the class number h is finite.

5. Reduced forms and the class group

Let $\Delta < 0$. In this case every form is equivalent to *exactly one* reduced form, see [14]. Hence the set R may be identified with the class group C. An efficient algorithm for the group multiplication $R \times R \to R$ is obtained by combining the formulae of sec. 2 with the reduction algorithm of sec. 4. The inverse of $(a, b, c) \in R$ is $(a, -b, c)$, except if $b = a$ or $a = c$, in which cases $(a, b, c)^{-1} = (a, b, c)$. This provides us with an explicit model for the class group.

Next let $\Delta > 0$. In this case it is not true that every form is equivalent to exactly one reduced form. Let $\rho: R \to R$ describe the effect of performing a reduction step on a form that is already reduced. More precisely, put $\rho((a, b, c)) = (c, b')$, where $b' \in J_c$, $b' \equiv -b \bmod 2c$; this form is equivalent to (a, b, c), and it belongs to R. It can be proved that ρ is a *permutation* of R, see [14, sec. 77]. The inverse of ρ is given by $\rho^{-1} = \tau\rho\tau$, where $\tau((a, b, c)) = (c, b, a)$. By a *cycle* of R we mean an orbit of R under the action of the powers of ρ. Since the leading coefficients alternate in sign, every cycle contains an *even* number of reduced forms.

It is a fundamental theorem that two reduced forms are equivalent if and only if they belong to the same cycle [14, sec. 82]. Hence C may be identified with the set of cycles of R. The cycle corresponding to the neutral element of C is called the *principal cycle*, notation: P; this is the cycle containing the form $(1, b_0)$, with $b_0 \in J_1$, $b_0 \equiv \Delta \bmod 2$.

The number of reduced forms in a cycle is $O(\Delta^{\frac{1}{2}+\varepsilon})$ for every $\varepsilon > 0$, by (6.2) and (11.4), and the exponent $\frac{1}{2}$ is best possible [8]. If Δ is large, it may be very difficult to decide whether two reduced forms are equivalent (see sec. 13 for an $O(\Delta^{\frac{1}{4}+\varepsilon})$-algorithm). Thus, while we can still do calculations in C using R, we have no efficient equality test. The way out of this difficulty is that, for the purposes of computation, we abandon the group C in favour of a group F, which resembles R more closely. The group F is defined in sec. 8; here we describe the phenomena that it is meant to explain.

We can define a multiplication $*: R \times R \to R$ as follows. Let $(a_1, b_1), (a_2, b_2) \in R$, and let (a_3, b_3) be defined by the formulae of sec. 2. Let $(a_4, b_4) \in R$ be the form obtained by reducing (a_3, b_3) using the algorithm of sec. 4. Then we put $(a_1, b_1) * (a_2, b_2) = (a_4, b_4)$. This multiplication satisfies the commutative law, the form $(1, b_0)$ defined above is a neutral element, and every $(a, b) \in R$ has an inverse (a, b'), with $b' \equiv -b \bmod 2a$, $b' \in J_a$. If the associative law were satisfied, then R would be a finite abelian group with subgroup $P \subset R$, and there would be an exact sequence

$$0 \to P \to R \to C \to 0.$$

It would follow that the cycles are the cosets of P, and that they all have the same cardinality. It is easy to find examples where this is not true, e.g. $\Delta = 40$. It can in fact be shown that $*$ makes R into a group if and only if all $(a, b) \in R$ are *ambiguous*, i.e. satisfy $b \equiv 0 \bmod a$. This occurs for only finitely many Δ, like 5, 8, ..., 5180, which can be effectively determined if the generalized Riemann hypothesis for the L-functions $L(s, (\frac{\Delta}{\cdot}))$ is assumed.

Even if R is no group, it exhibits a certain group-like behaviour. We have, for example, an approximate associative law:

$$F * (G * H) = \rho^n((F * G) * H), \qquad \text{with } n \in \mathbb{Z}, \; |n| \text{ "small"}, \tag{5.1}$$

for $F, G, H \in R$. Also, the cycles behave as the cosets of a cyclic subgroup:

$$(\rho^k F) * (\rho^\ell G) = \rho^{m(k,\ell)}(F * G) \qquad \text{for } k, \ell \in \mathbb{Z}, \tag{5.2}$$

where $m(k,\ell)$ is a function of k and ℓ that exhibits certain monotonicity properties in both variables, like $k+\ell$ does. These observations are basically due to Shanks [24].

The group F to be defined in sec. 8 can be used to analyze the above situation, and in particular to prove precise versions of (5.1) and (5.2); e.g., "small" in (5.2) can be replaced by $O(\log \Delta)$, as we shall see in sec. 12.

6. The analytic class number formula

Denote by χ the Kronecker symbol $\left(\frac{\Delta}{\cdot}\right)$, and let $L(s, \chi) = \Sigma_{n=1}^{\infty} \chi(n)n^{-s}$ for $s \in \mathbb{C}$, $\mathrm{Re}(s) > 0$. First let $\Delta < 0$. Then we have

$$h = \frac{w\sqrt{|\Delta|}}{2\pi} \cdot L(1, \chi)$$

(see [14]) where w is the number of roots of unity in A; so $w = 2$ for $\Delta < -4$. The number $L(1, \chi)$ may be expressed by the slowly converging product

$$L(1, \chi) = \Pi_{p \text{ prime}} \left(1 - \frac{\chi(p)}{p}\right)^{-1},$$

see [13, sec. 109]. The class number formula can be rewritten as a finite sum

$$h = \frac{w}{2} \cdot \frac{1}{2 - \chi(2)} \cdot \Sigma_{n=1}^{[\frac{1}{2}|\Delta|]} \chi(n)$$

if $\Delta = \Delta_0$. However, the number of terms is so large that for practical purposes the sum may be said to converge even slower than the non-absolutely converging product for $L(1, \chi)$.

Next let $\Delta > 0$. Let η be the smallest unit of A for which $\eta > 1$ and $N(\eta) = 1$. The *regulator* R (in the strict sense) of A is defined by $R = \log \eta$. The class number formula now reads

$$hR = \sqrt{\Delta} \cdot L(1, \chi)$$

(see [14]) where $L(1, \chi)$ is given by the same infinite product as above. The finite sum

$$hR = -2 \cdot \Sigma_{n=1}^{[\frac{1}{2}\Delta]} \chi(n) \log|1 - e^{2\pi i n/\Delta}|$$

(for $\Delta = \Delta_0$) is again useless for our purpose.

Satisfactory estimates for the rate of convergence of the

infinite product can be given if the Riemann-hypothesis for the zeta function of K is assumed. Then we have, both for $\Delta > 0$ and $\Delta < 0$:

$$\prod_{p \geq x} \left(1 - \frac{\chi(p)}{p}\right) = 1 + O(x^{-\frac{1}{2}} \cdot (\log|\Delta| + \log x)) \qquad (6.1)$$

for $x \geq 2$, the constant implied by the O-symbol being absolute and effectively computable. This can be deduced from [11, theorem 1.1]; cf. [18, théorème 3].

Schur [22] proved that

$$|L(1, \chi)| < \tfrac{1}{2}\log|\Delta| + \log\log|\Delta| + 1. \qquad (6.2)$$

If $\Delta > 0$, the term $\log\log|\Delta|$ can be omitted [5].

7. Shanks's algorithm for negative discriminants

Shanks described in [23] an algorithm to calculate h in the case that $\Delta < 0$. We indicate the main points of this algorithm.

Let X be some "large" integer, specified below. Calculate an integer $\tilde{h}$ that differs by at most 1 from

$$\frac{w\sqrt{|\Delta|}}{2\pi} \cdot \prod_{p \text{ prime},\ p < X} \left(1 - \frac{\chi(p)}{p}\right)^{-1}.$$

Then we expect that

$$h \text{ is "close" to } \tilde{h}. \qquad (7.1)$$

Select a form $F \in R$. By Lagrange's theorem in group theory, we have

$$F^h = 1, \qquad (7.2)$$

where 1 denotes the unit element of R. We try to determine h by combining (7.1) and (7.2). More specifically, we calculate $F^{\tilde{h}}$ and search for an integer n with

$$F^{\tilde{h}} = F^n, \qquad |n| \text{ "small"}. \qquad (7.3)$$

Then $\tilde{h} - n$ is a likely value for h. Searching among the divisors of $\tilde{h} - n$, we can determine the order e of F in the group R. If e is large, which it usually is, then $\tilde{h} - n$ is the only multiple of e that is sufficiently close to $\tilde{h}$, and we must have $h = \tilde{h} - n$. In that case we are done. If, on the other hand, e is small, then we select a second form $G \in R$ and determine the order of the subgroup of R generated by F and G in a like manner. We proceed until a subgroup $S \subset R$ has been found for which only one multiple of $\#S$ is sufficiently close to $\tilde{h}$ to be equal to h.

The exact meaning of "large", "close", "small" in the above

algorithm depends on how well one is able to estimate the convergence of the infinite product in sec. 6. Let us assume that (6.1) holds. Then we take for X an integer of order of magnitude $|\Delta|^{1/5}$. Let $\varepsilon > 0$ be an arbitrary real number. The calculation of $\tilde{h}$ can then be done in $O(|\Delta|^{(1/5)+\varepsilon})$ steps. From (6.1) and (6.2) we get

$$|h - \tilde{h}| \le Y \qquad \text{with } Y = O(|\Delta|^{(2/5)+\varepsilon}),$$

and this inequality can be made completely explicit. The calculation of $F^{\tilde{h}}$, for $F \in \mathcal{R}$, can be done in $O(|\Delta|^{\varepsilon})$ steps, by repeated squarings and multiplications using the binary representation of $\tilde{h}$. Searching for n as in (7.3), with "small" now meaning "$\le Y$", requires $O(|\Delta|^{(2/5)+\varepsilon})$ steps if one proceeds in the naive way. A significant improvement is made possible by using Shanks's "baby step - giant step" technique: if we write $n = iy + j$, where y has order of magnitude $\sqrt{2Y}$ and $|i|, |j| \le \frac{1}{2}y = O(|\Delta|^{(1/5)+\varepsilon})$, then (7.3) can be rewritten as

$$F^{\tilde{h}} \cdot F^{-iy} = F^{j}.$$

So we just have to multiply $F^{\tilde{h}}$ by small powers of $F^{\pm y}$, and wait until a small power of F appears; here the small powers of F are assumed to be calculated beforehand. In this way, determining n as in (7.3) can be done in $O(|\Delta|^{(1/5)+\varepsilon})$ steps. Factoring $\tilde{h} - n$ can be done in $O(|\Delta|^{(1/8)+\varepsilon})$ steps, see [19]. If $e = \text{order}(F)$ is larger than $|n| + Y$ then we must have $h = \tilde{h} - n$, and we are done. So let e be smaller; then $e = O(|\Delta|^{(2/5)+\varepsilon})$, and we have to proceed with a second form G. We have to determine the earliest power of G that is in the subgroup generated by F. By a strategy similar to the baby step - giant step technique this can be done in $O(|\Delta|^{(1/5)+\varepsilon})$ steps. In the same way we proceed with further forms, if necessary.

Assuming some extra Riemann hypotheses, besides those needed for (6.1), one can show that the selection of the forms $F, G, \ldots$ can be done in such a way that no more than $O((\log|\Delta|)^2)$ forms need be considered, see [10, Cor. 1.3].

We conclude that, modulo the Riemann hypotheses, the above method determines h in $O(|\Delta|^{(1/5)+\varepsilon})$ steps, for every $\varepsilon > 0$. If one does not stop before $F, G, \ldots$ generate the entire class group, one obtains an algorithm that determines the structure of the class group which runs in $O(|\Delta|^{(1/4)+\varepsilon})$ steps. In the present case of negative discriminants there is an additional technique, employing the decomposition of the class group in its p-primary subgroups, that reduces the exponent 1/4 to 1/5 in many cases; cf. [23, sec. 3]. This technique is, however, far less useful in the case of positive discriminants.

8. The group $\mathcal{F}$

Let Γ denote the subgroup $\{\begin{pmatrix} 1 & m \\ 0 & 1 \end{pmatrix} : m \in \mathbb{Z}\}$ of $SL_2(\mathbb{Z})$. It is easy to see that two forms (a, b) and (a', b') are in the same orbit under Γ if and only if

$$a = a', \qquad b \equiv b' \bmod 2a.$$

We denote the orbit space {forms of discriminant $\Delta\}/\Gamma$ by $\mathcal{F}$. Each orbit contains exactly one form (a, b) with b belonging to the interval J_a defined in sec. 4. It will be convenient to identify $\mathcal{F}$ with the set of such forms.

From $\Gamma \subset SL_2(\mathbb{Z})$ we see that there is a natural surjective map $\mathcal{F} \to \{\text{forms}\}/SL_2(\mathbb{Z}) \cong C$. We claim that there is a natural group law on $\mathcal{F}$ that makes this map into a group homomorphism. The easiest way to see this is, again, to use the connection with invertible ideals.

Consider the group $I \oplus (K^*/\mathbb{Q}^*_{>0})$ with I as in sec. 2. Elements of this group are pairs $(M, \alpha\mathbb{Q}^*_{>0})$ with $M \in I$ and $\alpha \in K^*$, and α can, in its coset mod $\mathbb{Q}^*_{>0}$, be uniquely chosen such that it is a primitive element of M. Choose $\beta \in M$ such that (3.1) holds; it is unique up to translation by $\mathbb{Z}\alpha$. Mapping the pair $(M, \alpha\mathbb{Q}^*_{>0})$ to the Γ-orbit of the form $N(X\alpha + Y\beta)/N(M)$, as in sec. 3, now defines a surjective map

$$I \oplus (K^*/\mathbb{Q}^*_{>0}) \to \mathcal{F}.$$

Using that the form $N(X\alpha + Y\beta)/N(M)$ equals (a, b), where $\beta/\alpha = (b + \sqrt{\Delta})/(2a)$, one checks that two pairs $(M, \alpha\mathbb{Q}^*_{>0})$ and $(M', \alpha'\mathbb{Q}^*_{>0})$ have the same image in $\mathcal{F}$ if and only if there exists $\gamma \in K^*$ such that

$$\gamma M = M', \qquad \gamma\alpha\mathbb{Q}^*_{>0} = \alpha'\mathbb{Q}^*_{>0}, \qquad N(\gamma) > 0.$$

So if we embed $K^*_{N>0} = \{\gamma \in K^* : N(\gamma) > 0\}$ in $I \oplus (K^*/\mathbb{Q}^*_{>0})$ by mapping γ to $(A\gamma, \gamma\mathbb{Q}^*_{>0})$, then we get a bijection

$$(I \oplus (K^*/\mathbb{Q}^*_{>0}))/K^*_{N>0} \to F.$$

The left hand side is a group, hence so is the right hand side, by transport of structure. Multiplication and inversion in F can be done by the formulae of sec. 2. We shall denote the unit element of F by 1; it is (the Γ-orbit of) the form $(1, b_0)$ with $b_0 \in J_1$, $b_0 \equiv \Delta \bmod 2$. It is obvious that the natural map $F \to C$ is a group homomorphism.

9. The algebraic structure of F

Some easy diagram chasing gives rise to an exact sequence

$$0 \to I/\mathbb{Q}^*_{>0} \to F \xrightarrow{\psi} K^*/K^*_{N>0} \to 0.$$

Here $\mathbb{Q}^*_{>0}$ is embedded in I by mapping x to Ax. To describe ψ, we first note that

$$K^*/K^*_{N>0} \cong \begin{cases} 0 & \text{if } \Delta < 0, \\ \{\pm 1\} & \text{if } \Delta > 0. \end{cases} \tag{9.1}$$

So ψ is trivial if $\Delta < 0$. If $\Delta > 0$, then ψ corresponds to the map sending (a, b) to $\operatorname{sign}(a)$.

We claim that the above exact sequence splits. This is clear if $\Delta < 0$, and if $\Delta > 0$ we can map the non-trivial element of $K^*/K^*_{N>0}$ to the element E of F corresponding to $(A, \sqrt{\Delta}\mathbb{Q}^*_{>0}) \in I \oplus (K^*/\mathbb{Q}^*_{>0})$; explicitly, E is the Γ-orbit of

$(-\Delta, \Delta, (1 - \Delta)/4)$ if Δ is odd,

$(-\Delta/4, 0, 1)$ if Δ is even.

(We could also have used the form $(-1, b_0)$ to split the sequence, but E is more convenient in the sequel.) We have proved

$$F \cong (K^*/K^*_{N>0}) \oplus (I/\mathbb{Q}^*_{>0}). \tag{9.2}$$

The group $I/\mathbb{Q}^*_{>0}$ can be analyzed by standard techniques from commutative algebra. Let A_p denote the semilocal ring $\{r/s: r \in A, s \in \mathbb{Z}, s \not\equiv 0 \bmod p\}$. Then we have $I \cong \oplus_{p \text{ prime}} (K^*/A^*_p)$, and

$$I/\mathbb{Q}^*_{>0} \cong \oplus_{p \text{ prime}} (K^*/\langle p\rangle A^*_p) \tag{9.3}$$

where $\langle p\rangle$ denotes the subgroup of K^* generated by p. The groups $K^*/\langle p\rangle A^*_p$ can be calculated explicitly. The result, which will not be used in the sequel, is as follows.

Write $\Delta = f^2 \cdot \Delta_0$ as in sec. 1, and let k be the number of factors p in f. The character χ is as in sec. 6, and χ_0 is the corresponding character for Δ_0. If $k = 0$ we have

$$\begin{aligned} K^*/\langle p\rangle A_p^* &\cong \mathbb{Z} && \text{if } \chi(p) = 1, \\ &\cong 0 && \text{if } \chi(p) = -1, \\ &\cong \mathbb{Z}/2\mathbb{Z} && \text{if } \chi(p) = 0. \end{aligned}$$

Next let $k > 0$. In most cases we have

$$\begin{aligned} K^*/\langle p\rangle A_p^* &\cong \mathbb{Z} \oplus (\mathbb{Z}/(p-1)p^{k-1}\mathbb{Z}) && \text{if } \chi_0(p) = 1, \\ &\cong \mathbb{Z}/(p+1)p^{k-1}\mathbb{Z} && \text{if } \chi_0(p) = -1, \\ &\cong (\mathbb{Z}/2\mathbb{Z}) \oplus (\mathbb{Z}/p^k\mathbb{Z}) && \text{if } \chi_0(p) = 0. \end{aligned}$$

The precise list of exceptions is as follows. The group $K^*/\langle p\rangle A_p^*$ is isomorphic to

$$\begin{aligned} &\mathbb{Z} \oplus (\mathbb{Z}/2\mathbb{Z}) \oplus (\mathbb{Z}/2^{k-2}\mathbb{Z}) && \text{if } p = 2,\ k > 2 \text{ and } \chi_0(2) = 1; \\ &(\mathbb{Z}/2\mathbb{Z}) \oplus (\mathbb{Z}/3\cdot 2^{k-2}\mathbb{Z}) && \text{if } p = 2,\ k > 2 \text{ and } \chi_0(2) = -1; \\ &\mathbb{Z}/4\mathbb{Z} && \text{if } p = 2,\ k = 1 \text{ and } \Delta_0 \equiv -4 \bmod 16; \\ &(\mathbb{Z}/4\mathbb{Z}) \oplus (\mathbb{Z}/2^{k-1}\mathbb{Z}) && \text{if } p = 2,\ k > 2 \text{ and } \Delta_0 \equiv -4 \bmod 32; \\ &(\mathbb{Z}/3\mathbb{Z}) \oplus (\mathbb{Z}/2\cdot 3^{k-1}\mathbb{Z}) && \text{if } p = 3,\ k > 1 \text{ and } \Delta_0 \equiv -3 \bmod 9. \end{aligned}$$

Combining this description of $K^*/\langle p\rangle A_p^*$ with (9.2), (9.1) and (9.3) we obtain an algebraic description of F. In particular, we see that F is the direct sum of a finite group and a free abelian group of countably infinite rank. The natural action of σ (see sec. 1) on F is given by $\sigma(F) = F^{-1}$, for $F \in F$.

10. The topological structure of F

From this point onward we assume that Δ is *positive*. The case of negative Δ is similar, but will not be needed in the sequel.

The group homomorphism $F \to C$ defined in sec. 8 maps the coset $(M, \alpha\mathbb{Q}^*_{>0})K^*_{N>0}$ to the ideal class of M. We denote by G the kernel of this homomorphism. The coset of $(M, \alpha\mathbb{Q}^*_{>0})$ belongs to G if and only if $M = A\beta$ for some $\beta \in K^*_{N>0}$, so, dividing by β:

$$G = \{(A, \gamma\mathbb{Q}^*_{>0})K^*_{N>0} : \gamma \in K^*\}.$$

For $\gamma_1, \gamma_2 \in K^*$ we have $(A, \gamma_1\mathbb{Q}^*_{>0})K^*_{N>0} = (A, \gamma_2\mathbb{Q}^*_{>0})K^*_{N>0}$ if and only if $\gamma_1\mathbb{Q}^*_{>0} = \zeta\gamma_2\mathbb{Q}^*_{>0}$ for some $\zeta \in A^*$ with $N(\zeta) = +1$. From this it follows that the map

$$d: G \to \mathbb{R}/R\mathbb{Z}$$

$$d((A, \gamma\mathbb{Q}^*_{>0})K^*_{N>0}) = (\tfrac{1}{2}\log|\gamma/\sigma(\gamma)| \bmod R)$$

is a well defined group homomorphism; here R is the regulator of A, defined in sec. 6. The map d is a small modification of the "distance" defined by Shanks [24]. We have $\ker(d) = \{1, E\}$, with E as defined in sec. 9. It follows that the map

$$G \to (\mathbb{R}/R\mathbb{Z}) \oplus \{\pm 1\}$$

obtained by combining d with the map ψ from sec. 9 is an *injective* group homomorphism. For cardinality reasons it is not surjective. However, its image is *dense* in $(\mathbb{R}/R\mathbb{Z}) \oplus \{\pm 1\}$; this follows from the fact that G is infinite (sec. 9), and it can also be seen directly.

We conclude that G may be considered as a dense subgroup of the product of a circle group of 'circumference' R and a group of order two. The h cosets of G in F may be considered as the cosets of such a subgroup. A coset of G in F will be called a *cycle* of F, and G itself is the *principal cycle*. The agreement with the terminology introduced in sec. 5 is intentional, and will be justified in sec. 11. Every cycle consists of two *circles*, a *positive* and a *negative* circle, containing forms with positive and negative leading coefficients, respectively; cf. figure 1.

If $F_1, F_2 \in F$ belong to the same cycle, the *distance* from F_1 to F_2 is defined to be $d(F_2F_1^{-1})$, which is a real number modulo R. The distance is zero if and only if $F_1 = F_2$ or $F_1 = F_2 \cdot E$. If $F_1, F_2 \in F$ do not belong to the same cycle, the distance from F_1 to F_2 is not defined.

Replacing G by the full group $(\mathbb{R}/R\mathbb{Z}) \oplus \{\pm 1\}$, and similarly with the cosets, we obtain an embedding of F as a dense

<u>Figure 1.</u> F.

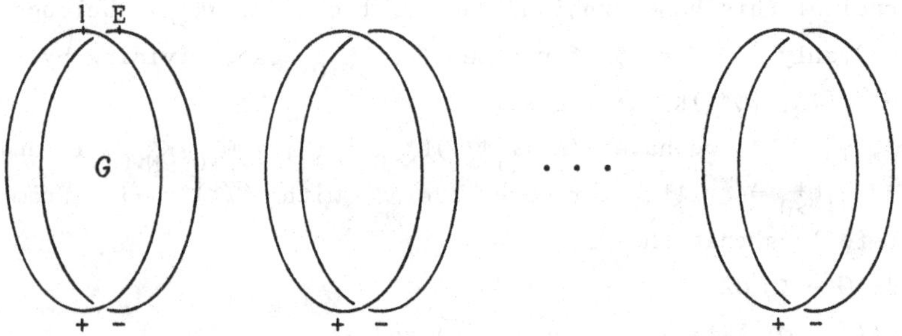

subset in a compact topological space $\bar{F}$. It is not difficult to see that the group multiplication of F can be extended to $\bar{F}$, making it into a topological group. This can be done using fibred sums, or by defining $\bar{F} = (I \oplus ((K \otimes_{\mathbb{Q}} \mathbb{R})^* / \mathbb{R}^*_{>0}))/K^*_{N>0}$. It is of interest to notice that the group $\bar{F}$ can also be described as a certain group of idele classes of K, as follows. For background, see [2].

Let $\hat{A} = \varprojlim A/nA$ be the profinite completion of A, with n ranging over the positive integers. We may consider $\hat{A}$ as a subring of the restricted product $\prod'_v K_v$, with v ranging over the finite places of K and K_v denoting the completion of K at v. Hence $\hat{A}^*$ may be considered as a subgroup of $\prod'_v K_v^*$; for example, if $A = A_0$ (see sec. 1) then $\hat{A}^* = \prod_v U_v$, where U_v consists of the local units at v. Adding 1's at the infinite places, we may consider $\hat{A}^*$ as a subgroup of the group J_K^1 of ideles of K satisfying the product formula. Now we have

$$\bar{F} \cong J_K^1/(K^*_{N>0} \cdot \hat{A}^*). \tag{10.1}$$

This group is very similar to the group $J_K^1/(K^* \cdot \prod_v U_v)$, the compactness of which is equivalent to the conjunction of the Dirichlet unit theorem and the finiteness of the class number. The isomorphism (10.1), which will not be used in the sequel, indicates what is the right generalization of F for algebraic number fields of higher degrees.

11. Reduced forms in F

Since no two forms in R are in the same orbit under Γ, we may consider R as a subset of F. By the fundamental theorem quoted in sec. 5, the cycles of R are precisely the intersections of the cycles of F with R; in particular, we have $P = G \cap R$. In fact, the cyclical structure of each cycle of R is reflected by the way it is sitting in the corresponding cycle of F, as suggested by fig. 2. More precisely, if $F \in R$, then $\rho(F)$ is the first element of R that is

Figure 2.

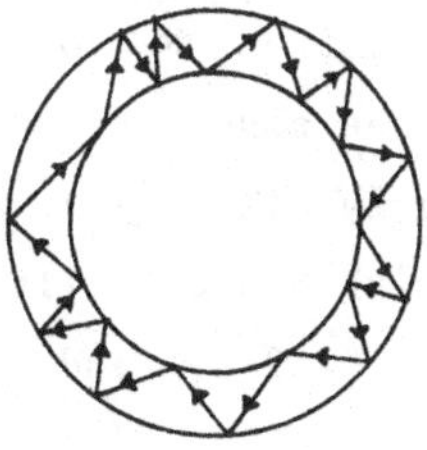

encountered if the two circles are simultaneously traversed in the positive direction, starting from F; this fixes $\rho(F)$ uniquely in the sense that for no $G \in \mathcal{R}$ one also has $G \cdot E \in \mathcal{R}$; and, finally, it is automatic that F and $\rho(F)$ are on different circles. The last statement reflects the fact that the sign of the leading coefficient is changed if ρ is applied.

The proof of these statements can most conveniently be given by interpreting $\mathcal{R}$ and ρ in terms of lattice points on the boundary of the convex hull of the totally positive part of a lattice in $\mathbb{R}^2$. We do not go into the details. The fundamental theorem quoted in sec. 5 is a consequence of the above results.

We calculate the distance from $F = (a, b) \in \mathcal{R}$ to $\rho(F)$. Let F correspond to the coset $(M, \alpha\mathbb{Q}^*_{>0})K^*_{N>0}$. Choosing α primitive in M we then have

$$M = \mathbb{Z}\alpha + \mathbb{Z}\beta, \qquad (\beta\sigma(\alpha) - \alpha\sigma(\beta))/\sqrt{\Delta} > 0,$$
$$aX^2 + bXY + cY^2 = N(X\alpha + Y\beta)/N(M).$$

Applying ρ means first applying the element $\begin{pmatrix} 0 & -1 \\ 1 & 0 \end{pmatrix}$ of $SL_2(\mathbb{Z})$ and next an element of Γ. The latter element does not change the Γ-orbit, so we only have to investigate the effect of $\begin{pmatrix} 0 & -1 \\ 1 & 0 \end{pmatrix}$. This changes the form into $N(X\beta - Y\alpha)/N(M)$, corresponding to the coset $(M, \beta\mathbb{Q}^*_{>0})K^*_{N>0}$. Since $(M, \beta\mathbb{Q}^*_{>0})(M, \alpha\mathbb{Q}^*_{>0})^{-1} = (A, (\beta/\alpha)\mathbb{Q}^*_{>0})$ and $\beta/\alpha = (b + \sqrt{\Delta})/(2a)$, we find that the distance from F to $\rho(F)$ is given by

$$d(\rho(F)F^{-1}) = \tfrac{1}{2}\log\left|\frac{\beta/\alpha}{\sigma(\beta/\alpha)}\right| = \tfrac{1}{2}\log\left|\frac{b + \sqrt{\Delta}}{b - \sqrt{\Delta}}\right|,$$

taken modulo R.

It is of interest to determine upper and lower bounds for this quantity. Since $0 < b < \sqrt{\Delta}$ for a reduced form (a, b), we have

$$\tfrac{1}{2}\log\left|\frac{b + \sqrt{\Delta}}{b - \sqrt{\Delta}}\right| = \tfrac{1}{2}\log\left|\frac{(b + \sqrt{\Delta})^2}{4ac}\right| < \tfrac{1}{2}\log \Delta. \tag{11.1}$$

Using that $b \geq 1$ one can prove the lower bound $\Delta^{-\frac{1}{2}}$, but this is useless. A more satisfactory lower bound is obtained by considering the distance traversed if ρ is applied *twice*, i.e. from F to $\rho^2(F)$. Let, with the notation as before, ρ map the coset of $(M, \alpha\mathbb{Q}^*_{>0})$ to the coset of $(M, \beta\mathbb{Q}^*_{>0})$, and similarly $(M, \beta\mathbb{Q}^*_{>0})$ to $(M, \gamma\mathbb{Q}^*_{>0})$. Using the geometrical interpretation with convex hulls that we suppressed it is quite easy to see that $|\gamma| > 2|\alpha|$ and

$|\sigma(\gamma)| < \frac{1}{2}|\sigma(\alpha)|$, so $\frac{1}{2}\log\left|\frac{\gamma/\alpha}{\sigma(\gamma/\alpha)}\right| > \log 2$. This gives the following lower bound for the distance traversed if ρ is applied twice:

$$\tfrac{1}{2}\log\left|\frac{b+\sqrt{\Delta}}{b-\sqrt{\Delta}}\right| + \tfrac{1}{2}\log\left|\frac{b'+\sqrt{\Delta}}{b'-\sqrt{\Delta}}\right| > \log 2, \tag{11.2}$$

where $\rho((a, b)) = (c, b')$. A heuristic argument suggests that the average of $\frac{1}{2}\log|(b+\sqrt{\Delta})/(b-\sqrt{\Delta})|$ over all reduced forms should be somewhere near Lévy's constant $\pi^2/(12\cdot\log 2) = 1.18656911\ldots$.

Since the circumference of the whole cycle is R, we have

$$R = \Sigma\ \tfrac{1}{2}\log\left|\frac{b+\sqrt{\Delta}}{b-\sqrt{\Delta}}\right|, \tag{11.3}$$

the sum ranging over the reduced forms (a, b) belonging to a fixed cycle. If there are ℓ reduced forms in the cycle, the above inequalities yield

$$\tfrac{1}{2}\ell\cdot\log 2 < R < \tfrac{1}{2}\ell\cdot\log\Delta. \tag{11.4}$$

Therefore, if two cycles of $\mathcal{R}$ contain ℓ_1 and ℓ_2 forms, respectively, we have

$$\ell_1/\ell_2 < \frac{\log\Delta}{\log 2}.$$

This is an explicit version of a theorem of Skubenko, asserting that $\ell_1/\ell_2 = O(\log\Delta)$, see [27; 15, pp. 558, 586]. I am indebted to A. Schinzel for mentioning this theorem to me.

12. Reduction in $\mathcal{F}$

The reduction algorithm of sec. 4 can be formulated as follows. Extend the map $\rho: \mathcal{R} \to \mathcal{R}$ to a map $\rho: \mathcal{F} \to \mathcal{F}$ by

$$\rho((a, b, c)) = (c, b'), \qquad b' \equiv -b \bmod 2c, \qquad b' \in J_c,$$

where we assumed that $b \in J_a$. As in the previous section, one shows that applying ρ comes down to moving along the cycle over a distance of $\frac{1}{2}\log|(b+\sqrt{\Delta})/(b-\sqrt{\Delta})| = \log|(b+\sqrt{\Delta})/\sqrt{|4ac|}|$; also, if $|b| < \sqrt{\Delta}$, one changes to the companion circle. The reduction map $\rho_0: \mathcal{F} \to \mathcal{R}$ is defined by $\rho_0(F) = \rho^k(F)$, where k is the least non-negative integer for which $\rho^k(F)$ is reduced. Clearly, ρ_0 is the identity on $\mathcal{R}$.

Figure 3.

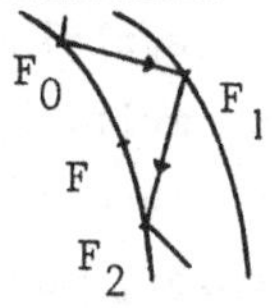

The map ρ_0 assigns to every form in $\mathcal{F}$ a form in $\mathcal{R}$ that is "not too far away" from it. More precisely, let F_0, F_1, F_2 be three consecutive forms on a cycle of $\mathcal{R}$ (possibly $F_0 = F_2$), and let $F \in \mathcal{F}$ be in the interval between F_0 and F_2 that is opposite to F_1. Then it can be shown that $\rho_0(F)$ is one of F_0, F_1, F_2. By (11.1) it follows from this, that the distance from F to $\rho_0(F)$ is at most $\log\Delta$ in absolute value. A more detailed analysis shows, in fact, that

$$|d(\rho_0(F)F^{-1})| < \tfrac{1}{2}\log(1+\theta\sqrt{\Delta}) \qquad \text{for all } F \in \mathcal{F}, \qquad (12.1)$$

where $\theta = (1+\sqrt{5})/2$ and $|x| = \min\{|y| : y \in x\}$ for $x \in \mathbb{R}/R\mathbb{Z}$. This is usually very small with respect to R, the circumference of the cycle, which may have order of magnitude $\Delta^{\frac{1}{2}}$.

The multiplication $*$ on $\mathcal{R}$ defined in sec. 5 is just multiplication in $\mathcal{F}$ followed by the reduction map ρ_0. This remark, and the inequalities (11.2) and (12.1), easily imply the approximate associative law (5.1), with $|n| < 1 + 4\log(1+\theta\sqrt{\Delta})/\log 2$. We leave the pleasure of investigating the properties of $m(k,\ell)$ in (5.2) to the reader.

13. The algorithm for positive discriminants

We shall mainly be concerned with the calculation of the regulator R, which is the circumference of each circle. It can be determined by applying the powers of ρ to a fixed form $F \in \mathcal{R}$, until we find $\rho^{\ell}(F) = F$, and then using (11.3). This is essentially the classical algorithm, which is often phrased in terms of continued fractions. It has running time $O(\Delta^{\frac{1}{2}+\varepsilon})$ for every $\varepsilon > 0$.

We describe two more efficient methods, which make use of the function d defined in sec. 10. The calculations are all done in the principal cycle $\mathcal{G}$, and mostly in $\mathcal{P} = \mathcal{G} \cap \mathcal{R}$. A form $F \in \mathcal{G}$ is not only specified by its coefficients a, b, but also by a real parameter δ which is such that $d(F) = (\delta \bmod R)$. It is not easy to read δ directly from the coefficients, but one can keep track of δ under all operations built up from ρ and multiplication and inversion in $\mathcal{F}$, by the following rules:

$1 = (1, b_0)$ has $\delta = 0$;

when applying ρ to (a, b), add $\frac{1}{2}\log\left|\frac{b+\sqrt{\Delta}}{b-\sqrt{\Delta}}\right|$ to δ;

when multiplying in $\mathcal{F}$, add up both δ's;

when inverting in $\mathcal{F}$, change the sign of δ.

In particular, we can keep track of δ under the composition

$*: P \times P \to P$ from sec. 5.

The inequality $R < \sqrt{\Delta}\cdot\log\Delta$ (see (6.2)) and the baby step - giant step technique now lead to an $O(\Delta^{(1/4)+\varepsilon})$-determination of R, as follows. Starting from the unit form $(1, b_0)$ we build up a stock of forms by successive applications of ρ ("baby steps"), until one of two things happens. It may happen, that (i) a form (a, b) $(\neq (1, b_0))$ is encountered that is its own inverse, i.e. for which a divides b; in that case, R is twice the current δ, and we stop. But for most large Δ it happens sooner, that (ii) one finds a form with $\delta \geq \delta_0 = (\sqrt{\Delta}\cdot\log\Delta)^{\frac{1}{2}}$. By (11.2), this happens after at most $1 + (2\delta_0/\log 2) = O(\Delta^{(1/4)+\varepsilon})$ applications of ρ. At this moment, we have a stock of forms that, together with their inverses, cover an interval of length $\geq 2\delta_0$ along the principal cycle. Now we start taking "giant steps", with step length a little bit less than $2\delta_0$. More precisely, by $*$-squaring the current form, and applying a small power of ρ^{-1}, one determines a form $F \in P$ whose δ satisfies

$$2\delta_0 - \tfrac{1}{2}\log(1 + \theta\sqrt{\Delta}) - \tfrac{1}{2}\log\Delta < \delta \leq 2\delta_0 - \tfrac{1}{2}\log(1 + \theta\sqrt{\Delta}).$$

The giant steps are taken by calculating $F^{*1} = F$, $F^{*2} = F * F$, $\ldots$, $F^{*(i+1)} = F * (F^{*i})$, $\ldots$. Our inequalities guarantee that the "step length", i.e. the distance from F^{*i} to $F^{*(i+1)}$, is for all i between δ_0 and $2\delta_0$. Hence after $O(R/\delta_0) = O(\Delta^{\frac{1}{4}+\varepsilon})$ giant steps we have traversed the entire cycle, and we will discover F^{*i} among our "baby" forms and their inverses. Then we have two values of δ for the same form, and the difference of these values is the regulator.

The above algorithm calculates the regulator to any prescribed precision in $O(\Delta^{(1/4)+\varepsilon})$ steps. The fundamental unit $\eta = e^R = (u + v\sqrt{\Delta})/2$ cannot be calculated in $O(\Delta^{(1/4)+\varepsilon})$ steps; in fact, since R ($\approx$ number of decimal digits of u and v) is often of order of magnitude $\Delta^{\frac{1}{2}}$, one cannot even write down η in time less than that, let alone calculate it. It is, however, possible to calculate u and v modulo any fixed positive integer m in time $O(\Delta^{(1/4)+\varepsilon})$, the implied constant depending on m, by a procedure similar to the above one, cf. [9]. The same remarks apply to the algorithm described below.

If the generalized Riemann hypothesis is assumed, we can give an $O(\Delta^{(1/5)+\varepsilon})$-algorithm for the calculation of R. The procedure is analogous to the determination of the order of F in the case $\Delta < 0$, see sec. 7, so we only sketch the main points. Using the class number formula, we find a number

$$\tilde{R} = \sqrt{\Delta}\cdot\prod_{p \text{ prime},\, p<X}\left(1 - \frac{\chi(p)}{p}\right)^{-1}, \qquad X \approx \Delta^{1/5}, \tag{13.1}$$

that is close to an integer multiple hR of R, the difference being $O(\Delta^{(2/5)+\varepsilon})$. The baby forms are now made as above, but with $\delta_0 \approx \Delta^{(1/5)+\varepsilon}$. Next, by repeated squarings and multiplications in P, we jump to a form F whose δ is close to $\tilde{R}$. Taking giant steps from this F, in both directions, we encounter a form that is already in the "baby" stock. That gives two δ's for the same form, and the difference $R^{\#}$ is an unknown integer multiple $\tilde{h}R$ of the regulator; here $\tilde{h}$ is supposedly not far from h. If $\tilde{h}$ is large ($\gtrsim \Delta^{1/10}$), this is discovered by finding another match after taking some more giant steps. The remaining cases $\tilde{h} \lesssim \Delta^{1/10}$ are checked by looking if the unit form $(1, b_0)$ is found at distance $\frac{1}{m}R^{\#}$ from itself, for $1 < m \lesssim \Delta^{1/10}$. We notice that the latter technique can also be applied in the case $\Delta < 0$, to avoid factoring.

This finishes our sketchy description of the algorithm to determine R. We notice that the Riemann hypothesis is only needed to guarantee the efficiency of the algorithm; once the answer is found, its correctness does not depend on any unproved assumptions.

The determination of the class number h now runs exactly as in the case $\Delta < 0$, with P and R playing the role of the subgroup generated by F, in sec. 7, and its order. If R is sufficiently large, h is determined by the class number formula. Otherwise, select a form $G \in R$, and determine its order in F/G. In this fashion one proceeds until a large enough subgroup of F/G has been determined to fix h uniquely.

In this procedure one needs an algorithm that tests if a given reduced form belongs to the principal cycle. By the baby step - giant step technique this can be done in $O(R^{\frac{1}{2}}\Delta^{\varepsilon})$ steps. In particular, equivalence of two reduced forms can be tested in $O(\Delta^{(1/4)+\varepsilon})$ steps.

The conclusion is exactly as in the case $\Delta < 0$. Modulo the Riemann hypotheses, h can be determined in $O(\Delta^{(1/5)+\varepsilon})$ steps, but the structure of the class group may take $O(\Delta^{(1/4)+\varepsilon})$ steps.

We have only considered the regulator, class number and class group in the *strict* sense. To obtain the regulator R', class number h' and class group C' in the *ordinary* sense, one has to look halfway the principal cycle, i.e. at distance $\frac{1}{2}R$ from the unit form $(1, b_0)$. If at this point the form $(-1, b_0)$ is found, then

$$R' = \tfrac{1}{2}R, \qquad h' = h, \qquad C' = C.$$

Otherwise, one finds halfway P a form $F = (a, b)$ with $|a| > 1$ and $b \equiv 0 \bmod a$. Then $|a|$ is a non-trivial factor of Δ, and one has

$$R' = R, \qquad h' = \tfrac{1}{2}h, \qquad C' = C/C_0$$

where $C_0 \subset C$ is the subgroup of order two generated by the class of the form $(-1, b_0)$.

The distance of two reduced froms (a, b) and (a', b') is an integer multiple of R' if and only if $|a| = |a'|$ and $b = b'$. This implies that the role of R in the above algorithm can also be played by R'. In particular, we can replace $\tilde{R}$ by $\frac{1}{2}\tilde{R}$, which is close to the integer multiple $h'R'$ of R'. I am indebted to R. Tijdeman for this observation.

14. A numerical example

The algorithms described in sections 7 and 13 have been programmed in Amsterdam by R.J. Schoof on the CDC Cyber 750 computer system, for discriminants of up to 28 digits [21]. Using only a hand held calculator like the HP67 one can deal with discriminants of up to 10 digits. For much smaller discriminants - up to 6 digits, roughly - it is often faster to apply the classical algorithm (see sec. 13).

We give an example which was calculated using an HP67. Let $\Delta = 40919537$. In table 1 one finds forms lying on the principal cycle P belonging to this discriminant. The first column gives an identification number to each form. In the text below, form #n is indicated by F_n. The second column shows how the form is obtained from previous forms in the table. Here ρ and the multiplication * are as in sec. 5, and ÷ is multiplication with the inverse. The next two columns contain the coefficients a, b of the form. The final column gives δ, the distance from F_1 to the form, rounded to five decimals from the value given by the calculator.

Table 1. Δ = 40919537.

# def.	a	b	δ
1 = unit	1	6395	0
2 = ρ(1)	-5878	5361	4.42393
3 = ρ(2)	518	6035	5.63858
4 = ρ(3)	-2171	2649	7.40699
5 = ρ(4)	3904	5159	7.84756
6 = ρ(5)	-916	5833	8.96447
7 = ρ(6)	1882	5459	10.50290
8 = ρ(7)	-1477	6357	11.77140
9 = ρ(8)	86	6371	14.65578
10 = ρ(9)	-959	5137	17.75720
11 = ρ(10)	3788	2439	18.86435
12 = ρ(11)	-2308	2177	19.26591
13 = ρ(12)	3919	5661	19.62037
14 = ρ(13)	-566	5659	21.01860
15 = ρ(14)	3929	2199	22.41539
16 = ρ(15)	-2296	2393	22.77375
17 = ρ(16)	3832	5271	23.16692
18 = ρ(17)	-857	5013	24.33607
19 = ρ(18)	4606	4199	25.39088
20 = ρ(19)	-1264	5913	26.17737
21 = ρ(20)	1178	5867	27.79557
22 = 19*19	7	6385	51.50454
23 = 22*22	49	6385	103.00908
24 = 23*23	2401	2465	206.01816
25 = 23*24	-157	6151	308.07526
26 = 25*25	-172	6113	617.15922

# def.	a	b	δ
27 = 26*26	2654	2391	1234.67199
28 = 27*27	-364	6159	2469.19812
29 = 28*28	-137	6371	4936.94461
30 = 29*29	-512	5671	9873.63784
31 = 30÷22	-3584	4647	9822.13330
32 = 31÷22	1586	3695	9770.79649
33 = 32÷22	-614	6129	9719.95084
34 = 33÷22	2294	3371	9668.63890
35 = 34÷22	2857	3553	9616.67814
36 = 35÷22	562	5345	9566.02209
37 = 36÷22	3934	1973	9514.51755
38 = 30*22	-3584	5671	9925.14238
39 = 38*22	-3581	1479	9976.97826
40 = 39*22	86	6371	10027.06848
41 = 29*22	-959	6371	4988.44915
42 = 28*21	-842	5735	2496.66310
43 = 42*3	794	5003	2502.05241
44 = ρ(43)	-5003	5003	2503.10318
45 = 36÷27	-1477	5459	8331.90585
46 = 37÷22	-56	6343	9461.41380
47 = 46÷22	-8	6391	9409.90926

Taking $X = 100$ in (13.1) we find $\tilde{R} = 9839.22$. Baby steps are taken from F_1 to F_{21}. Then we jump to F_{30}, which has $\delta \approx \tilde{R}$. Taking giant steps backward (F_{30} to F_{37}) we find no baby form, but going forward (F_{38} to F_{40}) we find one after three steps: $F_{40} = F_9$. Therefore R divides $R^\# = \delta(40) - \delta(9) = 10012.41270 = \tilde{h}R$, say. Since no other baby form, or inverse baby form, is found in the interval from F_{37} to F_{40}, we must have $R > 10012.41270 - \delta(37) + \delta(21) > 525$, so $\tilde{h} < 20$.

Looking halfway $R^\#$ we find another match: $F_{41} = F_{10}^{-1}$, since $6371 \equiv -5137 \bmod 2 \cdot 959$. Notice that $\delta(41) + \delta(10) = \frac{1}{2}R^\#$. Hence $\tilde{h}$ is *even*. Looking again halfway we find F_{42} with δ close to $\frac{1}{4}R^\#$ and $a = -842$. Since ± 842 is not in the baby list, this means that 4 does not divide $\tilde{h}$, and that exactly at $\frac{1}{4}R^\#$ a non-trivial factorization of Δ will be found. Looking there, out of curiosity, we find the ambiguous form F_{44}, yielding $\Delta = 5003 \cdot 8179$.

To test if 3 divides $\tilde{h}$, we look near $(5/6)R^\#$ and find the match $F_{45} = F_8^{-1}$. Therefore 6 divides $\tilde{h}$, and $\tilde{h} = 6$ or 18. We exclude the latter possibility by taking one more giant step (F_{46}) to improve the above upper bound to $R > 578$, $\tilde{h} < 18$. We have now proved that $R = (1/6)R^\# = 1668.73545$.

The most likely value for the strict class number h is $h = \tilde{h} = 6$. We show that in any case 6 *divides* h. By sec. 13, end, h is even. To see that 3 divides h we search for a form that is obviously a cube: e.g., $F_{47} = F_{46} \div F_{22}$ has $a = -8$, and it is, in F, the cube of $F = (-2, 6395)$ (we could also have used $F_{26} \div F_9$). We have $\delta(47) = 9409.90926 \equiv -602.50344 \bmod R$, so if F were on the principal cycle it would have $\delta \equiv (-602.50344)/3 \bmod R/3$, so $\delta \equiv -200.83448, 355.41067$ or $911.65582 \bmod R$. Multiplying F by F_{24} or by F_{34}, or raising it to the 11-th power, we derive in each of the three cases a contradiction. We conclude that F has order 3 in the class group, and that 6 divides h.

If one checks that 5003 and 8179 are primes, it is not difficult to prove that $h \equiv 2 \bmod 4$. So if $h \neq 6$ then $h \geq 18$, and

$$\prod_{p \text{ prime},\, p > 100} \left(1 - \frac{\chi(p)}{p}\right)^{-1} > 3.05,$$

which is very unlikely.

We leave to the reader the pleasure to find out how multiplicative relations between the a's can be exploited to shorten the above calculations.

15. Concluding remarks

(i) The algorithms described in this lecture can be used for an experimental approach to Gauss's class number problems [4, secs 302-307]. Thus, they have been employed in the search for fields with irregular class groups, see [20] for references. It would also be interesting to investigate the decreasing density of fields with class number one among the real quadratic fields with prime discriminants, cf. [25, sec. 5; 12; 16, sec. 1].

(ii) The connection between the factorizations of the discriminant and the elements of order two in the class group gives rise to interesting factorization algorithms. Using negative discriminants, as Shanks does in [23], one obtains an algorithm factoring any positive integer n in $O(n^{(1/5)+\varepsilon})$ steps, if we assume the Riemann hypotheses. Positive discriminants can be used in several ways. We can look halfway the principal cycle (cf. the end of sec. 13), for discriminants that are small multiples of n. Modulo the Riemann hypotheses it can be shown that this also leads to an $O(n^{(1/5)+\varepsilon})$-algorithm. A second factoring method employing positive discriminants will be described by Shanks [26], cf. [28; 17]. This method has expected running time $O(n^{(1/4)+\varepsilon})$, for composite n. It is so simple that it can be programmed for a pocket calculator like the HP67 for numbers of up to twenty digits.

(iii) As Shanks suggested in [25, sec. 1; 29, sec. 4.4], it should be possible to adapt his techniques for number fields of higher degrees, like complex cubic fields. From sec. 10 we know that the "right" group to consider is a group whose "size" is essentially the product of the class number and the regulator. The main complication is that the circles are replaced by higher dimensional tori.

References

1. Z.I. Borevič, I.R. Šafarevič, Teorija čisel, Moscow 1964. Translated into German, English and French.
2. J.W.S. Cassels, Global fields, pp. 42-84 in: J.W.S. Cassels, A. Fröhlich (eds), Algebraic number theory, Academic Press, London 1967.
3. J.W.S. Cassels, Rational quadratic forms, Academic Press, London 1978.
4. C.F. Gauss, Disquisitiones arithmeticae, Fleischer, Leipzig 1801.
5. L.-K. Hua, On the least solution of Pell's equation, Bull. Amer. Math. Soc. 48 (1942),731-735.
6. I. Kaplansky, Composition of binary quadratic forms, Studia Math. 31 (1968), 523-530.
7. J.C. Lagarias, Worst-case complexity bounds for algorithms in the theory of integral quadratic forms, J. Algorithms 1 (1980) 142-186.
8. J.C. Lagarias, On the computational complexity of determining the solvability or unsolvability of the equation $X^2 - DY^2 = -1$, Trans. Amer. Math. Soc. 260 (1980), 485-508.
9. J.C. Lagarias, Succinct certificates for the solvability of binary quadratic diophantine equations, Proc. 20th IEEE Symp. foundations comp. sci., 1979, 47-56.
10. J.C. Lagarias, H.L. Montgomery, A.M. Odlyzko, A bound for the least prime ideal in the Chebotarev density theorem, Inventiones math. 54 (1979), 271-296.
11. J.C. Lagarias, A.M. Odlyzko, Effective versions of the Chebotarev density theorem, pp. 409-464 in: A. Fröhlich (ed.), Algebraic number fields, Academic Press, London 1977.
12. R.B. Lakein, Computation of the ideal class group of certain complex quartic fields, II, Math. Comp. 29 (1975), 137-144.
13. E. Landau, Handbuch der Lehre von der Verteilung der Primzahlen, 2 Bände, Teubner, Leipzig 1909; 2nd ed., Chelsea, New York 1953.
14. P.G. Lejeune Dirichlet, R. Dedekind, Vorlesungen über Zahlentheorie, Braunschweig 1893^4; reprint, New York 1968.
15. A.V. Malyshev, Yu.V. Linnik's ergodic method in number theory, Acta Arith. 27 (1975), 555-598.
16. J.M. Masley, Where are number fields with small class number?, pp. 221-242 in: M.B. Nathanson (ed.), Number Theory Carbondale 1979, Lecture Notes in Mathematics 751, Springer, Berlin 1979.
17. L. Monier, Algorithmes de factorisation d'entiers, Thèse de 3° cycle, Orsay 1980.
18. J. Oesterlé, Versions effectives du théorème de Chebotarev sous l'hypothèse de Riemann généralisée, pp. 165-167 in: Astérisque 61 (Journées arithmétiques de Luminy), Soc. Math. de France 1979.
19. J.M. Pollard, Theorems on factorization and primality testing, Proc. Cambridge Philos. Soc. 76 (1974), 521-528.
20. R.J. Schoof, Quadratic fields and factorization, in: Number theory and computers, Mathematisch Centrum, Amsterdam, to appear.
21. R.J. Schoof, Two algorithms for determining class groups of quadratic fields, Mathematisch Instituut, Universiteit van Amsterdam, to appear.
22. I. Schur, Einige Bemerkungen zu der vorstehenden Arbeit des Herrn G. Pólya: Über die Verteilung der quadratischen Reste und Nichtreste, Nachr. Kön. Ges. Wiss. Göttingen, Math.-phys. Kl.

(1918), 30-36; pp. 239-245 in: Gesammelte Abhandlungen, vol. II, Springer, Berlin 1973.

23. D. Shanks, Class number, a theory of factorization, and genera, pp. 415-440 in: Proc. Symp. Pure Math. 20 (1969 Institute on number theory), Amer. Math. Soc., Providence 1971.
24. D. Shanks, The infrastructure of a real quadratic field and its applications, Proc. 1972 number theory conference, Boulder, 1972.
25. D. Shanks, A survey of quadratic, cubic and quartic algebraic number fields (from a computational point of view), pp. 15-40 in: Congressus Numerantium 17 (Proc. 7th S-E Conf. combinatorics, graph theory, and computing, Baton Rouge 1976), Utilitas Mathematica, Winnipeg 1976.
26. D. Shanks, Square-form factorization, a simple $O(N^{1/4})$ algorithm, unpublished manuscript.
27. B.F. Skubenko, The asymptotic distribution of integers on a hyperboloid of one sheet and ergodic theorems (Russian), Izv. Akad. Nauk SSSR Ser. Mat. 26 (1962), 721-752.
28. S.S. Wagstaff, Jr., M.C. Wunderlich, A comparison of two factorization methods, to appear.
29. H.C. Williams, D. Shanks, A note on class number one in pure cubic fields, Math. Comp. 33 (1979), 1317-1320.

H.W. Lenstra, Jr.

Mathematisch Instituut

Universiteit van Amsterdam

Roetersstraat 15

1018 WB Amsterdam

Netherlands.

PETITS DISCRIMINANTS DES CORPS DE NOMBRES

Jacques MARTINET *

Soit K un corps de nombres, de degré n et de signature (r, s) (i.e., K possède r places réelles et s places imaginaires). Les travaux récents de Odlyzko, Poitou et Serre, reposant sur des méthodes analytiques, ont conduit à des minorations de la valeur absolue du discriminant de K bien meilleures que celles qui avaient été obtenues jusqu'à présent à l'aide des méthodes géométriques issues des travaux de Minkowski. C'est d'abord de ces travaux que nous rendrons compte. Les lettres n, r, s, éventuellement indexées par K ont le sens ci-dessus dans la suite ; on note d_K le discriminant du corps K : c'est la valeur du déterminant $\det(Tr(\omega_i \omega_j))$ où $\omega_1, \ldots, \omega_n$ désigne une base de K sur $\mathbb{Q}$ et Tr désigne la trace de K sur $\mathbb{Q}$. Etant donnée une extension L/K, on dispose également d'une notion de discriminant relatif, qui est un idéal de K. Pour d'autres notions de discriminant, voir l'appendice 1.

---

* Laboratoire associé au C.N.R.S. n° 226

Le problème fondamental concernant les discriminants est le suivant : étant donnés un couple (r, s) d'entiers $(r \geq 0$, $s \geq 0$, $n = r+2s > 0)$, et un entier d , à quelle condition existe-t-il un corps de nombres K de signature (r, s) tel que $d_K = d$? L'entier d est soumis à un certain nombre de contraintes :

(i) le signe de d est lié au couple (r, s) par l'égalité $|d| = (-1)^s d$,

(ii) l'entier d vérifie la congruence de Stickelberger : $d \equiv 0$ ou $1 \bmod 4$ (voir l'appendice 2) ,

(iii) pour chaque nombre premier p , l'exposant $v_p(d)$ de p dans d ne peut prendre qu'un nombre fini de valeurs (voir l'appendice 3) ,

(iv) enfin, et c'est là l'essentiel, on dispose de minorations de $|d|$ en fonction du couple (r, s) .

La façon dont ces minorations s'obtiennent par une étude des fonctions zêta des corps de nombres est expliquée dans le paragraphe 1. On trouvera en fin d'article, grâce à l'obligeance de Odlyzko d'une part, de Diaz y Diaz et de Poitou d'autre part, des tables de minorations.

Le paragraphe 2 est consacré à des applications, essentiellement à des problèmes de nombres de classes, des minorations exposées dans le paragraphe 1 .

Enfin, dans le paragraphe 3, nous étudions le problème de construire des corps de discriminants (en valeur absolue) petits, c'est-à-dire aussi voisins que possible des bornes inférieures provenant du paragraphe 1. L'expérience montre que la recherche de petits discriminants est liée à la recherche de corps euclidiens par la méthode de Lenstra ; pour cette raison, nous exposons avec quelques détails en quoi consiste cette méthode. Deux tables illustrent le paragraphe 3 : une table de petits discriminants de corps de degré ≤ 8 et une table de petits discriminants de corps totalement imaginaires.

1. - Minorations

1.1. - Formules explicites. - Serre s'est rendu compte que les premières minorations obtenues par Odlyzko s'interprétaient comme cas particuliers des "formules explicites" dues à Weil [W1, W2] ; nous renvoyons le lecteur à l'exposé [P 1] de Poitou pour l'aspect historique de la question. Nous utilisons les formules explicites sous la forme donnée par Poitou dans [P 2] (les formules énoncées et démontrées dans [P 2] concernent les fonctions zêta des corps de nombres, ce qui suffit pour les applications que nous avons en vue ; l'extension aux fonctions L ne présente pas de difficulté particulière).

On considère un corps de nombres K , de degré $n = r+2s$, et une fonction réelle paire F , vérifiant certaines conditions de régularité et assez rapidement décroissante à l'infini. Voici les conditions données par Poitou :

(i) Les fonctions $x \mapsto F(x)$ et $x \mapsto (F(0)-F(x))/x$ sont à variation bornée sur toute la droite, et leur valeur en chaque point est la moyenne de leurs limites à gauche et à droite.

(ii) A l'infini, la fonction F est $O(e^{-(\frac{1}{2}+\delta)|x|})$ pour un $\delta > 0$.

On considère la fonction Φ définie pour s complexe appartenant à la bande critique $0 < \mathrm{Re}(s) < 1$ par la formule :

$$\Phi(s) = \int_{-\infty}^{+\infty} F(x)\, \exp((s-\tfrac{1}{2})x)\, dx$$

et les trois nombres réels A_F, B_F, C_F définis par les intégrales :

$$A_F = \int_0^{+\infty} \frac{F(0) - F(x)}{2\,\mathrm{ch}(x/2)}\, dx$$

$$B_F = \int_0^{+\infty} \frac{F(0) - F(x)}{2\,\mathrm{sh}(x/2)}\, dx$$

$$C_F = 4\int_0^{+\infty} F(x)\, \mathrm{ch}(x/2)\, dx \, .$$

THÉORÈME. - <u>Quelle que soit la fonction</u> F <u>paire vérifiant les conditions</u> (i) <u>et</u> (ii), <u>on a l'égalité</u> :

$$F(0)\left[\log|d| - n(\gamma + \log 8\pi) - r\frac{\pi}{2}\right] + r\,A_F + n\,B_F + C_F =$$

$$\lim_{T\to\infty} \Big(\sum_{|\mathrm{Im}(\rho)| \leq T} \Phi(\rho)\Big) + 2 \sum_{\mathfrak{p},\, m} \frac{\log N\mathfrak{p}}{(N\mathfrak{p})^{m/2}}\, F(m \log N\mathfrak{p})\, .$$

Dans le membre de gauche, d est le discriminant d'un corps de nombre K de degré n possédant r places réelles et s places complexes ; γ désigne la constante d'Euler ($\gamma = 0,577\,215\ldots$). Le membre de droite est la somme de deux termes. Dans le premier, la sommation est étendue aux zéros de la fonction zêta de K de partie réelle positive ; dans le second, la sommation est étendue aux entiers $m > 0$ et aux idéaux premiers $\mathfrak{p}$ de K.

Les convergences des séries et intégrales qui interviennent (N désigne la norme) sont clairement assurées par les hypothèses faites sur la fonction F.

1. 2. - <u>Principe d'obtention des inégalités</u>. - Pour tirer du théorème ci-dessus des inégalités sur les discriminants, on choisit F de façon que le second membre de l'égalité soit ≥ 0, en prenant

une fonction F positive dont la transformée de Mellin Φ prend des valeurs de parties réelles positives aux zéros de la fonction zêta de K (la partie réelle suffit par symétrie). Faute de pouvoir localiser ces zéros avec précision, le mieux que l'on puisse faire dans l'état actuel de nos connaissances est de s'arranger pour que Φ soit de partie réelle positive dans la bande critique $0<\mathrm{Re}(s)<1$. Si l'on veut bien admettre l'hypothèse de Riemann généralisée (GRH), il suffit de prendre la partie réelle de Φ positive sur la droite $s=\frac{1}{2}$; alors, lorsque l'on pose $s=\frac{1}{2}+it$, Φ devient une fonction φ de la variable réelle t, définie par la formule

$$\varphi(t)=\int_{-\infty}^{+\infty} F(x)\, e^{itx}\, dx .$$

On reconnaît la transformée de Fourier de F, et il suffit alors d'utiliser des fonctions F positives dont la partie réelle de la transformée de Fourier (transformée de Fourier en cosinus) soit également positive.

1.3. - Minorations asymptotiques. - On se donne un réel $\rho\in[0,1]$, et l'on considère une suite de corps K dont les degrés n tendent vers l'infini pour laquelle le rapport r/n tend vers ρ ; il s'agit de minorer la limite inférieure de la suite $|d_K|^{1/n}$.

a) Minoration sous GRH (Serre)

On considère une fonction F positive à transformée de Fourier positive ; un exemple de telle fonction est donné par la figure de gauche ci-après, les fonctions de ce type étant les carrés de convolution des fonctions représentées par la figure de droite.

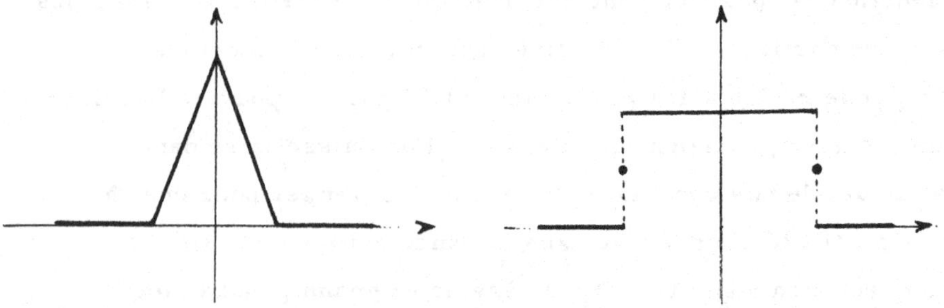

On considère ensuite une suite de fonctions de la forme $x \to aF(x/b)$ dans laquelle on ajuste les nombres réels a et b de façon à obtenir la convergence de la suite (au sens des distributions) vers la mesure de Dirac en 0. Après division par la valeur en 0 des fonctions, on constate que les contributions dues aux termes contenant A_F, B_F et C_F tendent vers 0. On en déduit la minoration asymptotique

$$\liminf |d_K|^{1/n} \geq 8\pi e^{\gamma+(\pi/2)\rho} .$$

Numériquement, on obtient les inégalités $\liminf |d_K|^{1/n} \geq 44,7$ (resp. $\liminf |d_K|^{1/n} \geq 215,3$) pour une suite de corps totalement imaginaires (resp. totalement réels).

b) <u>Minorations inconditionnelles(Odlyzko)</u>

Le procédé ci-dessus n'assure pas la positivité de Φ dans toute la bande critique. Odlyzko s'est aperçu que l'on y parvenait en remplaçant les fonctions $x \mapsto aF(x/b)$ par les fonctions $x \mapsto aF(x/b)/\mathrm{ch}(x/2)$. La contribution des termes en A_F et B_F intervient alors dans le résultat, et l'on trouve la minoration asymptotique

$$\liminf |d_K|^{1/n} \geq 4\pi\, e^{\gamma+\rho} .$$

Par rapport à la minoration trouvée par Serre sous GRH, on a perdu un facteur croissant en fonction de ρ de 2 dans le cas totalement imaginaire à $2e^{(\pi/2)-1} = 3,539\ldots$ dans le cas totalement réel.

1.4. - Minorations pour une signature donnée. - Par rapport aux minorations données de 1.3, que ce soit sous GRH ou inconditionnellement, les termes A_F, B_F et C_F introduisent une correction affaiblissant les minorations de $|d_K|^{1/n}$. Il est possible de construire des tables ; la qualité des résultats dépend du choix de la fonction F. Grâce à l'obligeance de Odlyzko (resp. Poitou et Diaz y Diaz), on trouvera de telles tables à la fin de cet article sous GRH (resp. inconditionnelles).

Ces tables partagent les deux propriétés suivantes :

(a) pour un rapport r/n donné, la minoration de $|d_K|^{1/n}$ est une fonction croissante de n ;

(b) pour un degré donné, la minoration de $|d_K|^{1/n}$ est une fonction croissante de r/n.

Outre les tables citées ci-dessus, Poitou a donné diverses formules faciles à programmer sur une calculatrice de poche, concernant le cas "inconditionnel". Nous reproduisons avec son autorisation la formule suivante, qui est un bon compromis entre la simplicité et la qualité du résultat ; elle conduit à une minoration de $\frac{1}{n}\log|d|$ en fonction homographique de r/n :

$$\frac{1}{n}\log|d| \geq \gamma + \log 4\pi + \frac{r}{n} - (\tfrac{3}{5})(12\pi)^{2/3} b_3^{1/3}/(n^{2/3} + (\tfrac{1}{4})(12\pi)^{2/3} c_2 b_3^{1/3}).$$

Dans cette formule, γ est la constante d'Euler. Les coefficients b_3 et c_2, qui dépendent de r/n, sont définis ainsi : pour k entier positif, on pose

$\lambda(k) \quad = 1 + 1/3^k + 1/5^k + \dots$

$\eta(k) \quad = 1 - 1/2^k + 1/3^k - 1/4^k + \dots$

$b_{2k+1} = \lambda(2k+1) + \eta(2k)\, r/n$

$c_2 \quad = (12\, b_5)/(35\, b_3)$

Voici les valeurs numériques utiles :

$\lambda(3) \quad = 1{,}051\ 799\ 790 \dots$ $\qquad \eta(2) = 0{,}822\ 467\ 033 \dots$

$\lambda(5) \quad = 1{,}004\ 523\ 762 \dots$ $\qquad \eta(4) = 0{,}947\ 032\ 829 \dots$

$\gamma + \log 4\pi = 3{,}108\ 239\ 911 \dots$

$(12\pi)^{2/3} \quad = 11{,}243\ 134\ 665 \dots$

Exemple. - Pour les corps totalement imaginaires, on peut utiliser la formule

$$\frac{1}{n} \log |d| \geq 3{,}108 - 6{,}861 / (n^{2/3} + 0{,}936)\ .$$

Pour $n \geq 30$, cette formule donne des résultats meilleurs que les tables inconditionnelles de Odlyzko (qui ne sont pas reproduites ici, celles de Diaz y Diaz et Poitou étant un petit peu meilleures).

1.5. - Le second membre des formules explicites. - Dans les paragraphes précédents, le second membre des formules explicites n'intervenait que par la condition qu'il soit ≥ 0 . La contribution du premier terme semble difficile à estimer. On peut en revanche tenir compte du second terme, et cela est particulièrement facile à faire lorsque l'on utilise des fonctions nulles en dehors d'un compact, ce qui est le cas des fonctions utilisées par Odlyzko pour construire ses tables. Voici un exemple d'application : considérons le corps de degré 8 , extension cyclique de degré 4 de $\mathbf{Q}(i)$ de conducteur un idéal premier au-dessus de 17 . Son discriminant est $2^8.17^3$; c'est le plus petit discriminant connu parmi les discriminants de corps totalement imaginaires de degré 8 , et , dans ce corps , 2 est le carré d'un idéal premier de degré 4 . En tenant compte du second terme, on peut montrer inconditionnellement que, si un corps totalement imaginaire de degré 8 a un discriminant au plus égal à $2^8.17^3$, alors la décomposition de 2 dans K est de l'une des trois formes suivantes : 2 est inerte, $(2) = \mathfrak{P}^2$ ou $(2) = \mathfrak{P}\mathfrak{P}'$, où $\mathfrak{P}$ et $\mathfrak{P}'$ désignent des idéaux premiers de degré 4 . De façon générale, si le discriminant d'un corps est proche de la limite inférieure donnée par les tables, ce corps ne contient pas d'idéaux premiers de petite norme. En outre, dans ces conditions, le premier terme du second membre des formules explicites est lui aussi petit.

1. 6. - Grands discriminants. - Fixons un degré n et une signature (r, s) . Les corps soumis à ces conditions peuvent avoir des discriminants arbitrairement grands. Au premier membre de la formule explicite, le corps n'intervient que par le terme en $\frac{1}{n}\log|d|$. Le second terme du second membre est borné : une borne supérieure se détermine en calculant la somme $\sum_{m,\mathfrak{p}} \frac{F(\log N\mathfrak{p})}{(N\mathfrak{p})^{\frac{1}{2}}} F(m \log \mathfrak{p})$ dans la situation (impossible en fait à cause du théorème de Tchebotarev) où la décomposition des nombres premiers est la plus défavorable. Il s'en suit que, pour une suite de corps dont les discriminants tendent vers l'infini, le terme $\lim_{T\to\infty} \left(\sum_{|\mathrm{Im}(\rho)|\leq T} \Phi(\rho)\right)$ tend vers l'infini. Cela traduit une tendance des zéros de la fonction zêta à se rapprocher de l'axe réel. Il y a peut-être là (remarque de Poitou) une possibilité d'obtenir des renseignements sur la répartition statistique des zéros des fonctions zêta en fonction des discriminants en utilisant des familles convenables de fonctions F .

1.7. - Peut-on faire mieux ? - La question posée est de savoir si l'on peut espérer améliorer les minorations ci-dessus. Nous verrons au paragraphe 3 que pour de petits degrés, on connaît des exemples de corps K pour lesquels $|d_K|^{1/n}$ est proche des minorations en question. Il n'en est pas de même quand on considère de grands degrés, mais cela est sans doute dû au fait que l'on ne connaît pas de bonnes méthodes pour décrire des corps de grands degrés.

A défaut de disposer d'exemples, on peut tenter d'analyser de plus près la méthode elle même. Faute de renseignements précis sur la localisation des zéros des fonctions zêta, et si l'on ne désire pas introduire de conditions sur la décomposition des nombres premiers, on ne peut que considérer des fonctions F positives telles que Φ soit positive dans la bande critique (ou simplement sur la droite $s = 1/2$ si l'on accepte d'utiliser GRH). Lorsque l'on recherche des minorations asymptotiques, les minorations de Serre (sous GRH) et de Odlyzko (sans hypothèse) apparaissent être les meilleures possibles ; Odlyzko m'a écrit qu'il avait étudié la situation générale, et qu'il disposait de résultats montrant que l'on ne pouvait pas espérer obtenir d'améliorations substantielles des minorations en s'imposant les règles ci-dessus sur F et Φ .

1.8. - Classes d'idéaux. - Dans les méthodes géométriques (Minkowski et ses continuateurs, jusqu'à Mulholland), les minorations des discriminants apparaissent comme corollaire à un résultat prouvant la finitude du nombre de classes d'idéaux des corps de nombres : il existe une constante $c_{r,s}$ telle que toute classe d'idéaux contient un idéal $\mathfrak{A}$ de norme $\leq c_{r,s}\sqrt{|d|}$; en minorant la norme de $\mathfrak{A}$ par 1 , on obtient une minoration de $|d|$.

L'emploi des formules explicites ne conduit pas à des résultats de ce type. Il convient néanmoins de signaler que Zimmert [Z] a obtenu récemment des résultats de la forme ci-dessus par des méthodes analytiques, desquels il est possible de déduire la minoration asymptotique trouvée par Odlyzko.

2. - Applications des minorations des discriminants

2.1. - Extensions non ramifiées. - Soit K un corps de nombres de degré n et soit L une extension non ramifiée de K. On a entre les discriminants de K et L la relation $|d_K|^{1/[K:\mathbb{Q}]} = |d_L|^{1/[L:\mathbb{Q}]}$. En conséquence, si $|d_K|^{1/n}$ est inférieur à la minoration asymptotique du paragraphe 1, l'extension maximale non ramifiée de K est de degré fini ; si $|d_K|^{1/n}$ est même inférieur à la minoration correspondant au degré $2n$, le corps K ne possède aucune extension non ramifiée autre que lui-même. Les résultats de Minkowski permettaient déjà de prouver cette propriété pour certains corps K, en particulier pour $K = \mathbb{Q}$, permettant à Minkowski de résoudre un des problèmes ouverts célèbres de la théorie des nombres ; les progrès récents des minorations des discriminants ont considérablement augmenté le champ d'application de la méthode.

Bien entendu, nous devons faire des démonstrations, et par conséquent utiliser les minorations inconditionnelles. Ceux qui croient en la validité de l'hypothèse de Riemann généralisée en concluront que le procédé ci-dessus ne s'applique qu'à un nombre fini de corps.

2.2. - Corps de nombre de classes 1. - On note h_K le nombre de classes du corps de nombres K. La théorie du corps de classes montre qu'un corps de nombres K a un nombre de classes 1 si et seulement si K ne possède pas d'extension abélienne non ramifiée autre que lui-même. Le procédé indiqué en 2.1. fournit donc des exemples de corps dont le nombre de classes est 1. L'égalité $h_K = 1$ équivaut aussi à l'absence d'extension résoluble non ramifiée de K autre que K lui-même. Mais rien n'empêche qu'un corps de nombres avec $h_K = 1$ possède des extensions non ramifiées à groupe de Galois simple non abélien ; nous n'avons pas d'interprétation de l'existence de telles extensions.

Comme l'a remarqué Masley (cf [Ms] et sa bibliographie), on peut, en utilisant l'action des groupes de Galois sur les groupes de classes, prouver l'égalité $h_K = 1$ dans le cas de corps pour lesquels les minorations des discriminants n'entraînent pas immédiatement l'absence d'extension non ramifiée. Le principe est le suivant : soit L une extension galoisienne non ramifiée d'un corps K, de groupe de Galois G ; le groupe G opère sur le groupe des classes Cl_L de L. Si l'on peut prouver que l'action de G est fidèle, on a l'inégalité $h_L \geq [L:K]$, d'où l'existence d'une extension non ramifiée M/K de K de degré relatif $\geq [L:K]^2$, ce qui peut être contradictoire avec les minorations des discriminants pour le degré $[M:\mathbb{Q}]$. Voici comment on peut utiliser cette remarque dans le cas où G est cyclique d'ordre premier ℓ. Soit p un nombre premier, et soit H_p le sous-groupe de Cl_L formé des classes vérifiant l'égalité $x^p = 1$; H_p fournit une représentation de G sur $\mathbb{F}_p$. Si $p \neq \ell$, cette représentation est semi-simple ; par conséquent, si H_p est non trivial, son ordre est divisible par p^m, où m est le plus petit entier pour lequel $\mathbb{F}_{p^m}$ contient une racine d'ordre ℓ de l'unité, i.e. $p^m \equiv 1 \bmod \ell$. Si $p = \ell$, et si H_p est non trivial, le sous-groupe de H_p fixé par G est également non trivial ; des formules classiques de classes invariantes remontant à Furtwängler, Takagi et Chevalley fournissent un calcul explicite du nombre de classes invariantes permettant éventuellement de prouver que h_L est d'ordre premier à ℓ . Un cas particulier important est le suivant : si h_K est premier à ℓ , et si une unique place se ramifie dans L/K , alors h_L est également premier à ℓ .

2. 3. - Exemples

2. 3. 1. - Masley a prouvé que, pour $n \leq 50$, le sous-corps réel maximal du corps des racines n-èmes de l'unité a pour nombre de classes 1 . Ces résultats ont été étendus par van der Linden [vL] à un grand nombre de corps abéliens réels.

2. 3. 2. - On peut à l'aide de ces méthodes, et en utilisant les résultats connus sur les classes des corps de degré 3 ou 4 , étudier les extensions maximales non ramifiées des corps quadratiques imaginaires ; pour $|d|^{\frac{1}{2}} < 4\pi e^{\gamma}$, i. e. $|d| \leq 499$, les corps quadratiques imaginaires de discriminant d ont une extension maximale non ramifiée de degré fini. Pour $|d| \leq 250$, on trouve que l'extension maximale non ramifiée d'un corps quadratique de discriminant d est son corps de classes de Hilbert, aux sept exceptions près suivantes, pour lesquelles nous donnons la structure du groupe de Galois relatif de l'extension maximale non ramifiée, le symbole D_n (resp. H_n) désignant un groupe diédral (resp. quaternionien) d'ordre n

$d = -115 : D_6$; $d = -120 : H_8$; $d = -155 : H_{12}$;
$d = -184 : H_{12}$; $d = -195 : H_{16}$; $d = -235 : D_{10}$;
$d = -248 : H_{24}$.

Voici deux autres exemples intéressants :

<u>d = -283</u> : l'extension maximale non ramifiée est de degré 48 ; son groupe de Galois relatif (resp. absolu) est le groupe $\widetilde{A}_4$

(resp. $\tilde{S}_4$), obtenu comme image réciproque dans le revêtement $S^3 \to SO_3(\mathbb{R})$ d'un sous-groupe de $SO_3(\mathbb{R})$ isomorphe à A_4 (resp. S_4).

d = -420 : le groupe des classes du corps quadratique est de type (2, 2, 2), et c'est le seul exemple dans lequel le groupe des classes contient un groupe à trois générateurs pour lequel on sache montrer que l'extension maximale non ramifiée est de degré fini. On montre sans difficulté que l'extension maximale non ramifiée K a un degré sur $\mathbb{Q}$ multiple de 64 ; si l'on accepte l'hypothèse de Riemann généralisée, il est immédiat que $[K:\mathbb{Q}] = 64$; démontrer ce résultat sans utiliser l'hypothèse de Riemann généralisée est beaucoup plus difficile ; une démonstration m'a été communiquée par René Schoof pendant la conférence d'Exeter [Sch].

2. 3. 3. - On trouve dans [Mr 2] un exemple de corps totalement imaginaire de degré 116 et de nombres de classes 1 ; c'est le plus grand degré connu pour un coips avec $h_K = 1$. Un autre exemple intéressant, signalé par Lenstra, est celui du corps de classe de Hilbert du corps des racines 23-èmes de l'unité, de degré 66.

3. - Recherche de petits discriminants

Dans ce paragraphe, nous cherchons des exemples de corps ayant un petit discriminant relativement aux minorations exposées dans le paragraphe 1. Nous avons systématiquement comparé, pour un corps K de degré n, $|d_K|^{1/n}$ avec la minoration donnée par Odlyzko sous GRH pour la signature de K ; il semble en effet plus raisonnable d'admettre GRH pour avoir une idée de ce que l'on peut espérer ; de toute façon, pour les degrés assez petits, les minorations inconditionnelles sont proches des minorations obtenues sous GRH.

Les corps de discriminant petit ont pour nombre de classes 1 . Souvent, on peut montrer qu'ils sont euclidiens. Inversement, la recherche de corps euclidiens par la méthode de Lenstra conduit à des exemples de petits discriminants ; pour cette raison, nous avons donné quelques détails sur cette méthode.

3.1. - Tours de corps de classes. - Dans ce sous-paragraphe, nous cherchons à tester les minorations asymptotiques de Serre. On a vu (cf §.1.3), que pour tout réel $\rho \in [0,1]$, on avait pour une suite de corps K de degré n tendant vers l'infini pour lesquels r/n tend vers ρ la minoration asymptotique $|d_K|^{1/n} \geq 8\pi e^{\gamma+(\pi/2)\rho}$ (resp. $4\pi e^{\gamma+\rho}$) selon que l'on accepte ou non d'utiliser l'hypothèse de Riemann généralisée. Un résultat important de Golod et Šafarevič est que, pour tout ρ rationnel $\in [0,1]$, il existe une suite de corps de degré tendant vers l'infini avec $\lim \frac{r}{n} = \rho$ et $\lim |d_K|^{1/n} < +\infty$ (autrement dit, la limite inférieure de $|d_K|^{1/n}$ pour $n \to \infty$ et $\frac{r}{n} \to \rho$ est finie). Le principe est d'utiliser des tours de corps de classes, c'est-à-dire "d'empiler" des extensions abéliennes non ramifiées ; on ne sait pas ce qui se passe si $\lim \frac{r}{n} = \rho$ n'est pas un nombre rationnel. Voici brièvement comment on procède.

a) On se fixe un nombre premier p . Etant donné un pro-p-groupe G , on sait définir le nombre minimum de générateurs $d_1(G)$ de G et le nombre de relations $d_2(G)$ entre ces générateurs ; une formulation de ces définitions est $d_i(G) = \dim_{\mathbb{F}_p} H^i(G, \mathbb{Z}/p\mathbb{Z})$. Pour un groupe abélien fini A , on note $d(A)$ le p-rang de A : $d(A) = \dim_{\mathbb{F}_p} \mathbb{Z}/p\mathbb{Z} \otimes A$.

b) Soit K un corps de nombres, et soit L/K une extension finie, non ramifiée, galoisienne, dont le groupe de Galois G est un p-groupe ; soit E_K (resp. E_L) le groupe des unités de K (resp. L). Alors, si p ne divise pas h_L, on a l'égalité (cf [Se]):

$$d_2(G) - d_1(G) = d(E_K/N_{L/K}(E_L)).$$

c) Soit G un p-groupe fini ; Golod et Šafarevič (cf [Ro]) ont montré l'inégalité

$$d_2(G) > d_1(G)^2/4 .$$

d) Il résulte de b) et c) que si le p-rang t du groupe des classes de K est assez grand, alors la p-tour de corps de classes de K est infinie : en effet, soit G le groupe de Galois de la p-extension maximale non ramifiée de K ; la théorie du corps de classes montre que l'on a l'égalité $d_1(G) = t$; en utilisant b) et le théorème de Dirichlet sur les unités, on voit tout de suite que, si G est fini, on a l'inégalité $d_2(G) - d_1(G) \leq r+s$. Cette inégalité est contradictoire avec c) si t dépasse $2 + 2\sqrt{r+s+1}$.

e) Soit alors K_o un corps de nombre de degré $n_o = r_o + 2s_o$ avec $r_o/n_o = \rho$. Un calcul de classes invariantes montre que le 2-rang du groupe des classes du corps $K_o(\sqrt{m})$, m entier positif, tend vers l'infini avec le nombre de facteurs premiers de m . Pour un choix convenable de m , le corps $K = K_o(\sqrt{m})$ a donc une 2-tour de corps de classes infinie, et le rapport r/n ainsi que le discriminant élevé à la puissance 1/n sont constants, égaux respectivement à ρ et à $|d_K|^{1/n}$, lorsque l'on considère les différents étages de la tour de corps de classes, C.Q.F.D.

Les exemples de tours avec une valeur pas trop grande pour $|d_K|^{1/n}$ ont été obtenues en rendant d) explicite . Pour $\rho = 0$, le meilleur exemple connu (cf [Mr 1]) est celui du corps $K = \mathbb{Q}(\cos 2\pi/11 , \sqrt{-46})$, de degré 10 , pour lequel $|d_K|^{1/10} = 11^{4/5} . 2^{3/2} . 23^{1/2} = 92,368 \ldots$, ce qui dépasse considérablement la borne inférieure asymptotique $8\pi e^{\gamma} = 44,763\ldots$.

Il est extrêmement probable qu'il existe des tours donnant de meilleurs résultats. Ainsi, considérons un corps quadratique imaginaire K avec cinq nombres premiers ramifiés, si bien que le 2-rang du groupe des classes de K est égal à 4 . Si la 2-tour de corps de classes de K est finie, alors d'après b) et c), le groupe G (en tant que pro-2-groupe) peut être défini par quatre générateurs et cinq relations. On ignore s'il existe de tels 2-groupes ; les spécialistes conjecturent qu'un tel groupe n'existe pas. Une démonstration de ce fait permettrait alors de remplacer 92, 368... par $2.(3.5.7.13)^{\frac{1}{2}} = 73,891\ldots$, comme on le voit en considérant le corps $\mathbb{Q}(\sqrt{-3.5.7.13})$.

A défaut d'améliorer le théorème de Golod et Šafarevič, ce qui est probablement difficile, on peut essayer d'utiliser d'autres propriétés des groupes de Galois. L'exemple suivant dû à Koch et Venkov ([K - V]) est encourageant, bien que n'améliorant pas les estimations de $|d_K|^{1/n}$: ces deux auteurs ont montré que pour p impair, l'éventualité $d_1(G) = d_2(G) = 3$ était impossible pour le groupe de Galois (supposé fini) de la tour de corps de classes d'un corps quadratique imaginaire, bien qu'il existe des p-groupes finis ayant ces valeurs pour d_1 et d_2 ; leur démonstration est en fait un théorème de théorie des groupes, que l'on applique aux corps quadratiques imaginaires en utilisant l'existence de l'automorphisme d'ordre 2 induit par la conjugaison complexe.

A mon avis, l'obtention de résultats significativement meilleurs passe par des progrès en théorie des groupes.

3.2. - Corps totalement imaginaires. - La théorie du corps de classes permet de décrire les extensions abéliennes d'un corps de nombres k à partir de données provenant exclusivement de k :

groupes des classes d'idéaux et des unités de k , et décomposition dans k des nombres premiers. L'itération du procédé est difficile, car la théorie du corps de classes, si elle fournit la décomposition des idéaux premiers dans les extensions abéliennes de k , n'indique rien au sujet des unités ni des classes d'idéaux de ces extensions. On n'a alors pas d'autre recours que la détermination explicite de polynômes, ce qui conduit à des calculs d'une complexité croissant rapidement avec le degré.

L'utilisation de la théorie du corps de classes est d'autant plus facile que le groupe des unités est de rang petit. La table I, construite d'après des calculs faits dans [Mr 2], donne, pour 22 entiers n compris entre 2 et 80 , des exemples de corps totalement imaginaires K de degré n pour lesquels $|d_K|^{1/n}$ n'excède que de 3 % au plus la minoration obtenue sous GRH par Odlyzko ; tous les corps sont des corps de classes de rayon sur un corps de degré 2 , 3 ou 4 possédant au plus une unité fondamentale ; ils sont tous de nombre de classes 1 .

3. 3. - <u>Corps euclidiens</u>. - L'étude des corps de nombres euclidiens (pour la norme) a été complètement renouvelée par Lenstra. Introduisons, pour tout corps K , la constante L(K) , borne inférieure des normes des idéaux premiers de K , et la "constante de Lenstra" M(K) , borne supérieure des entiers ℓ tels qu'il existe une suite $\omega_1, \ldots, \omega_\ell$ d'éléments de K dont les différences sont des unités (on a clairement $\ell \leq L(K)$ pour toute suite du type ci-dessus, d'où l'existence de M(K) ; en outre, on peut toujours prendre $\omega_1 = 0$ et $\omega_2 = 1$, d'où les inégalités $2 \leq M(K) \leq L(K) \leq 2^n$). Dans [Ln] , Lenstra détermine des constantes $\alpha_{r,s}$ telles que, si K est un corps de nombres possédant r places réelles et s places complexes, l'inégalité $M(K) \geq \alpha_{r,s}\sqrt{|d_K|}$ entraîne que K

est euclidien. En cherchant parmi les corps de petits discriminants, on a donc de bonnes chances de découvrir des corps euclidiens. Il se trouve que, réciproquement, on trouve des corps de petits discriminants en cherchant des corps euclidiens par la méthode de Lenstra : c'est en effet un fait d'expérience que les corps pour lesquels $|d_K|^{1/n}$ est très proche des minorations que l'on obtient à l'aide des formules explicites ont une grande constante $M(K)$. Par exemple, pour le corps $K_{4,3}$ de la table 2, on a $M = 9$, alors que l'inégalité triviale $M \geq 2$ suffit à prouver que ce corps est euclidien. Un autre exemple est le corps $K_{4,4}$ pour lequel on sait que M vaut 10 ou 11 alors que l'inégalité facile $M \geq 3$ suffit à prouver qu'il est euclidien. On peut noter à ce propos que Lenstra établit dans [Ln] une minoration de $M(K)$ en fonction de $L(K)$. Or, il se trouve qu'un corps pour lequel $|d_K|^{1/n}$ est proche des minorations obtenues par les formules explicites a nécessairement une grande constante $L(K)$: cela se voit immédiatement en examinant le deuxième terme du second membre des formules explicites. Il y a peut être dans cette remarque une explication au lien qui semble exister entre les constantes M et les discriminants.

3.4. - Petits discriminants jusqu'au degré 8. - On trouve dans la table II pour chaque couple $(n, r+s)$ avec $2 \leq n \leq 8$ la racine n-ème du discriminant d'un corps, noté $K_{n,r+s}$; ces discriminants sont en valeur absolue les plus petits que je connaisse dans leur catégorie. Voici comment ont été choisis les 23 corps de la table II :

a) pour 13 d'entre eux, il a été prouvé qu'ils réalisent le minimum en valeur absolue des discriminants des corps correspondant à ce couple $(n, r+s)$; ce sont les corps avec $n \leq 5$, $n = 6$ et $r+s = 3$ ou 6, et enfin $n = r+s = 7$. Le résultat était connus de Gauss pour les degrés 2 et 3 en termes de minima de formes quadratiques ou cubiques. Les résultats pour $n=4$ sont dûs à Mayer ([My], 1929) et ceux pour $n=5$ à Hunter ([H], 1957). Les trois autres résultats sont beaucoup plus récents, et ont été obtenus par Kaur ([Ka], 1970) pour $n=r+s=6$, par Liang et Zassenhaus ([L-Z], 1977) pour $n=6$, $r+s=3$ et par Pohst ([Po], à paraître) pour $n=r+s=7$.

b) Les autres corps de la table de degré 6 ou 7 et le corps totalement imaginaire de degré 8 (6 corps) ont été choisis à cause de la valeur de leur constante M. Pour prouver que les corps $K_{p,q}$ sont euclidiens il suffit de prouver l'inégalité $M \geq 5$ si $(p,q)=(6,4)$, $(7,4)$ ou $(8,4)$ et l'inégalité $M \geq 6$ si $(p,q)=(6,5)$. Des recherches très étendues de corps vérifiant l'une des inégalités $M \geq 5$ ou $M \geq 6$ ayant été faites, par Lenstra en particulier, on peut conjecturer, compte tenu des remarques du paragraphe précédent, que ces quatre corps fournissent les plus petits discriminants parmi ceux correspondant à l'un des couples $(6,4)$, $(6,5)$, $(7,4)$ et $(8,4)$; les deux corps de degré 6 sont du reste bien connus des spécialistes.

On ne peut pas être aussi affirmatif en ce qui concerne les deux autres corps, faute d'expérimentation suffisante. Le corps $K_{7,5}$ a été trouvé lors d'une recherche de corps vérifiant $M \geq 7$ en considérant les polynômes unitaires f pour lesquels $f(0)$, $f(1)$, $f(-1)$, $f(i)$, $f(-j)$ $(i^2+1=0,\ j^2+j+1=0)$ sont des unités, l'inégalité $M \geq 7$ se voyant alors sur la suite $0, 1, x, 1/(1-x)$, $(x-1)/x$, $x+1$, x^2+1 où x désigne une racine de f. Leutbecher ([Lt], cf [Mr 3]) a montré que l'on avait l'inégalité $M \geq 7$ pour les corps définis par une racine d'un polynôme unitaire f tel que $f(0)$, $f(1)$, $f(-1)$, $f(2)$, $f(\theta)$ $(\theta^2-\theta-1=0)$ sont des unités, à l'aide de la suite $0, 1, x, x+1, x^2, x/(x-1), 1/(2-x)$. C'est cette suite que nous avons utilisée pour trouver le corps $K_{7,6}$: nous avons déterminé les polynômes vérifiant $f(-1) = f(2) = 1$, $f(0) = 1$, $f(1) = f(\theta) = -1$, ce qui assure l'existence d'au moins 5 racines réelles pour f ; on trouve une famille de polynômes dépendant d'un paramètre, que l'on choisit pour que les coefficients de f ne soient pas trop grands. Le premier essai a conduit au corps $K_{7,6}$.

c) Enfin, les quatre corps non totalement imaginaires de degré 8 ont été cherchés comme extensions quadratiques de corps de degré 4 au moyen de la théorie du corps de classes ; le corps $K_{8,8}$ m'a été signalé il y a longtemps par Lenstra.

Tous les corps de la table II sont de nombre de classes 1 ; on sait que dix-huit d'entre eux sont euclidiens, et cela peut se démontrer à l'aide de minorations de M ; nous en dressons la liste ci-dessous en indiquant l'inégalité à démontrer pour M en fonction du couple (n , r+s) :

$M \geq 2$: (2, 1) , (2, 2) , (3, 2) , (3, 3) , (4, 2) , (4, 3) .
$M \geq 3$: (4, 4) , (5, 3) , (6, 3)
$M \geq 4$: (5, 4)
$M \geq 5$: (5, 5) , (6, 4) , (7, 4) , (8, 4)
$M \geq 6$: (6, 5)
$M \geq 8$: (7, 5)
$M \geq 9$: (6, 6) , (8, 5) .

A cette liste, on pourrait adjoindre le corps totament imaginaire de degré 10 de la table I ; il est euclidien, l'inégalité $M \geq 15$ ayant été démontrée par Mestre ([Me]) en utilisant la multiplication complexe ($M \geq 10$ suffit).

4. - Tables

Nous expliquons dans ce paragraphe l'usage des tables.

Table I. - La table I décrit des corps totalement imaginaires K. La première colonne contient le degré du corps, obtenu comme corps de classes de rayon sur un corps k dont le degré (2, 3 ou 4) est donné dans la colonne 2 et le discriminant dans la colonne 3 ; la colonne 4 contient le conducteur de K sur k, la notation $\mathfrak{P}_p$ (resp. $\mathfrak{p}_p$) désignant un idéal premier convenable de k au-dessus du nombre premier p, de degré 1 (resp. 2) ; les colonnes 5 et 6 contiennent la valeur de $|d_K|^{1/n}$, arrondie aux trois décimales les plus proches dans la colonne 6 ; la colonne 7 contient la minoration Odl_n pour le degré n obtenue sous GRH par Odlyzko (cf. table 3) ; enfin, la colonne 8 contient l'expression $(|d_K|^{1/n}/Odl_n)-1$, exprimée en pourcentage et arrondie aux deux décimales les plus proches.

Table II. - A l'intersection de la colonne n et de la ligne r+s figurent trois nombres, relatifs à un corps $K_{n,r+s}$ de degré n $(2 \leq n \leq 8)$ et de signature (r, s) : $|d_K|^{1/n}$, arrondi aux trois décimales les plus proches, $Odl_{n,r+s}$ minoration de Odlyzko sous GRH pour le degré n et la signature (r, s), et enfin l'expression $(|d_K|^{1/n}/Odl_{n,r+s})-1$ exprimée en pourcentage, le résultat étant arrondi aux deux décimales les plus proches. Les corps $K_{n,r+s}$ sont décrits ci-dessous au moyen de l'un des procédés suivants :

a) Si $K_{n,r+s}$ est abélien sur $\mathbb{Q}$, on le définit par un générateur mis sous forme trigonométrique.

b) Si $K_{n,r+s}$ n'est pas abélien sur $\mathbb{Q}$ mais est abélien sur un sous-corps k, on le définit par k et son conducteur, la notation $\mathfrak{P}_p$ désignant un idéal premier convenable de degré 1 de k au-dessus du nombre premier p .

c) Lorsque les procédés ci-dessus ne sont pas applicables, on définit $K_{n,r+s}$ par le polynôme minimal de l'un de ses éléments

$K_{2,1}$	$d = -3$	, premier	$\mathbb{Q}(e^{2i\pi/3})$
$K_{2,2}$	$d = +5$	, premier	$\mathbb{Q}(2\cos 2\pi/5)$
$K_{3,2}$	$d = -23$	, premier	X^3-X-1 ; $K_{3,2} \subset \text{Hilbert}(\mathbb{Q}(\sqrt{-23}))$
$K_{3,3}$	$d = +49 =$	7^2	$\mathbb{Q}(2\cos 2\pi/7)$
$K_{4,2}$	$d = +117 =$	$3^2.13$	conducteur $\mathfrak{P}_{13}$ sur $K_{2,1}$
$K_{4,3}$	$d = -275 =$	$-5^2.11$	conducteur $\mathfrak{P}_{11}$ sur $K_{2,2}$
$K_{4,4}$	$d = +725 =$	$+5^2.29$	conducteur $\mathfrak{P}_{29}$ sur $K_{2,2}$
$K_{5,3}$	$d = +1\,609$,	premier	$X^5 - X^3 + X^2 + X - 1$
$K_{5,4}$	$d = -4\,511$,	premier	$X^5 - 2X^3 + X^2 - 1$
$K_{5,5}$	$d = +14\,641 =$	11^4	$\mathbb{Q}(2\cos 2\pi/11)$
$K_{6,3}$	$d = -9\,747 =$	$-3^3.19^2$	conducteur $\mathfrak{P}_{19}$ sur $K_{2,1}$
$K_{6,4}$	$d = +28\,037 =$	$23^2.53$	conducteur $\mathfrak{P}_{53}$ sur $K_{3,2}$
$K_{6,5}$	$d = -92\,779$,	premier	$X^6+X^5-2X^4-3X^3-X^2+2X+1$
$K_{6,6}$	$d = +300\,125 =$	$5^3.7^4$	$\mathbb{Q}(2\cos 2\pi/5 + 2\cos 2\pi/7)$
$K_{7,4}$	$d = -184\,607$,	premier	$X^7 - X^6 - X^5 + X^3 + X^2 - X - 1$
$K_{7,5}$	$d = -612\,233 =$	$71.8\,623$	$X^7-2X^6+3X^5-2X^4-2X^3+2X^2-3X+1$
$K_{7,6}$	$d = -2\,306\,599 =$	$-107.21\,557$	$X^7-3X^6-X^5+9X^4-3X^3-7X^2+2X+1$
$K_{7,7}$	$d = +20\,134\,393 =$	$71.283\,583$	$X^7+X^6-6X^5-5X^4+8X^3+5X^2-2X-1$
$K_{8,4}$	$d = +1\,257\,728 =$	$2^8.17^3$	conducteur $\mathfrak{P}_{17}$ sur $\mathbb{Q}(i)$
$K_{8,5}$	$d = -4\,461\,875 =$	$-5^4.11^2.59$	conducteur $\mathfrak{P}_{59}$ sur $K_{4,3}$
$K_{8,6}$	$d = +15\,243\,125 =$	$5^4.29^3$	conducteur $\mathfrak{P}_{29}$ sur $K_{2,2}$
$K_{8,7}$	$d = -68\,856\,875 =$	$-5^4.29^2.131$	conducteur $\mathfrak{P}_{131}$ sur $K_{4,4}$
$K_{8,8}$	$d = +282\,300\,416 =$	$2^{12}.41^3$	conducteur $\mathfrak{P}_{41}$ sur $\mathbb{Q}(\sqrt{2})$

Table III. - Cette table est extraite de tables multigraphiées dues à Odlyzko, datées du 29 novembre 1976 ([Odl]). Elle donne, sous l'hypothèse de Riemann généralisée, une minoration de la valeur absolue du discriminant d d'un corps K en fonction du degré n et de la signature (r, s) .

La première partie de la table concerne les corps totalement réels ou totalement imaginaires ; une minoration de $|d|^{1/n}$ s'obtient par lecture directe en fonction de n ; l'usage du paramètre b figurant dans les colonnes 2 et 4 est expliqué ci-dessous.

La seconde partie de la table concerne des corps de signature (r, s) arbitraire ; des fonctions A , B et E du paramètre b donné dans la première colonne sont calculées, et l'on a, quel que soit le choix de b , $|d| \geq A^r B^{2s} e^{-E}$. Il faut optimiser b ; la valeur à chercher est comprise entre les deux valeurs données dans la première partie de la table qui concerne les cas extrêmes $r=0$ et $r=n$.

Si l'on tient compte du deuxième terme du second membre des formules explicites, on obtient des minorations de $|d|$ de la forme $|d| > A^r B^{2s} e^{f-E}$, où $f = 2 \sum_{\mathfrak{p},m} (\log(N\mathfrak{p})/N(\mathfrak{p})^{m/2}) F(\log N(\mathfrak{p})^m)$ (cf. § . 1) , pour un choix convenable de la fonction F . Le choix à faire est le suivant : on prend $F(x) = G(x/b)$, où G est la fonction paire, nulle en dehors de l'intervalle $[-2, +2]$, et définie, pour $0 \leq x \leq 2$, par la formule

$$G(x) = (1-x/2)\cos(\pi/2)x + (1/\pi)\sin(\pi/2)x \ .$$

Odlyzko, en serrant de plus près les minorations, peut améliorer quelque peu les minorations de la table III. Il m'a signalé les trois exemples suivants concernant des corps totalement imaginaires : on peut remplacer 1,721 par 1,725 en degré 2 , 5,734 par 5,743 en degré 8 et 15,225 par 15,238 en degré 48 .

Les pourcentages indiqués dans la table I seraient alors 0,41% au lieu de 0,64% en degré 2, 0,76% au lieu de 0,92% en degré 8 et 1,54% au lieu de 1,62% en degré 48.

Table IV. - Nous donnons des extraits des tables calculées par Diaz y Diaz ([Dy D]), donnant, pour un corps K de degré n, une minoration de $|d_K|^{1/n}$ en fonction de la signature du corps, indépendante de GRH (minorations inconditionnelles).

Nous donnons les résultats jusqu'au degré 10 pour toutes les signatures possibles, et de larges extraits des tables concernant les corps totalement réels ou totalement imaginaires. Nous n'avons reproduit que quatre des huit décimales calculées par Diaz y Diaz.

TABLE I

n	m	d_k	f	$\lvert d_K\rvert^{1/n}$	$\lvert d_K\rvert^{1/n}$	Odl	%
2	2	-3	(1)	$3^{1/2}$	1,732	1,721	0,64
4	2	-3	$\mathfrak{P}_{13}$	$3^{1/2}.13^{1/4}$	3,289	3,263	0,79
6	2	-3	$\mathfrak{P}_{19}$	$3^{1/2}.19^{1/3}$	4,622	4,592	0,65
8	2	-2^2	$\mathfrak{P}_{17}$	$2.17^{3/8}$	5,787	5,734	0,92
10	2	-3	$\mathfrak{P}_{31}$	$3^{1/2}.31^{2/5}$	6,841	6,726	1,70
12	4	$3^2.13$	$\mathfrak{P}_{163}$	$3^{1/2}.13^{1/4}.163^{1/6}$	7,687	7,598	1,17
14	2	-71	(1)	$71^{1/2}$	8,426	8,371	0,66
16	4	$3^2.13$	$\mathfrak{P}_{241}$	$3^{1/2}.13^{1/4}.241^{3/16}$	9,198	9,068	1,43
18	3	-23	$\mathfrak{P}_7\mathfrak{P}_{19}$	$7^{2/9}.19^{5/18}.23^{1/3}$	9,929	9,697	2,40
20	4	2^8	$\mathfrak{p}_{11}$	$2^2.11^{2/5}$	10,438	10,270	1,64
22	2	-7	$\mathfrak{P}_{23}$	$7^{1/2}.23^{5/11}$	11,003	10,797	1,91
24	4	$3^2.13$	$\mathfrak{P}_{397}$	$3^{1/2}.13^{1/4}.397^{5/24}$	11,441	11,283	1,40
32	2	-15	$\mathfrak{P}_2^2\,\mathfrak{P}_3\,\mathfrak{P}_5$	$2^{1/2}.3^{3/4}.5^{7/8}$	13,181	12,912	2,08
36	4	229	$\mathfrak{P}_{307}$	$229^{1/4}.307^{2/9}$	13,889	13,581	2,26
40	4	$3^2.17^3$	(1)	$3^{1/2}.17^{3/4}$	14,501	14,183	2,24
44	4	$3^3.7$	$\mathfrak{P}_{463}$	$3^{3/4}.7^{1/4}.463^{5/22}$	14,960	14,728	1,57
48	4	$5^2.19^2$	$\mathfrak{p}_2\mathfrak{p}'_2$	$2^{2/3}.5^{1/2}.19^{1/2}$	15,472	15,225	1,62
52	4	5^3	$\mathfrak{P}_{911}$	$5^{3/4}.911^{3/13}$	16,114	15,680	2,77
56	4	257	$\mathfrak{P}_2^2\,\mathfrak{P}_{211}$	$2^{1/4}.211^{13/56}.257^{1/4}$	16,493	16,097	2,46
60	4	$3^2.37$	$\mathfrak{p}_{19}$	$3^{1/2}.37^{1/4}.19^{7/15}$	16,880	16,482	2,41
72	4	2^8	$\mathfrak{P}_{577}$	$2^2.577^{17/72}$	17,948	17,497	2,57
80	4	257	$\mathfrak{P}_{641}$	$257^{1/4}.641^{19/80}$	18,583	18,073	2,82

TABLE II

r+s \ n	2	3	4	5	6	7	8
1	1,732 1,721 0,64 %						
2	2,236 2,225 0,50 %	2,844 2,820 0,85 %	3,289 3,263 0,79 %				
3		3,659 3,639 0,56%	4,072 4,036 0,90 %	4,378 4,345 0,77 %	4,622 4,592 0,65 %		
4			5,189 5,124 1,27 %	5,381 5,322 1,11%	5,512 5,484 0,51%	5,654 5,619 0,61%	5,787 5,734 0,92%
5				6,809 6,640 2,55%	6,728 6,638 1,36%	6,710 6,653 0,85%	6,779 6,675 1,56%
6					8,182 8,143 0,48%	8,110 7,960 1,88%	7,905 7,834 0,90 %
7						11,051 9,611 14,99%	9,544 9,266 3,00 %
8							11,385 11,036 3,16%

TABLE III - Minorations sous GRH

Première Partie

corps totalement réels			corps totalement imaginaires	
n	b	$\|d\|^{1/n}$	b	$\|d\|^{1/n}$
1	0.340	0.996	0.300	0.874
2	0.700	2.225	0.580	1.721
3	1.050	3.630	0.800	2.519
4	1.350	5.124	1.050	3.263
5	1.550	6.640	1.200	3.954
6	1.750	8.143	1.350	4.592
7	1.900	9.611	1.500	5.185
8	2.050	11.036	1.600	5.734
9	2.200	12.410	1.700	6.247
10	2.300	13.736	1.800	6.726
11	2.400	15.012	1.900	7.176
12	2.500	16.238	2.000	7.598
13	2.550	17.422	2.050	7.997
14	2.650	18.559	2.100	8.371
15	2.700	19.657	2.200	8.730
16	2.800	20.711	2.250	9.068
17	2.850	21.734	2.300	9.390
18	2.900	22.720	2.400	9.697
19	2.950	23.672	2.450	9.990
20	3.000	24.594	2.500	10.270
21	3.100	25.474	2.550	10.539
22	3.100	26.351	2.550	10.797
23	3.100	27.178	2.600	11.045
24	3.200	28.001	2.650	11.283
25	3.200	28.787	2.700	11.512
26	3.300	29.554	2.750	11.733
28	3.300	31.020	2.800	12.153
30	3.400	32.425	2.850	12.545
32	3.500	33.750	2.950	12.912
34	3.500	35.005	3.000	13.258
36	3.600	36.219	3.000	13.581
38	3.700	37.356	3.100	13.894
40	3.700	38.471	3.100	14.183
42	3.800	39.514	3.200	14.465
44	3.800	40.542	3.200	14.728
46	3.800	41.504	3.300	14.984
48	3.900	42.456	3.300	15.225
50	3.900	43.356	3.400	15.456

TABLE III - Minorations sous GRH

Première Partie (suite)

corps totalement réels			corps totalement imaginaires	
n	b	$\|d\|^{1/n}$	b	$\|d\|^{1/n}$
52	4.000	44.230	3.400	15.680
56	4.000	45.884	3.500	16.097
60	4.100	47.452	3.500	16.482
64	4.200	48.913	3.600	16.846
68	4.200	50.285	3.700	17.180
72	4.300	51.601	3.700	17.497
76	4.400	52.822	3.800	17.793
80	4.400	54.014	3.800	18.073
84	4.500	55.119	3.900	18.338
88	4.500	56.204	3.900	18.589
92	4.500	57.214	4.000	18.826
96	4.600	58.205	4.000	19.055
100	4.600	59.141	4.000	19.268
110	4.700	61.335	4.100	19.770
120	4.800	63.335	4.200	20.221
130	4.900	65.169	4.300	20.631
140	5.000	66.853	4.400	21.003
150	5.000	68.426	4.400	21.345
160	5.100	69.897	4.500	21.666
170	5.200	71.255	4.600	21.959
180	5.200	72.553	4.600	22.236
190	5.300	73.760	4.700	22.493
200	5.300	74.909	4.700	22.735
220	5.400	77.026	4.800	23.178
240	5.500	78.943	4.900	23.575
260	5.600	80.689	5.000	23.934
280	5.700	82.283	5.100	24.258
300	5.700	83.775	5.100	24.560
320	5.800	85.155	5.200	24.838
340	5.900	86.424	5.200	25.091
360	5.900	87.642	5.300	25.332
380	6.000	88.760	5.400	25.552
400	6.000	89.833	5.400	25.763
480	6.200	93.555	5.600	26.485
600	6.400	97.979	5.800	27.328
720	6.600	101.488	6.000	27.984
840	6.800	104.361	6.100	28.515
960	6.900	106.815	6.300	28.961

TABLE III - Minorations sous GRH

Première Partie (suite)

corps totalement réels			corps totalement imaginaires	
n	b	$\|d\|^{1/n}$	b	$\|d\|^{1/n}$
1 000	6.900	107.548	6.300	29.094
1 200	7.200	110.728	6.500	29.673
1 332	7.200	112.575	6.600	29.992
2 400	7.800	122.112	7.200	31.645
4 800	8.400	132.020	7.800	33.298
4 840	8.600	132.126	7.800	33.315
8 862	9.200	139.766	8.400	34.541
10 000	9.200	141.218	8.600	34.768
31 970	10.400	153.252	9.800	36.613
100 000	11.600	162.651	10.800	37.994
254 228	12.500	168.971	11.800	38.895
1 000 000	14.000	176.415	13.000	39.923
2 391 978	14.500	180.319	14.000	40.458
10 000 000	16.000	185.655	15.000	41.122

Table III - Minorations sous GRH

Deuxième Partie

b	A	B	E
0.300	2.623	2.324	0.9769
0.320	2.820	2.478	1.0426
0.340	3.020	2.633	1.1085
0.360	3.222	2.787	1.1745
0.380	3.427	2.941	1.2406
0.400	3.636	3.095	1.3068
0.420	3.847	3.249	1.3732
0.580	5.640	4.475	1.9107
0.600	5.878	4.628	1.9788
0.625	6.179	4.818	2.0643
0.650	6.485	5.008	2.1501
0.675	6.795	5.198	2.2363
0.700	7.110	5.387	2.3229
0.750	7.753	5.765	2.4973
0.800	8.416	6.142	2.6735
0.850	9.096	6.517	2.8517
0.900	9.795	6.891	3.0319
0.950	10.513	7.262	3.2143
1.000	11.248	7.633	3.3991
1.050	12.001	8.001	3.5863
1.100	12.772	8.368	3.7761
1.150	13.561	8.732	3.9686
1.200	14.366	9.095	4.1641
1.250	15.189	9.455	4.3626
1.300	16.028	9.814	4.5643
1.350	16.883	10.170	4.7693
1.400	17.754	10.524	4.9779
1.450	18.641	10.876	5.1902
1.500	19.543	11.225	5.4062
1.550	20.459	11.572	5.6264
1.600	21.390	11.916	5.8507
1.650	22.335	12.258	6.0793
1.700	23.293	12.598	6.3126
1.750	24.264	12.935	6.5505
1.800	25.248	13.269	6.7935
1.850	26.244	13.600	7.0415
1.900	27.251	13.929	7.2949
1.950	28.269	14.256	7.5539

TABLE III - Minorations sous GRH

Deuxième Partie (suite)

b	A	B	E
2.000	29.298	14.579	7.8187
2.050	30.338	14.900	8.0894
2.100	31.386	15.218	8.3664
2.200	33.511	15.845	8.9400
2.250	34.585	16.154	9.2371
2.300	35.667	16.461	9.5414
2.350	36.757	16.764	9.8532
2.400	37.853	17.065	10.173
2.450	38.955	17.363	10.501
2.500	40.063	17.658	10.837
2.550	41.176	17.950	11.181
2.600	42.295	18.239	11.535
2.650	43.417	18.525	11.898
2.700	44.543	18.808	12.270
2.750	45.673	19.089	12.653
2.800	46.806	19.366	13.045
2.850	47.941	19.641	13.448
2.900	49.079	19.912	13.863
2.950	50.218	20.181	14.289
3.000	51.359	20.446	14.726
3.100	53.643	20.969	15.638
3.200	55.928	21.480	16.603
3.300	58.211	21.980	17.624
3.400	60.490	22.469	18.706
3.500	62.762	22.946	19.851
3.600	65.024	23.413	21.066
3.700	67.275	23.868	22.355
3.800	69.513	24.313	23.723
3.900	71.735	24.747	25.176
4.000	73.940	25.171	26.719
4.100	76.126	25.585	28.360
4.200	78.292	25.988	30.105
4.300	80.436	26.382	31.962
4.400	82.558	26.766	33.939
4.500	84.656	27.141	36.044
4.600	86.730	27.507	38.286
4.700	88.778	27.863	40.676
4.800	90.799	28.211	43.224
4.900	92.794	28.550	45.941

TABLE III - Minorations sous GRH

Deuxième Partie (suite)

b	A	B	E
5.000	94.761	28.881	48.840
5.100	96.701	29.203	51.934
5.200	98.612	29.518	55.237
5.300	100.495	29.825	58.764
5.400	102.349	30.123	62.532
5.500	104.174	30.415	66.559
5.600	105.970	30.699	70.863
5.700	107.737	30.976	75.465
5.800	109.475	31.247	80.387
5.900	111.184	31.510	85.652
6.000	112.863	31.767	91.287
6.100	114.514	32.018	97.319
6.200	116.137	32.262	103.78
6.300	117.731	32.501	110.70
6.400	119.296	32.733	118.11
6.500	120.834	32.960	126.05
6.600	122.344	33.181	134.56
6.700	123.826	33.397	143.68
6.800	125.282	33.608	153.47
6.900	126.710	33.813	163.96
7.000	128.112	34.013	175.22
7.200	130.839	34.400	200.26
7.400	133.464	34.768	229.13
7.600	135.991	35.119	262.43
7.800	138.423	35.453	300.88
8.000	140.764	35.772	345.31
8.200	143.015	36.076	396.69
8.400	145.182	36.366	456.15
8.600	147.266	36.643	525.04
8.800	149.272	36.908	604.89
9.000	151.201	37.161	697.52
9.200	153.058	37.403	805.06
9.400	154.845	37.634	929.98
9.600	156.565	37.855	1.0753 e 03
9.800	158.220	38.067	1.2442 e 03
10.000	159.814	38.270	1.4409 e 03
10.200	161.348	38.464	1.6700 e 03
10.400	162.826	38.650	1.9371 e 03
10.600	164.250	38.828	2.2485 e 03

TABLE III - Minorations sous GRH

Deuxième Partie (suite)

b	A	B	E
10.800	165.622	39.000	2.6120 e 03
11.000	166.944	39.164	3.0365 e 03
11.200	168.219	39.322	3.5324 e 03
11.400	169.447	39.473	4.1122 e 03
11.600	170.633	39.619	4.7903 e 03
11.800	171.776	39.759	5.5840 e 03
12.000	172.879	39.893	6.5134 e 03
12.500	175.472	40.207	9.5972 e 03
13.000	177.848	40.494	1.4194 e 04
13.500	180.031	40.754	2.1065 e 04
14.000	182.037	40.993	3.1366 e 04
14.500	183.886	41.211	4.6851 e 04
15.000	185.592	41.412	7.0186 e 04
16.000	188.628	41.766	1.5882 e 05
17.000	191.237	42.069	3.6298 e 05
18.000	193.493	42.328	8.3717 e 05
19.000	195.455	42.553	1.9467 e 06
20.000	197.170	42.748	4.5606 e 06
22.000	200.009	43.069	2.5532 e 07
24.000	202.240	43.320	1.4625 e 08
26.000	204.024	43.520	8.5416 e 08
28.000	205.471	43.681	5.0734 e 09
30.000	206.660	43.813	3.0577 e 10
32.500	207.868	43.946	2.9393 e 11
35.000	208.842	44.054	2.8743 e 12
37.500	209.639	44.141	2.8528 e 13
40.000	210.298	44.214	2.8685 e 14
42.500	210.849	44.274	2.9175 e 15
45.000	211.315	44.325	2.9977 e 16
47.500	211.712	44.369	3.1082 e 17
50.000	212.053	44.406	3.2493 e 18
52.500	212.348	44.438	3.4220 e 19
55.000	212.605	44.466	3.6281 e 20
57.500	212.830	44.491	3.8702 e 21
60.000	213.029	44.512	4.1518 e 22
65.000	213.361	44.549	4.8503 e 24
70.000	213.626	44.577	5.7572 e 26

TABLE IV - Minorations inconditionnelles

Première Partie : $n \leq 10$

Dans chaque ligne, on lit, pour un degré n donné, $2 \leq n \leq 10$, la suite des minorations de $|d_K|^{1/n}$ pour r+s croissant de $[(n+1)/2]$ à n ; ces minorations tiennent compte de la correction la plus défavorable que donne le second membre des formules explicites en fonction de la décomposition dans K des petits nombres premiers ; pour n = 1 , le résultat avec ses 8 décimales est 0,99 999 524.

n = 2 : 1.7297 ; 2.2280

n = 3 : 2.8185 ; 3.6128

n = 4 : 3.2584 ; 4.0143 ; 5.0674

n = 5 : 4.3175 ; 5.2638 ; 6.5240

n = 6 : 4.5577 ; 5.4194 ; 6.5239 ; 7.9424

n = 7 : 5.5485 ; 6.5355 ; 7.7664 ; 9.3027

n = 8 : 5.6593 ; 6.5540 ; 7.6448 ; 8.9749 ; 10.5972

n = 9 : 6.5763 ; 7.5583 ; 8.7337 ; 10.1404 ; 11.8242

n = 10 : 6.6003 ; 7.4952 ; 8.5502 ; 9.7934 ; 11.2585 ; 12.9853 .

TABLE IV - Minorations inconditionnelles

Deuxième Partie

Corps totalement réels

N	Minoration	N	Minoration	N	Minoration
1	0.9979	38	30.4117	240	49.2319
2	2.2234	40	31.0865	260	49.7675
3	3.6108	42	31.7232	280	50.2442
4	5.0670	44	32.3252	300	50.6717
5	6.5235	46	32.8954	320	51.0579
6	7.9414	48	33.4365	340	51.4087
7	9.3017	50	33.9508	360	51.7292
8	10.5964	52	34.4405	380	52.0233
9	11.8238	56	35.3532	400	52.2945
10	12.9850	60	36.1874	450	52.8888
11	14.0831	64	36.9536	480	53.1982
12	15.1217	68	37.6605	500	53.3882
13	16.1047	72	38.3151	600	54.1850
14	17.0359	76	38.9235	700	54.7962
15	17.9192	80	39.4908	720	54.9021
16	18.7580	84	40.0213	800	55.2829
17	19.5555	88	40.5188	840	55.4513
18	20.3148	92	40.9865	900	55.6813
19	21.0386	96	41.4272	960	55.8880
20	21.7294	100	41.8434	1 000	56.0147
21	22.3896	110	42.7899	1 100	56.2986
22	23.0212	120	43.6232	1 200	56.5438
23	23.6261	130	44.3638	1 300	56.7581
24	24.2061	140	45.0273	1 332	56.8212
25	24.7628	150	45.6260	1 400	56.9474
26	25.2976	160	46.1696	1 500	57.1161
28	26.3071	170	46.6658	1 600	57.2674
30	27.2440	180	47.1211	1 700	57.4042
32	28.1165	190	47.5406	1 800	57.5285
34	28.9315	200	47.9287	1 900	57.6421
36	29.6948	220	48.6246	2 000	57.7464

TABLE IV - Minorations inconditionnelles

Deuxième Partie

Corps totalement imaginaires

N	Minoration	N	Minoration	N	Minoration
2	1.7221	72	15.3591	360	19.5903
4	3.2545	76	15.5549	380	19.6813
6	4.5570	80	15.7371	400	19.7652
8	5.6593	84	15.9071	480	20.0443
10	6.6003	88	16.0663	500	20.1029
12	7.4128	92	16.2158	600	20.3483
14	8.1224	96	16.3563	700	20.5363
16	8.7484	100	16.4889	720	20.5688
18	9.3056	110	16.7898	800	20.6858
20	9.8057	120	17.0539	840	20.7375
22	10.2575	130	17.2880	900	20.8081
24	10.6683	140	17.4974	960	20.8715
26	11.0438	150	17.6859	1 000	20.9103
28	11.3889	160	17.8568	1 100	20.9973
30	11.7072	170	18.0126	1 200	21.0724
32	12.0022	180	18.1553	1 300	21.1380
34	12.2764	190	18.2867	1 332	21.1573
36	12.5322	200	18.4081	1 400	21.1959
38	12.7715	210	18.5207	1 500	21.2475
40	12.9960	220	18.6254	1 600	21.2937
42	13.2071	230	18.7232	1 700	21.3355
44	13.4061	240	18.8148	1 800	21.3735
46	13.5941	250	18.9007	1 900	21.4082
48	13.7721	260	18.9815	2 000	21.4401
50	13.9409	270	19.0577	2 200	21.4966
52	14.1013	280	19.1297	2 400	21.5453
56	14.3993	290	19.1979	2 600	21.5877
60	14.6707	300	19.2625	2 800	21.6252
64	14.9193	320	19.3823	3 000	21.6585
68	15.1479	340	19.4911	4 000	21.7825

APPENDICE I

Diverses notions de discriminant

Classiquement, on définit, pour toute extension K/k de corps de nombres, un discriminant $\delta_{K/k}$ qui est un idéal entier de k. Lorsque $k = \mathbb{Q}$, cela ne définit le discriminant habituel qu'au signe près. En considérant des bases de K sur k, on définit naturellement un discriminant $d_{K/k}$ qui est un élément de k^*/k^{*2}. Lorsque $k = \mathbb{Q}$, la connaissance du discriminant usuel d_K équivaut à la connaissance de $\delta_{K/\mathbb{Q}}$ et $d_{K/\mathbb{Q}}$.

Le discriminant $d_{K/k}$ partage avec le discriminant usuel lorsque $k = \mathbb{Q}$ les deux propriétés suivantes :

(i) Ce discriminant met en bijection les éléments de k^*/k^{*2} avec les extensions quadratiques de k, à condition de considérer k lui-même comme une extension quadratique de k.

(ii) Il permet d'associer à toute extension K/k une extension quadratique canonique K' de k caractérisée par l'égalité $d_{K/k} = d_{K'/k}$ dans k^*/k^{*2}.

Dans cette situation, soit $x \in K$ tel que $k(x) = K$. Le groupe de Galois G de la clôture galoisienne de K/k dans une clôture algébrique donnée $\bar{k}$ de k opère sur l'ensemble R des racines du polynôme minimal de x sur k, permettant d'identifier G à un sous-groupe du groupe symétrique $S(R)$. Alors, l'extension K'/k correspond à l'intersection de G et du sous-groupe alterné de $S(R)$; en particulier, $d_{K/k}$ est trivial dans k^*/k^{*2} si et seulement si G se plonge dans le sous-groupe alterné de $S(R)$.

Soit $\mathfrak{p}$ un idéal premier non nul de k. La clôture intégrale dans K de l'anneau local de k en $\mathfrak{p}$ est un $(\mathcal{O}_k)_{\mathfrak{p}}$-module libre, auquel on peut associer un discriminant dans $(\mathfrak{O}_k)_{\mathfrak{p}}^{*2} \backslash (\mathfrak{O}_k)_{\mathfrak{p}}$, qui définit le discriminant $d_{K/k}$. En remplaçant $(\mathfrak{O}_k)_{\mathfrak{p}}$ par son complété en $\mathfrak{p}$, et en introduisant les places à l'infini, on obtient le discriminant idélique introduit par Fröhlich (Discriminants of algebraic number fields, Math. Z. 74 (1960), 18-28 ; Ideals in an extension field as modules over the algebraic integers in a finite number field, Math. Z. 74 (1960), 29-38) ; c'est un élément de J_K/U_K^2 (idèles modulo les carrés des idèles unités), qui appartient plus précisément à $K^* J_K^2/U_K^2$.

La considération des discriminants définis dans $(\mathfrak{O}_k)_{\mathfrak{p}}^{*2} \backslash (\mathfrak{O}_k)_{\mathfrak{p}}$ permet de vérifier pour $\delta_{K/k}$ une égalité de la forme $\delta_{K/k} = \delta_{K'/k}\, \mathfrak{a}^2$, où $\mathfrak{a}$ est un idéal entier de k , ce qui entraîne un résultat analogue pour le discriminant idélique.

APPENDICE II

La congruence de Stickelberger

La congruence classique est le fait que, pour tout corps de nombre K , on a $d_K \equiv 0$ ou 1 mod. 4 ; une démontration très simple de ce théorème de Stickelberger a été donnée par Schur (Elementarer Beweis eines Satzes von L. Stickelberger, Math. Z 29 (1929), 464-465,et oeuvres, tome 3, p. 87-88). Comme l'a montré Fröhlich (cf. appendice I), la démonstration de Schur permet de démontrer un théorème analogue pour le discriminant idélique.

Voici une généralisation du théorème de Stickelberger aux discréminants "idéaux" $\delta_{K/k}$: soit K/k une extension finie de corps de nombres ; soit r_2 le nombre de places complexes de K au-dessus d'une place réelle de k ; alors, $N_{k/\mathbb{Q}}$ désignant la norme absolue, on a la congruence :

$$N_{K/\mathbb{Q}}(\delta_{K/k}) \equiv 0 \text{ ou } (-1)^{r_2} \mod 4 .$$

Je n'ai pas de référence à proposer pour cet énoncé ; on peut le déduire de la congruence "idélique" de Fröhlich. On peut également le démontrer en le vérifiant pour les extensions quadratiques, puis en passant au cas général au moyen de la formule $\delta_{K/k} = \delta_{K'/k}\,\mathfrak{a}^2$ énoncée dans l'appendice I.

APPENDICE III

Valeurs de $v_p(d_K)$

Soit k un corps de nombres (ou une extension finie d'un corps p-adique $\mathbb{Q}_p$), soit n un entier et soit $\mathfrak{p}$ un idéal premier non nul de k. L'exposant $v_{\mathfrak{p}}(\delta_{K/k})$ de $\mathfrak{p}$ dans le discriminant relatif d'une extension K/k de degré n ne peut prendre qu'un nombre fini de valeurs, dont on peut dresser la liste en fonction de la caractéristique résiduelle p de $\mathfrak{p}$, du degré n de K/k et de l'indice absolu de ramification e_o de $\mathfrak{p}$. Cela résulte des travaux de Ore (Existenzbeweise für algebraische Körper mit vorgeschriebenen Eigenschaften, Math. Z. 25 (1926), 474-489) et de Thomson (On the possible forms of discriminants of algebraic fields, Amer. J. Math. 53 (1931), 81-90 et 55 (1933), 111-118).

Voici le principe de la méthode de Ore ; on commence par étudier le cas local totalement ramifié. Une uniformisante Π de K est racine d'un polynôme d'Eisenstein $f \in k[X]$:
$f(X) = X^n + a_{n-1} X^{n-1} + \ldots + a_1 X + \pi$ où $a_i \in \mathfrak{p}$ et π est une uniformisante de k. La valuation dans k du discriminant de K/k est

égale à la valuation dans K de la différente de K/k , qui est engendrée par $f'(\Pi)=n\,\Pi^{n-1}+(n-1)\,a_{n-1}\Pi^{n-2}+\ldots+a_1$; les différents termes de cette somme ont des valuations dans K deux à deux incongrues modulo n , si bien que la valuation cherchée est la valeur minimum des valuations des différents termes ci-dessus ; on trouve ainsi une liste de valeurs possibles pour $v_{\mathfrak{p}}(\delta_{K/k})$, qui se présentent toutes effectivement. La plus petite est n , ou n-1, que l'on obtient par exemple pour le polynôme $X^n+\pi X+\pi$, et la plus grande est $n\,v_{\mathfrak{p}}(n)+n-1$, que l'on obtient par exemple pour le polynôme $X^n-\pi$.

On passe delà au cas local général en écrivant K comme extension totalement ramifiée d'une extension de k non ramifiée, puis on globalise en approchant par le théorème d'approximation un produit de polynômes de $k_{\mathfrak{p}}[X]$ par un polynôme de $k[X]$ dont on peut assurer l'irréductibilité en lui imposant d'être un polynôme d'Eisenstein en un idéal premier de k autre que $\mathfrak{p}$; le théorème d'approximation montre également que l'on peut imposer le nombre de places réelles de k ramifiées dans K .

Voici les résultats pour le discriminant d'un corps de nombres de degré n $(e_0=1)$. On écrit le développement p-adique de n : $n=\sum_{i=0}^{q} a_i p^i$, avec $0\leq a_i<p$. La valeur maximum de $v_p(d)$ est

$$N_{n,p}=\sum_{i=0}^{q} a_i(i+1)p^i-r\ ,$$

où r désigne le nombre de coefficients a_i non nuls.

Les valeurs possibles pour $v_p(d)$ sont celles de l'intervalle $[0\ ,N_{n,p}]$ à l'exclusion des suivantes :

(i) ip^i-1 si $n=p^i$, $i\geq 1$

(ii) ip^i-1 si $n=p^i+1$, $i\geq 2$

(iii) 1 pour $p=2$ (cf. appendice II).

BIBLIOGRAPHIE

[Dy D] F. DIAZ Y DIAZ. - Tables minorant la racine n-ième du discriminant d'un corps de degré n, Publications Mathématiques d'Orsay, à paraître.

[H] J. HUNTER. - The minimum discriminant of quintic fields, Proc. Glasgow Math. Association 3 (1957), 57-67.

[Ka] G. KAUR. - The minimum discriminant of sixth degree totally real algebraic number fields, J. Indian Math. Soc. 34 (1970), 123-134.

[K-V] H. KOCH, B.B. VENKOV. - Über den p-Klassenkörperturm eines imaginär-quadratischen Zahlkörpers, Astérisque 24-25 (1975), 57-67.

[Ln] H.W. LENSTRA. - Euclidean number fields of large degree, Invent. Math. 36 (1977), 237-254.

[Lt] A. LEUTBECHER. - Communication privée.

[L-Z] J. LIANG, H. ZASSENHAUSS. - The minimum Discriminant of Sixth Degree Totally Complex Algebraic Number Fields, J. Number Theory 9 (1977), 16-35.

[v L] F.J. van der LINDEN. - Class Numbers of real abelian number fields of small conductors, exposé aux Journées Arithmétiques d'Exeter.

[Mr 1] J. MARTINET. - Tours de corps de classes et estimation de discriminants, Invent. Math. 44 (1978), 65-73.

[Mr 2] J. MARTINET. - Petits Discriminants, Ann. Inst. Fourier 29, 1 (1979), 159-170.

[Mr 3] J. MARTINET. - Sur la constante de Lenstra des corps de nombres, Séminaire de Théorie des Nombres, Bordeaux, 1979-80, exp. n°17.

[Ms] J.M. MASLEY. - Where are number fields with small class number ? in Number Theory, Carbondale 1979, Lecture Notes in Mathematics, n°751, Berlin-Heidelberg-New-York; Springer, 1979.

[My] J. MAYER. - Die Absolut kleinsten Discriminanten der biquadratischen Zahlkörper, S.B. Akad. Wiss. Wien, II a, 138 (1929), 733-742.

[Me] J.-F. MESTRE. - Corps euclidiens, unités exceptionnelles et courbes elliptiques, J. Number Theory (à paraître).

[Odl] A. ODLYZKO. - Discriminant Bounds, tables multigraphiées datées du 29 Novembre 1976.

[Po] M. POHST. - The minimum discriminant of seventh degree totally real algebraic number field, J. Number Theory (à paraître).

[P 1] G. POITOU. - Minorations de discriminants (d'après A.M. Odlyzko), Séminaire Bourbaki, exposé 479, Février 1976 ; Lecture Notes in Maths, n°567, Berlin-Heidelberg-New-York ; Springer, 1977.

[P 2] G. POITOU. - Sur les petits discriminants, Séminaire D.P.P., Paris, exposé n°6, 1976-77.

[Ro] P. ROQUETTE. - On class field towers, in J.W.S. Cassels and A. Fröhlich, Algebraic Number Theory, p.231-249 ; New-York, London, Academic Press, 1967.

[Sch] R. SCHOOF. - Communication privée.

[Se] J.-P. SERRE. - Cohomologie galoisienne, 4ème éd., Lecture Notes in Maths, n°5 ; Berlin - Heidelberg - New-York ; Springer, 1973.

[W 1] A. WEIL. - Sur les "formules explicites" de la théorie des nombres premiers, Comm. Sem. Math. Lund (1952), 252-265.

[W 2] A. WEIL. - Sur les formules explicites de la théorie des nombres, Izvestia Akad. Nauk. S.S.S.R., Ser. Math., 36 (1972), 3-18.

[Z] R. ZIMMERT. - Ideale kleiner Norm in Idealklassen und eine Regulatorabschätzung, à paraître.

STICKELBERGER RELATIONS IN CLASS GROUPS AND GALOIS MODULE STRUCTURE

Leon R. McCulloh*

Hilbert's [H] proof of the Stickelberger relations for the cyclotomic field $\mathbb{Q}(\mu_\ell)$ is based on the fact that tame cyclic extensions of $\mathbb{Q}$ of degree ℓ (prime) have normal integral bases. This connects two apparently distinct aspects of Galois module structure:

a) the structure of the ideal class group $\mathrm{Cl}(\mathbb{Z}[\mu_\ell])$ under the action of the Galois group C $(\cong(\mathbb{Z}/\ell\mathbb{Z})^\times)$ of $\mathbb{Q}(\mu_\ell)/\mathbb{Q}$ and

b) the structure of rings of integers O_L in tame cyclic extensions $L/\mathbb{Q}$ of degree ℓ under the action of their Galois groups G $(\cong\mathbb{Z}/\ell\mathbb{Z})$.

The connection is made by the resolvents:

$$(v|\chi) = \sum_{\sigma\in G} \sigma(v)\chi(\sigma) \qquad \text{for } v \in L\ ,\ \chi \in \mathrm{Hom}(G,\mu_\ell)$$

which belong to the composite fields $L(\mu_\ell)$, are acted on by both Galois groups C and G, and embody the Galois module structure of the rings O_L.

Using the same bridge, Fröhlich [F] was able to prove the Stickelberger relations for an arbitrary cyclotomic field $\mathbb{Q}(\mu_f)$, avoiding Stickelberger's explicit factorization of Gauss sums.

This approach can also be used to obtain relations on class groups associated to $\mathbb{Z}G$ for certain abelian groups G. These relations arise by applying general results on the Galois module structure of tame extensions L/K with Galois group isomorphic to G to the ground field $K = \mathbb{Q}$ where all such extensions have normal integral bases.

With an arbitrary number field K as ground field, the situation becomes more complicated in two ways:

1) L/K may have no (relative) integral basis and

2) even if it does and is tame (or even unramified), it may have no (relative) normal integral basis. (Example: $\mathbb{Q}(\sqrt{-5},\sqrt{-1})/\mathbb{Q}(\sqrt{-5})$.) (In fact, K has tame quadratic extensions without normal integral bases if and only if the ray class group (mod (2)) of K is non-

* This research was made possible by the enlightened sabbatical policy of the University of Illinois. The author wishes to express his gratitude to Universität Regensburg and to King's College, London for the generous hospitality extended to him while the major part of this work was being done.

trivial.) To deal with these difficulties, we introduce the class group.

Let

K = an algebraic number field,

G = a finite abelian group, and

$o = O_K$ = the ring of integers in K .

<u>Definition</u> The class group $Cl(oG)$ of the group ring oG is

$$Cl(oG) = I(oG)/((KG)^{\times})$$

where $I(oG)$ is the group of invertible fractional oG-ideals in KG and $((KG)^{\times})$ is the group of principal invertible oG-ideals.

If L/K is a tame extension with Galois group $Gal(L/K) \cong G$, the class (O_L) of O_L is an element of $Cl(oG)$ defined as follows: By the normal basis theorem of Galois theory, $L = KG{\cdot}v$ for some $v \in L$. Then $O_L = m{\cdot}v$ for some $m \in I(oG)$.

<u>Definition</u> (O_L) is the image of m in $Cl(oG)$.

Evidently, $(O_L) = 1$ iff m is principal iff $O_L = OG\ v'$ for some $v' \in L$ (i.e., L/K has a normal integral basis).

Notice that (O_L) depends on the choice of the isomorphism $Gal(L/K) \cong G$ and is unique only up to the action of $AutG$ on $Cl(oG)$.

<u>Definition</u> The set of 'realizable classes' is defined as

$$R(oG) = \{(O_L) \mid L/K \text{ tame, } Gal(L/K) \cong G\} \subseteq Cl(oG) \quad ,$$

and is a union of $AutG$ orbits in $Cl(oG)$. (Note: $Cl(oG)$ is a $\mathbb{Z}[AutG]$-module.)

One easily sees, in fact, that

$$R(oG) \subseteq Cl'(oG) = Kernel(Cl(oG) \xrightarrow{\varepsilon} Cl(o))$$

where ε is the map on class groups induced by the augmentation $oG \xrightarrow{\varepsilon} o$. The reason, in essence, is that if L/K is tame, then $Tr(O_L) = o$, the principal class in $Cl(o)$. Beyond this, for arbitrary ground field K , $R(oG)$ has been described precisely only for elementary abelian groups.

Let G be elementary abelian of order ℓ^k . Then $\mathrm{Aut}G \cong Gl_k(\mathbb{F}_\ell)$. If we identify G with the additive group of the finite field $\mathbb{F}_{\ell^k}$, then the multiplicative group $C = C_k = (\mathbb{F}_{\ell^k})^\times$ acts (by multiplication) on G and is embedded as a (Cartan) subgroup of $\mathrm{Aut}G$. Thus, $Cl(oG)$ becomes a $\mathbb{Z}C$ module. We can show that there is an ideal $J \subseteq \mathbb{Z}C$, analogous to the classical Stickelberger ideal which describes $R(oG)$ in the following sense:

THEOREM 1. $R(oG) = Cl'(oG)^J$ *for* G *elementary abelian of order* ℓ^k . *(We write the group* $Cl(oG)$ *multiplicatively and the action of* $\mathbb{Z}C$ *exponentially.*

The ideal J is a relative of one introduced by Kubert and Lang in their series on units in modular function fields. It is defined as follows: Let $tr: C \longrightarrow \mathbb{Z}$ be the composite of the trace map $Tr: \mathbb{F}_{\ell^k} \longrightarrow \mathbb{F}_\ell$ with the canonical lifting to the least non-negative residue $\mathbb{F}_\ell \longrightarrow [0,\ell) \cap \mathbb{Z}$. Then

<u>Definition</u> The Stickelberger element and ideal are defined, respectively as

$$\theta = \theta_k = \sum_{\delta \in C} tr(\delta)\delta^{-1} \in \mathbb{Z}C \quad \text{and}$$

$$J = J_k = \mathbb{Z}C \cdot (\theta/\ell) \cap \mathbb{Z}C \quad .$$

<u>Remarks</u>

1) If $k = 1$, $C_1 = \mathbb{F}_\ell^\times = (Z/\ell Z)^\times = Gal(Q(\mu_\ell)/Q)$ and J_1 is the classical Stickelberger ideal.

2) If $k = 1$ and $\ell = 2$ then $C_1 = 1$, $\mathbb{Z}C_1 = J_1 = \mathbb{Z}$ and the above theorem gives $R(oG) = Cl'(oG)$. One can also compute that $Cl'(oG)$ is isomorphic to the ray class group (mod (2)) of K from which it follows that the vanishing of this ray class group is necessary and sufficient for all tame quadratic extensions L/K to have normal integral bases.

3) If $k = 1$ and $K = Q$, then $o = \mathbb{Z}$ so $Cl'(\mathbb{Z}G) = Cl(\mathbb{Z}G)$. Moreover, $Cl(\mathbb{Z}G) = Cl(\mathbb{Z}[\mu_\ell])$ (see [R]) and the isomorphism preserves the C_1 action. So our theorem says $R(\mathbb{Z}G) = Cl(\mathbb{Z}[\mu_\ell])^{J_1}$. This points up explicitly the fact, implicit in Hilbert's proof, that the Stickelberger relations for $\mathbb{Q}(\mu_\ell)$ are equivalent to the normal integral basis theorem for tame cyclic extensions of $\mathbb{Q}$ of prime degree ℓ .

The theorem was obtained first [M] in the special case where $k = 1$ and $\mu_\ell \leq K$. In 1977, L.N. Childs, [C], still in the case $k = 1$ was able to obtain the inclusion $R(oG) \supseteq Cl'(oG)^J$ without assuming $\mu_\ell \leq K$, which is sufficient to recover the Stickelberger relations from the normal basis theorem. In 1978, C. Glass [G] found in the case $k = 2$ an explicit description of $R(oG)$ under the hypothesis $\mu_\ell \leq K$, based on the diagonal (split Cartan) subgroup $C_2^{(s)}$ of $Gl_2(\mathbb{F}_\ell)$. This work of Glass, in preliminary version, was instrumental in drawing my attention to the elementary abelian case.

Now, just as in Remark 3) above, if we put $K = \mathbb{Q}$ in the theorem with $k \geq 1$, we obtain Stickelberger relations on the class group of $\mathbb{Z}G$:

<u>Corollary</u> $Cl(\mathbb{Z}G)^J = 1$ for G elementary abelian of order ℓ^k.

This suggested looking for an analogue of Iwasawa's class number formula [I]. Let $\tau: G \longrightarrow G$ be the involution sending each element to its inverse. This allows us to consider the skew-symmetric part A^-, with respect to τ, of a $\mathbb{Z}C$ module A.

THEOREM 2. *For* G *elementary abelian of order* ℓ^k,

$|Cl(\mathbb{Z}G)^-| = [\mathbb{Z}C^- : J^-]$ *if* ℓ *is odd, and*

$|Cl(\mathbb{Z}G)| = [\mathbb{Z}G : J]$ *if* $\ell = 2$.

The proof of this theorem consists of computing both sides and comparing the results. Let M be the maximal order of $\mathbb{Q}G$. Iwasawa's class number formula allows us to express $Cl(M)^-$ as the index of the ideal of $\mathbb{Z}C$ generated by the classical Stickelberger ideal $J_1 \subseteq \mathbb{Z}C_1 \leq \mathbb{Z}C$, that is, $|Cl(M)^-| = [\mathbb{Z}C^- : (\mathbb{Z}C \cdot J_1)^-]$. Denoting by $D(\mathbb{Z}G)$ the kernel of the surjection $Cl(\mathbb{Z}G) \longrightarrow Cl(M)$, we get $|Cl(\mathbb{Z}G)^-| = |Cl(M)^-| \cdot |D(\mathbb{Z}G)^-|$. Fröhlich [F'] has computed $|D(\mathbb{Z}G)^-|$ for arbitrary abelian ℓ-groups and Kubert and Lang [KL] have developed techniques which can be applied to computing $[(\mathbb{Z}C \cdot J_1)^- : J^-]$. Details appear in [M'].

The proof of Theorem 1 follows the spirit of Hilbert's proof discussed at the beginning of this article. One attempts to connect two situations:

a) the structure of the class group $Cl(O_K G)$ under the action of the group C of automorphisms of KG over K and

b) the Galois module structure of rings of integers O_L in tame extensions L/K with Galois group isomorphic to G. The connection is made by the 'resolvends':

$$\tilde{v} = \sum_{\sigma \in G} \sigma(v)\sigma^{-1} \quad \text{for } v \in L,$$

which are elements of the group ring LG and whose images under the characters χ of G are the resolvents: $\chi(\tilde{v}) = (v \mid \bar{\chi})$. Both groups, C and G, act on LG. The action of C is through its action on the elements of G. The action of G is through its action on the coefficients which lie in L. Of course, G may also act on LG by left multiplication, but that action does not preserve the ring structure. However, both actions of G agree on the resolvends, and indeed, the resolvends are characterized by that property. From that fact, one deduces an important result on the behaviour of resolvends under the action of C: Let $A = \mathrm{Ann}_{\mathbb{Z}C}G$, the annihilator of G regarded as a $\mathbb{Z}C$ module. Then, for $\alpha \in \mathbb{Z}C$, $\tilde{L}^{\alpha} \subseteq KG$ if and only if $\alpha \in A$. (For the exponent to make sense in this relation, we regard $\tilde{L}$ as an element of the multiplicative group of all rank one, free KG modules in LG which are generated by units of LG.) The annihilator ideal A is also closely related to the Stickelberger ideal J. One shows that $J = A(\theta/\ell)$ and that $A + J = \mathbb{Z}C$. Now, let v be a normal basis generator of L/K, and $O_L = m \cdot v$ where $m \in I(oG)$ and represents the class $(O_L) \in Cl(oG)$. Raising the module of integral resolvends to the ℓth power (noting that $\ell \in A$) we find that $\tilde{O}_L^{\ell} = m^{\ell} \cdot w$ where $\tilde{v}^{\ell} = w \in KG$. The crucial step is to observe that, in an appropriate sense, $\tilde{O}_L^{\ell}$ is an integral ℓth power free ideal of oG and can be represented in the form

$$\tilde{O}_L^{\ell} = a^{\theta}$$

where a is a square-free integral ideal of oG, distinct from all of its C-conjugates. Then, one shows that $(O_L)^{\alpha} \in Cl(oG)^J$ for every $\alpha \in A$, and using the fact that $A + J = \mathbb{Z}C$, one obtains $(O_L) \in Cl(oG)^J$. (The details are actually somewhat more involved. It is more convenient to choose v to be a normal integral basis generator at all prime divisors of ℓ, and then to replace m by m', the $o'G$ ideal which it generates, where $o' = o[1/\ell]$. The class group can be regarded as a quotient of $I(o'G)$, and since $o'G$ is a maximal o'-order, the notions of ℓth power free and square free are meaningful.)

The most difficult part of the proof is to construct, for each class in $Cl'(oG)^J$, a tame extension L/K with Galois group G for which (O_L) is the given class. This is done by constructing, in effect, a 'Kummer extension' of the group algebra KG with the elements of G playing the role of the group of roots of unity. This algebra is then shown to be of form $L \otimes_K KG$ for a suitable abelian extension L/K which turns out to be the desired extension. From the construction, one sees that there are infinitely many such extensions, and that the discriminant may be chosen relatively prime to any preassigned ideal of K.

It is natural to look for generalizations of these theorems to other abelian groups. Now let G be an abelian group of type $(\ell^n,\ldots,\ell^n)$, rank k. Again we consider a certain Cartan subgroup of $\mathrm{Aut}\, G = Gl_k(\mathbb{Z}/\ell^n\mathbb{Z})$. Specifically, if R is the ring of integers in the unique unramified extension of $\mathbb{Q}_\ell$ of degree k, then $(R/\ell^nR)^+ \cong G$ and we take $C = C_{k,n} = (R/\ell^nR)^\times$, acting on G by multiplication. As before, let $tr: C \longrightarrow \mathbb{Z}$ denote the composite of the trace $Tr: C \longrightarrow \mathbb{Z}/\ell^n\mathbb{Z}$ with the canonical lifting to the least non-negative residue mod ℓ^n. Then

<u>Definition</u>

$$\theta = \theta_{k,n} = \sum_{\delta \in C} tr(\delta)\delta^{-1} \in \mathbb{Z}C \quad \text{and}$$

$$J = J_{k,n} = \mathbb{Z}C(\theta/\ell^n) \cap \mathbb{Z}C \quad .$$

In case $k = 1$, $G = \mathbb{Z}/\ell^n\mathbb{Z}$ and $J_{1,n}$ is again classical. However,

$$[\mathbb{Z}C^-_{1,n} : J^-_{1,n}] = |Cl(\mathbb{Z}[\mu_{\ell^n}])^-| = h^-_{\ell^n}$$

which, in general is much smaller than

$$|Cl(\mathbb{Z}G)^-| = |D(\mathbb{Z}G)^-| \prod_{r=1}^{n} h^-_{\ell^r} \quad .$$

To generalize the results, we need an analogue to the ring of cyclotomic integers $\mathbb{Z}[\mu_{\ell^n}]$ for $k > 1$.

Let H be the elementary abelian subgroup of G of rank k. Then the group algebra $\mathbb{Q}G$ decomposes as a product $\mathbb{Q}(G/H) \times A_{k,n}$ where $A_{k,n}$ is a product of copies of $\mathbb{Q}(\mu_{\ell^n})$. In fact, the algebra $A_{k,n}$ is a Galois extension of $\mathbb{Q}$ under the action of the Galois group $C_{k,n}$. For increasing n, these extensions can be fitted into a tower of Galois extensions producing an infinite Galois extension with Galois group $\varprojlim C_{k,n} \cong R^\times$. Denote by $\Lambda_{k,n}$

the image of $\mathbb{Z}G$ in $A_{k,n}$ and by $\mu_{\ell^n}^{(k)}$ the image of G . (Then $\Lambda_{k,n} = \mathbb{Z}[\mu_{\ell^n}^{(k)}]$.)

THEOREM 3.
i) $Cl(\Lambda_{k,n})^{J_{k,n}} = 1$.

ii) *If* ℓ *is odd,* $[\mathbb{Z}C_{k,n}^- : J_{k,n}^-] = \ell^E |Cl(\Lambda_{k,n})^-|$
where $E = (k-1)(\ell^{(n-1)k} - 1)/2$.

Remark We can enlarge $J_{k,n}$ to an ideal $S_{k,n} \subseteq \mathbb{Z}C_{k,n}$ analogous to the Stickelberger ideal used by Sinnott [S]. For the enlarged ideal, we obtain equality between the index of the ideal and the class number without the extra power of ℓ appearing in ii).

The proof of ii) parallels that of Theorem 2. Fröhlich's method for computing $|D(\mathbb{Z}G)^-|$ extends also to $|D(\Lambda_{k,n})^-|$. The only additional fact needed is that the units of finite order in $\Lambda_{k,n}$ are precisely $\pm 1 \cdot \mu_{\ell^n}^{(k)}$. The proof also shows that the major ingredient of the equality ii), in addition to Iwasawa'a formula for $h_{\ell^n}^-$ is the fact that the order $\Lambda_{k,n}$ satisfies a conductor-discriminant formula with respect to the characters of $C_{k,n}$. Taken together, these results show that the (non-maximal) orders $\Lambda_{k,n}$ generalize the rings of integers in cyclotomic fields in a significant number of ways.

The proof of i) comes by adapting Fröhlich's proof [F] of the Stickelberger relations in cyclotomic fields. As yet, it has not been possible to extend Theorem 1 to the group G in the case $n > 1$, and obtain i) as a corollary of such an extension. In the case where G is cyclic of order ℓ^n one could conceivably show $R(oG) \subseteq Cl\,(oG)^{J_{1,n}}$, by methods similar to the proof of Theorem 1. The converse is, however, false. For, if it were true, it would imply in the case $o = \mathbb{Z}$ that $Cl(\mathbb{Z}G)$ is annihilated by $J_{1,n}$ and in particular that $D(\mathbb{Z}G)^-$ is annihilated by $J_{1,n}^-$. But if ℓ is a regular prime, the index of $J_{1,n}^-$ in $\mathbb{Z}C_{1,n} \cdot (1-\tau)/2$ is relatively prime to ℓ , whereas Fröhlich's computation of $|D(\mathbb{Z}G)^-|$ shows it is a non-trivial power of ℓ when $n \geq 2$ (except for $\ell = 3$ when $n = 2$).

REFERENCES

[C] Childs, L.N., Stickelberger relations and tame extensions of prime degree, Illinois J.Math. (to appear).

[F] Fröhlich, A., Stickelberger without Gauss sums, in Algebraic Number Fields, Proc. Durham Symposium, Academic Press, London, 1977.

[F'] Fröhlich, A., On the classgroup of integral grouprings of finite abelian groups II, Mathematika 19 (1972), 51-56.

[G] Glass, C.A., Realizable classes in the class groups of integral group rings, Thesis, London University, 1980.

[H] Hilbert, D., Die Theorie der algebraischen Zahlkörper, in Gesammelte Abhandlungen, Chelsea, New York, 1965.

[I] Iwasawa, K., A class number formula for cyclotomic fields, Ann.Math. 76 (1962), 171-179.

[KL] Kubert, D. and Lang, S., Cartan-Bernoulli numbers as values of L-series, Math.Ann. 240 (1979), 21-26.

[M] McCulloh, L.R., A Stickelberger condition on Galois module structure for Kummer extensions of prime degree, in Algebraic Number Fields, Proc. Durham Symposium, Academic Press, London, 1977.

[M'] McCulloh, L.R., A class number formula for elementary-abelian-group rings, J. Algebra 68 (1981), 443-452.

[R] Rim, D.S., Modules over finite groups. Ann.Math. 69 (1959), 700-712.

[S] Sinnott, W., On the Stickelberger ideal and the circular units of a cyclotomic field, Ann.Math. 108 (1978), 107-134.

UNIFORM DISTRIBUTION OF SEQUENCES OF INTEGERS

W. Narkiewicz
Mathematical Institute
Wrocław University
PL-50-384 Wrocław, Plac Grunwaldzki 2/4
Poland

1. In this report we shall be concerned with uniform distribution of sequences of integers in residue classes and shall deal with two related notions: uniform distribution in all residue classes with respect to a given modulus N, which we shall denote by UD (mod N) and uniform distribution in residue classes (mod N), which are prime to N, which we shall call weakly uniform distribution (mod N) and denote by WUD (mod N).

Thus a sequence $a_1, a_2, \dots$ of integers is said to be UD (mod N) provided for all j one has

$$\lim_{x\to\infty} \frac{N\{n \le x : a_n \equiv j \pmod N\}}{x} = \frac{1}{N} \tag{1}$$

and it is said to be WUD (mod N), provided the set

$$\{n : (a_n, N) = 1\}$$

is infinite and moreover for all j prime to N one has

$$\lim_{x\to\infty} \frac{N\{n \le x : a_n \equiv j \pmod N\}}{N\{n \le x : (a_n, N) = 1\}} = \frac{1}{\phi(N)} \tag{2}$$

Although the first paper in which UD (mod N) was considered from a general point of view appeared in 1961 (I. Niven [27]), many results concerning this notion were known much earlier. The oldest dealt with uniform distribution of values of polynomials with integer coefficients in residue classes with respect to a prime. Dickson called them permutational polynomials and there is a huge literature concerning them. Here we want to point out the highlights of this theory.

Let f be a given polynomial over Z, the ring of rational integers and denote by $M(f)$ the set of all integers N such that the sequence $f(1), f(2), \dots$ is UD (mod N). I. Schur [30] stated in 1926 the conjecture, that if $M(f)$ contains infinitely many primes, then the degree of f is prime to 6 and moreover f can be written as a composition of cyclic polynomials $ax^n + b$ and Tchebycheff polynomials

$$T_n(x) = 2^{-1-n}\{(x+[x^2+4]^{\frac{1}{2}})^n + (x-[x^2+4]^{\frac{1}{2}})^n\} .$$

Schur himself proved its truth in the case, when the degree of f was equal to an odd prime and later V.A. Kurbatov [18] extended his result and covered the case when the degree of f is either divisible by two primes or is squarefree with at most four prime divisors. The long-awaited proof of Schur's conjecture was finally given in 1970 by M. Fried [10].

The following question seems to be still unanswered:

Problem 1 Let f be a given polynomial over Z and let p be a prime number. For any $j = 0,1,\dots,p-1$ denote by $d_j(f,p)$ the number of solutions of the congruence

$$f(X) \equiv j \pmod{p} ,$$

and denote by $A_p(f)$ the sequence

$$d_0(f,p), d_1(f,p), \dots, d_{p-1}(f,p) .$$

Let finally f_0 be a fixed polynomial over Z. Is it possible to characterize all those polynomials f with the property that for infinitely many primes p one has

$$A_p(f) = A_p(f_0)$$

in a way similar to the conjecture of Schur, which covers the case $f_0(X) = X$?

2. Another conjecture concerning permutational polynomials was put forward by L.E. Dickson. He noted namely that if the degree of f equals 4, then $M(f)$ cannot contain any prime exceeding 7 and if the degree equals 8 then the biggest prime which may belong to $M(f)$ is 11. These observations lead to the conjecture that for polynomials of even degree $M(f)$ is finite. This follows obviously from Fried's theorem but was established earlier by H. Davenport and D.J. Lewis [4], and D.R. Hayes [11].

It follows from the Chinese Remainder Theorem that if $(a,b) = 1$ then the product ab lies in $M(f)$ if and only if both factors a and b belong to $M(f)$ and the usual lifting argument gives that a prime power p^c $(c \geq 2)$ lies in $M(f)$ if and only if $p \in M(f)$ and the derivative of f has no zeros (mod p). This shows in particular that if $p^2 \in M(f)$ then all powers of p lie in $M(f)$. Denote by $P(f)$ the set of all primes in $M(f)$ and by $S(f)$ the set of all primes p such that $p^2 \in M(f)$. Then $M(f)$ consists of all integers of the form

$$p_1 \cdots p_m q_1^{a_1} \cdots q_m^{a_m}$$

where the p_i's are different and lie in $P(f)\setminus S(f)$, the q_i's lie in $S(f)$ and the exponents a_i are arbitrary. Thus the determination of $M(f)$ reduces to that of $P(f)$ and $S(f)$ (W. Nöbauer [28]). An interesting result concerning these sets was obtained in W. Nöbauer [29]: if $S \subset P$ are two finite sets of primes, then there exists an infinite sequence $f_1, f_2, \ldots$ of polynomials with the degree of f_n tending to infinity and such that for every n one has $S(f_n) = S$ and $P(f_n) = P$.

The following question is still open:

Problem 2 Determine all pairs $S \subset P$ of subsets of the set of all primes such that

a) there exists a polynomial f with $S(f) = S$ and $P(f) = P$;

b) there exists infinitely many such polynomials with arbitrarily large degree.

3. Uniform distribution of arithmetical functions f which are not polynomials was considered in its full generality by I. Niven [27], although certain particular functions were studied earlier (e.g. the UD property for $\omega(n)$ and $\Omega(n)$ was known long ago) Niven gave a necessary condition for UD (mod N) and later S. Uchiyama [34] gave the appropriate Weyl-type condition which is both necessary and sufficient and runs as follows:

A sequence $f(1), f(2), \ldots$ is UD (mod N) if and only if for $h = 1, \ldots, N-1$ one has

$$\sum_{k=1}^{T} \exp\{\frac{2\pi i}{N} hf(k)\} = o(T) .$$

Of course this is a special case of the known criterion for uniform distribution in compact abelian groups. See also A. Dijksma, H.G. Meijer [8], L. Kuipers, S. Uchiyama [17], S. Uchiyama [35] and M. Uchiyama, S. Uchiyama [33].

For certain classes of functions $f(n)$ one can state conditions for UD (mod N) which are easier to apply than the one stated above. In the case of integer-valued additive functions this was done by H. Delange [5] who showed that an additive function f is UD (mod N) if and only if either for $m = 1, 2, \ldots, N-1$ and

$r = 1,2,3,\ldots$ the ratio

$$2mf(2^r)/N$$

is integral and odd, or the series Σp^{-1} taken over all primes p such that $N \nmid mf(p)$ diverges.

A similar result ([6]) holds for systems of additive functions, provided one extends appropriately the notion of uniform distribution. Examples of multiplicative functions which are UD (mod N) for all N were given by H. Delange.

The question for which integers N a given linear recurrence sequence of the second order is UD (mod N) was solved by R.T. Bumby [1]. He showed that such a sequence, defined by

$$a_{n+2} = Aa_{n+1} + Ba_n$$

is UD (mod N) if and only if it is UD (mod p^m) for all prime powers p^m which divide exactly N and it is UD (mod p^m) for $m > 1$ if and only if it is UD (mod p) and one of the following cases holds:

(i) $p \geq 5$

(ii) $p = 3$, $A^2 + B \not\equiv 0 \pmod 9$

(iii) $p = 2$, $A \equiv 2 \pmod 4$, $B \equiv 3 \pmod 4$.

Finally the sequence $\{a_n\}$ is UD (mod p) if and only if we have in case odd p, $p \mid A^2 + 4B$, $p \nmid A$, $p \nmid 2a_2 - Aa_1$ and in the case of $p = 2$, $2 \nmid A$, $2 \nmid Ba_2 - a_1$.

This condition for primes was obtained also by M.B. Nathanson [24] and for prime powers by P. Bundschuh, J.S. Shiue [3] and W.A. Webb, C.T. Long [36].

Earlier the special cases of the Fibonacci and Lucas sequences were settled (P. Bundschuh [2], L. Kuipers, J.S. Shiue [13], [15], H. Niederreiter [26]).

If, as before, we define $M(f)$ the set of all those integers N for which the sequence $f(1), f(2), \ldots$ is UD (mod N) then one can ask which subsets M of the set of positive integers can serve as $M(f)$ for a suitably chosen f. An answer to that question was obtained by A. Zame [38] who proved that M has this property if and only if M contains all divisors of its elements. He obtained this as a corollary to a more general result concerning groups.

One considered also joint distribution of values of a finite set of arithmetic functions in residue classes. In this respect see

H. Delange [6], L. Kuipers, J.S. Shiue [16], L.Kuipers, H. Niederreiter [12] and M.B. Nathanson [25].

4. Now we turn to WUD (mod N) , as defined in (2) . Particular functions with this property were studied long ago, in fact, the quantitative form of the Dirichlet's prime number theorem expresses the fact that the sequence of all primes is WUD (mod N) for all N . One can easily formulate the Weyl-type conditions for WUD , namely for all nonprincipal characters X mod N one should have

$$\sum_{n\leq x} X(a_n) = o(\sum_{n\leq x} X_o(a_n))$$

where X_o denotes the principal character (mod N) . In certain cases however one may obtain much simpler criteria. Let $f(n)$ be an integer-valued multiplicative function, which is polynomial-like, i.e. for every $j = 1,2,...$ there exists a polynomial $V_j \in Z[x]$ such that for all primes p one has $f(p^j) = V_j(x)$. (This condition may be weakened without affecting most results but we do not want to complicate matters.) Denote by R_j the set

$$\{V_j(x) : (xV_j(x),N) = 1\}$$

and let Λ_j be the subgroup of $G(N)$, the multiplicative group of reduced residue classes (mod N) , generated by R_j . If we now assume that not all sets $R_1,R_2,...$ are empty, and M will denote the smallest index k with R_k non-empty, then the sequence $f(1),f(2),...$ will be WUD (mod N) if and only if for every non-principal character X (mod N) which is trivial on $_M$ there exists a prime p such that (W. Narkiewicz [19])

$$1 + \sum_{j=1}^{\infty} \frac{X(f(p^j))}{p^{j/M}} = 0 . \tag{3}$$

This implies that if $\Lambda_M = G(N)$, then our sequence is WUD (mod N) which was already observed in the case $M = 1$ by E. Wirsing [37]. Using this criterion one can find all integers N for which the Euler function, the divisor function or the sum of the divisors are WUD (mod N) (W. Narkiewicz [19], J. Śliwa [32]). Thus e.g. the values of $\phi(n)$ are WUD (mod N) if and only if $(6,N) = 1$ and the values of $\sigma(n)$ are WUD (mod N) if and only if $6 \nmid N$. For the divisor function $d(n)$ things are more complicated - its values are WUD (mod N) if and only if the smallest prime which does not divide N is a primitive root (mod N) .

One can reformulate this criterion so that it could be applied to arbitrary multiplicative integer-valued functions, not necessarily polynomial-like. Let f be such a function and assume that there exists an integer k such that the series

$$\sum_{\substack{p \\ (f(p^k),N)=1}} p^{-1}$$

diverges. Let the smallest such k be denoted by M. Now let Λ be the subgroup of $G(N)$ generated by those $j \pmod N$, $(j,N) = 1$ for which the series

$$\sum_{\substack{p \\ f(p^M)\equiv j \pmod N}} p^{-1}$$

diverges. Now we can repeat the condition involving (3) with Λ in place of Λ_M.

This condition makes sense for arbitrary multiplicative functions but it is not clear whether it is equivalent with WUD (mod N) except in two cases:

1) when for each $j \pmod N$, $(j,N) = 1$ the set A_j of primes p such that $f(p^M) \equiv j \pmod N$ is regular, i.e. for $\operatorname{Re} s > 1$ one has

$$\sum_{p \in A_j} \frac{1}{p^s} = c_j \log \frac{1}{s-1} + f_j(s)$$

where c_j are nonnegative integers and $g_j(s)$ is a function regular in the closed half-plane $\operatorname{Re} s \geq 1$, and

2) when the series

$$\sum_{\substack{p \\ (f(p),N)>1}} \frac{1}{p}$$

converges (H. Delange [7], W. Narkiewicz [20]).

In the general case it is equivalent with the following condition, which may be called WUD (mod N) in the sense of Dirichlet: for all j with $(j,N) = 1$

$$\lim_{s \to \frac{1}{M}+0} \frac{\sum\limits_{\substack{n \\ f(n)\equiv j \pmod N}} n^{-s}}{\sum\limits_{\substack{n \\ (f(n),N)=1}} n^{-s}} = \frac{1}{\phi(N)}$$

(W. Narkiewicz, J. Śliwa [23]). To obtain WUD from this one needs tauberian theorems and it seems that one should be able to construct a multiplicative function for which the last condition is satisfied for a certain N but nevertheless it is not WUD (mod N) .

5. If f(n) is a multiplicative function satisfying f(p) = V(p) with a non-constant polynomial V , about which we assume that it is not of the form $V = cW^k$ where W is a polynomial and $k \geq 2$, then it is possible to utilize evaluations of character sums resulting from A. Weil's proof of Riemann hypothesis for curves to deduce that there is an integer D , which can be given explicitly in terms of V such that if (N,D) = 1 , then the values of f are WUD (mod N) (W. Narkiewicz [21]). In particular they are WUD (mod p) for all sufficiently large primes p . This answers a question of P. Erdös asked at one of the meetings of the Oberwolfach Number Theory Conference.

An analogous result holds also for systems of multiplicative functions, provided one adapts appropriately the notion of WUD . This was recently applied to the study of joint distribution of values of $\phi(n)$ and $\sigma(n)$ (W. Narkiewicz [22]). This paper brings also an explicit procedure which in most cases leads to the determination of the set of all N's for which a given polynomial-like multiplicative function has its values WUD (mod N) .

No analogue of Zame's result for UD quoted in section 3 is known for WUD . In fact there are difficulties in determining when from WUD (mod M) one can deduce WUD (mod N) . This question was studied by E.J. Scourfield [31], who obtained certain sufficient conditions. So we have:

<u>Problem 3</u> Prove an analogue of Zame's result for WUD .

<u>Problem 4</u> Characterize all pairs of integers M,N such that WUD (mod M) implies WUD (mod N) .

Maybe these questions would be easier if one would consider only polynomial-like multiplicative functions.

<u>REFERENCES</u>

1. Bumby, R.T., A distribution property for linear recurrence of the second order, Proc.Amer.Math.Soc. <u>50</u> (1975), 101-106.

2. Bundschuh, P., On the distribution of Fibonacci numbers, Tamchang J.Math. <u>5</u> (1974), 75-79.

3. Bundschuh, P., Shiue, J.S., Solution of a problem on the uniform distribution of integers, Atti Accad.Naz. Lincei, Rend.Cl.Sci.

Fis.Mat.Nat. (8), 55 (1973), 172-177.

4. Davenport, H. and Lewis, D.J., Notes on congruences I, Quart. J.Math., Oxford Ser. 14 (1963), 51-60.

5. Delange, H., On integral-valued additive functions, J. Number Theory 1 (1969), 419-430.

6. Delange, H., On integral-valued additive functions II, ibidem 6 (1974), 161-170.

7. Delange, H., Sur les fonctions multiplicatives à valeurs entieres, C.R.Acad.Sci. Paris 283 (1976), A1065-A1067.

8. Dijksma, A. and Meijer, H.G., Note on uniformly distributed sequences of integers, Nieuw Arch.Wisk. 17 (1969), 210-213.

9. Dowidar, A.F., Summability methods and distribution of sequences on integers, J.Nat.Sci.Math. 12 (1972), 337-341.

10. Fried, M., On a conjecture of Schur, Michigan Math.J. 17 (1970), 41-55.

11. Hayes, D.R., A geometric approach to permutation polynomials over a finite field, Duke Math.J. 34 (1967), 293-305.

12. Kuipers, L. and Niederreiter, H., Asymptotic distribution (mod m) and independence of sequence of integers, I, II, Proc. Japan Acad.Sci. 50 (1974), 256-260, 261-265.

13. Kuipers, L. and Shiue, J.S., A distribution property of the sequence of Fibonacci numbers, Fibonacci Quart. 10 (1972), 375-376, 392.

14. Kuipers, L. and Shiue, J.S., A distribution property of a linear recurrence of the second order, Atti Accad.Naz.Lincei, Rend. Cl.Sci.Fis.Mat.Nat. (8), 52 (1972), 6-10.

15. Kuipers, L. and Shiue, J.S., A distribution property of the sequence of Lucas numbers, Elem.Math. 27 (1972), 10-11.

16. Kuipers, L. and Shiue, J.S., Asymptotic distribution modulo m of sequences of integers and the notion of independence, Atti Accad. Naz. Lincei, Mem.Cl.Sci.Fis.Mat.Nat. (8), 11 (1972), 63-90.

17. Kuipers, L. and Uchiyama, S., Notes on the uniform distribution of sequences of integers, Proc.Japan.Acad. 44 (1968), 608-613.

18. Kurbatov, V.A., on the monodromy group of an algebraic function (in Russian), Mat. Sbornik 25 (1949), 51-94.

19. Narkiewicz, W., On distribution of values of multiplicative functions in residue classes, Acta Arith. 12 (1966-67), 269-279.

20. Narkiewicz, W., Values of integer-valued multiplicative functions in residue classes, ibidem 32 (1977), 179-182.

21. Narkiewicz, W., On a kind of uniform distribution for systems of multiplicative functions, Litovskij Mat. Sbornik, to appear.

22. Narkiewicz, W., Euler function and the sum of divisors, J. Reine Angew.Math. 323 (1981), 200-212.

23. Narkiewicz, W. and Śliwa, J., On a kind of uniform distribution of values of multiplicative functions in residue classes, Acta Arith. 31 (1976), 291-294.

24. Nathanson, M.B., Linear recurrences and uniform distribution, Proc.Amer.Math.Soc. 48 (1975), 289-291.

25. Nathanson, M.B., Asymptotic distribution and asymptotic independence of sequences of integers, Acta Math.Hungar. 29 (1977), 207-218.

26. Niederreiter, H., Distribution of Fibonacci numbers mod 5^k, Fibonacci Quart. 10 (1972), 373-374.

27. Niven, I., Uniform distribution of sequences of integers, Trans.Amer.Math.Soc. 98 (1961), 52-61.

28. Nöbauer, W., Uber Permutationspolynome und Permutationsfunktionen für Primzahlpotenzen, Monatsh.f.Math. 69 (1965), 230-238.

29. Nöbauer, W., Polynome, welche für gegebene Zahlen Permutationspolynome sind, Acta Arith. 11 (1966), 437-442.

30. Schur, I., Über den Zusammenhang zwischen einem Problem der Zahlentheorie und einem Satz über algebraische Funktionen, SBer. Preuss.Akad.Wiss. (1923), 123-134.

31. Scourfield, E.J., On polynomial-like multiplicative functions weakly uniformly distributed (mod N), J. London Math.Soc. 9 (1974), 245-260.

32. Śliwa, J., On distribution of values of $\sigma(n)$ in residue classes, Colloq.Math. 27 (1973), 283-291, 332.

33. Uchiyama, M. and Uchiyama, S., A characterization of uniformly distributed sequences of integers, J.Fac.Sci. Hokkaido 16 (1962), 238-248.

34. Uchiyama, S., On the uniform distribution of sequences of integers, Proc.Japan.Acad. 37 (1961), 605-609.

35. Uchiyama, S., A note on the uniform distribution of sequences of integers, J.Fac.Sci. Shinshu Univ. 3 (1968), 163-169.

36. Webb, W.A. and Long, C.T., Distribution modulo p^h of the general linear second order recurrence, Atti Accad.Naz. Lincei, Rend. Cl.Sci.Fis.Mat.Nat. (8), 58 (1975), 92-100.

37. Wirsing, E., Das asymptotische Verhalten von Summen über multiplikativen Funktionen, Math. Annalen 143 (1967), 75-102.

38. Zame, A., On a problem of Narkiewicz concerning uniform distributions of sequences of integers, Colloq.Math. 24 (1972), 271-273.

DIOPHANTINE EQUATIONS WITH PARAMETERS

A. Schinzel

Introduction The starting point of the investigations to be presented here is:

Hilbert's Irreducibility Theorem (1892) [7]. Let $F_j(x_1,\ldots,x_s, t_1,\ldots,t_r)$ $(1 \leq j \leq k)$ be polynomials irreducible over $\mathbb{Q}$. Then for every polynomial $G \in \mathbb{Q}[t_1,\ldots,t_r]$, $G \neq 0$, there exist integers $t_1^*,\ldots,t_r^*$ such that $G(t_1^*,\ldots,t_r^*) \neq 0$ and for all $j \leq k$ the polynomials $F_j(x_1,\ldots,x_s, t_1^*,\ldots,t_r^*)$ are irreducible over $\mathbb{Q}$.

In what follows, we shall use the abbreviated vector notation $(x_1,\ldots,x_s) = x$ and $(t_1,\ldots,t_r) = t$; so that, for example,

$$F_j(x_1,\ldots,x_s, t_1,\ldots,t_r) = F_j(x,t).$$

Where $r = 1$ we write $t_1 = t$, but there should be no risk of confusion.

Hilbert's theorem easily implies:

THEOREM I. Let $F \in \mathbb{Q}[x,t]$, and suppose that, for all $t^* \in \mathbb{Z}^r$, there exists $x \in \mathbb{Q}^s$ such that $F(x,t^*) = 0$. Then there exists $x \in \mathbb{Q}(t)^s$ such that $F(x(t),t) = 0$.

For the proof it suffices to decompose F over Q into irreducible factors

$$F(x,t) = c \prod_{j=1}^{k} F_j(x,t)^{e_j}$$

and then in Hilbert's Irreducibility Theorem take G to be the product of leading coefficients of $F_j(x,t)$ with respect to x.

Closely related to Theorem I is:

THEOREM II (Kojima (1915) [8] - Skolem (1921) [18]). Let $F \in Q[x,t]$, and suppose that, for all $t^* \in \mathbb{Z}^r$, there exists $x \in \mathbb{Z}^s$ such that $F(x,t^*) = 0$. Then there exists $X \in Q[t]^s$ such that $F(X(t),t) = 0$.

Let us denote by I_r the set of all polynomials in $Q[t_1,\ldots,t_r]$ that take integer values for all $t^* \in \mathbb{Z}^r$. Theorem II implies:

COROLLARY 1. Let $G \in I_r$. Suppose that, for all $t^* \in \mathbb{Z}^r$, there exists $x \in \mathbb{Z}$ such that $x^k = G(t^*)$. Then there exists $X \in I_r$ such that $X(t)^k = G(t)$.

COROLLARY 2. Let $G \in \mathbb{Z}[t]$. Suppose that for all $t^* \in \mathbb{Z}^r$, there exists $x \in \mathbb{Z}$ such that $x^k = G(t^*)$. Then there exists $X \in \mathbb{Z}[t]$ such that $X(t)^k = G(t)$.

For $r = 1$, both Corollaries have been proved by many authors: Franel [4], Grosch |6| (only for $k = 2$), Fried and Surányi [5], Lovász [10] and see also Shapiro [17]. For $r > 1$ an extension to algebraic number fields has been given by Ribenboim [12].

Theorems I and II and Corollaries 1 and 2 suggest four types of statements to be investigated for Diophantine equations $F(x_1,\ldots,x_s, t_1,\ldots,t_r) = 0$, with s unknowns and r parameters. Namely the following four statements.

(i) If, for all $t^* \in \mathbb{Z}^r$, there exists $x \in Q^s$ such that $F(x,t^*) = 0$, then there exists $x \in Q(t)^s$ such that $F(x(t),t) = 0$.

(ii) If, for all $t^* \in \mathbb{Z}^r$, there exists $x \in \mathbb{Z}^s$ such that $F(x,t^*) = 0$, then there exists $X \in Q[t]^s$ such that $F(X(t),t) = 0$.

(iii) If, for all $t^* \in \mathbb{Z}^r$, there exists $x \in \mathbb{Z}^s$ such that $F(x,t^*) = 0$, then there exists $X \in I_r^{\ s}$ such that $F(X(t),t) = 0$.

(iv) If for all $t^* \in \mathbb{Z}^r$, there exists $x \in \mathbb{Z}^s$ such that $F(x,t^*) = 0$, then there exists $X \in \mathbb{Z}[t]^s$ such that $F(X(t),t) = 0$.

In addition to the examples afforded by Corollaries 1 and 2, the statements (iii) and (iv) occur in the following theorems and examples.

THEOREM III. (Skolem (1937), [19]). Statement (iii) holds if $F = A_o(t) + \sum_{i=1}^{s} A_i(t)\, x_i$, where the polynomials $A_i(t)$ $(1 \leqslant i \leqslant s)$ have no common zero.

Example 1. (Skolem (1940) [19]). Statement (iii) fails if $F = t_1^2 + (t_1^2 + t_2^2)\, x_1 + t_1 t_2 x_2$.

THEOREM IV. (Davenport, Lewis and Schinzel (1964) [2] - Chowla (1966) [1]). Statement (iv) holds if $F = x_1^2 + x_2^2 - C(t)$, where $C(t) \in \mathbb{Z}[t]$.

The results and conjectures given in the sequel have been obtained or proposed during the last ten years and many of them are not yet published. In their formulation capital letters (except $\mathbb{Q}$ and $\mathbb{Z}$) denote polynomials with integral coefficients.

1. Concerning Statement (i)

THEOREM 1. Statement (i) holds if $s = 2$ and $F(x,t) = 0$ represents a finite union of curves of genus 0 over the algebraic closure of $\mathbb{Q}(t)$.

The crucial case when F is quadratic over $\mathbb{Q}(t)$ was settled in [3] for $r = 1$ and in [9] for $r > 1$. The case of one curve of genus 0 reduces to the former due to a theorem of Poincaré (see [21], p 71) as has been pointed out by M.Fried.

CONJECTURE 1. Statement (i) fails for $F = x_1^4 - x_2^2 - (8\, t^2 + 5)^2$.

It is shown in [9] how via a result of Stephens [22], the conjecture follows from the so-called Selmer's conjecture in the theory of rational points on elliptic curves.

2. Concerning statement (ii)

THEOREM 2. Statement (ii) holds if either:
1) $F = G(x_1,x_2) - C(t)$, where G is a quadratic form; or $r = 1$ and F satisfies one of the conditions:
2) $F = 0$ represents a parabola over $\mathbb{Q}(t)$,
3) $F = L(x_1,t) - M(t)\, x_2$, where L is a polynomial of degree at most 4 in x_1,
4) $F = A(t)\, x_1^n + B(t) - M(t)\, x_2$, where $n \not\equiv 0 \pmod 8$.

The proof of (ii) in case 1) is implicit in [2] for $r = 1$ and in [12] for $r > 1$. The proof in cases 2) to 4) will appear in [15].

The assumption in 1) that G is a quadratic form seems to be due to the imperfections of the method. In fact we quote:

CONJECTURE 2. (cf [14]). Statement (ii) holds for $F = G(x_1,x_2) - C(t)$, where G is an arbitrary form, C an arbitrary polynomial.

On the other hand, assumption 2) to 4) are natural, as is shown by the following examples.

Example 2 ([15]). Statement (ii) fails for $F = x_1^2 - (4t^2+1)^3 x_2^2+1$.
Example 3 ([15]). Statement (ii) fails for $F = (x_1^2 + 3)(x_1^3 + 3) - (3t + 1)\, x_2$.
Example 4 ([15]). Statement (ii) fails for $F = x_1^8 - 16 - (2t + 1)x_2$.
Example 5 ([15]). Statement (ii) fails for $F = x_1^2 + 1 - ((4t_1^2 + 1)^2 + t_2^2)\, x_2$.

In particular, Example 5 shows that in the assumptions 2) to 4) one parameter t cannot be replaced by two parameters. Nevertheless we have the following:

THEOREM 2a. ([16]). Let $F = L(x_1,\tau,t) - M(\tau,t)x_2$, where L is of degree at most 4 in x_1 or $L(x_1,\tau,t) = A(\tau,t)x_1{}^n + B(\tau,t)$, where A,B,M are arbitrary and $n \not\equiv 0 \pmod 8$. Suppose that, for all $\tau^* \in \mathbb{Z}^{r-1}$ and all $t^* \in \mathbb{Z}$, there exists $x \in \mathbb{Z}^2$ such that $F(x,\tau^*,t^*) = 0$. Then there exists $x \in \mathbb{Q}(\tau)[t]^2$ such that $F(x,\tau,t) = 0$.

3. Concerning statement (iii)

THEOREM 3 ([15]). Statement (iii) holds if r = 1, $F = A(t)x_1{}^n + B(t) - M(t)x_2$ and $n \not\equiv 0 \pmod 8$.

Example 6 ([3]). Statement (iii) fails for $F = (2x_1 + t)(2x_1 + t+1)$

Example 7 ([15]). Statement (iii) fails for $F = (2x_1 + 1)(3x_1 + 1) - (5t + 1)x_2$.

4. Concerning statement (iv)

THEOREM 4 ([14]). Statement (iv) holds if $F = G(x_1,x_2) - C(t)$, $C \in \mathbb{Z}[t]$ and either G is an integral quadratic form with fundamental discriminant equivalent (properly or improperly) to every form in its genus or $G = x_1{}^k x_2{}^\ell$ and the greatest common divisor of the values of C (the fixed divisor of C) equals the greatest common divisor of the coefficients of C (the content of C).

Theorem 4 suggests the following:

CONJECTURE 3. Statement (iv) holds if $F = G(x_1,x_2) - C(t)$, G is an arbitrary integral form and the following conditions are satisfied.

1) $G(x_1,x_2) = H(a_{11}x_1 + a_{12}x_2,\ a_{21}x_1 + a_{22}x_2)$, $H \in \mathbb{Z}[x,y]$, $a_{ij} \in \mathbb{Z}$ implies $\det(a_{ij}) = \pm 1$.

2) the fixed divisor of C equals the content of C.

The following examples show that neither condition can be dispensed with.

Example 8 ([11], see also [23]). Statement (iv) fails for
$F = x_1^2 + 3x_2^2 - (t_1^2 + t_1t_2 + t_2^2)$.

Example 9 ([14]). Statement (iv) fails for $F = x_1^2x_2^3 - 2t^2(t+1)^2$.

5. *The case $s > 2$.*

Theorems 1 to 4 all referred to the case $s = 2$. Besides Theorem III there are a few affirmative results concerning the case $s > 2$. Some of the simpler ones are quoted below.

THEOREM 5 ([13]). Statement (i) holds if $F = A(t)\,N_K(x) + B(t)$, where $N_K(x)$ is the norm form of a field K of prime degree s.

THEOREM 6 ([2] for r = 1, [13] for r > 1). Statement (ii) holds if $F = N_K(x) + C(t)$, where $N_K(x)$ is the norm form of a cyclic field K.

The assumptions about the field K made in Theorems 5 and 6 are essential, as the following example shows.

Example 10 ([2]). Statements (i) and (ii) fail for
$F = N_K(x) - t^2$, where $K = \mathbb{Q}(\zeta_8)$.

References

1. Chowla S, Some problems of elementary number theory, J.Reine Angew.Math.222 (1966), pp 71-74.
2. Davenport H., Lewis D.J. and Schinzel A., Polynomials of certain special types, Acta Arith.9 (1964) pp 107-116.
3. Davenport H., Lewis D.J. and Schinzel A., Quadratic diophantine equations with a parameter, Acta Arith.11 (1966) pp 353-358
4. Franel J., Sixième réponse à la question 37, Intermed. Math.2 (1985) pp94-96
5. Fried E. and Surányi J., Neuer Beweis eines zahlentheoretischen Satzes über Polynome (Hungarian), Math.Lapok 11 (1960) pp 75-84.
6. Grosch W., Lösung zu Aufgabe 402, Arch.Math.Phys. (3)21 (1913) pp 368-369.
7. Hilbert D. Über die Irreduzibilität ganzer rationalen Functionen mit ganzzahligen Coeffizienten, J.Reine Angew.Math.110(1892) pp 104-129 Ges.Abh.Bd.II, Springer 1970 pp 264-286.

8. Kojima T. Note on Number-theoretic properties of algebraic functions, Tohôku Math.J.8 (1915) pp 24-37.
9. Lewis D.J. and Schinzel A., Quadratic diophantine equations with parameters, Acta Arith. 37 (1980) pp 133-141.
10. Lovasz L., Connections between number theoretic properties of polynomials and their substitutional values (Hungarian), Mat.Lapok 20 (1969) pp 129-132.
11. Perlis R. and Schinzel A., Zeta functions and the equivalence of integral forms, J.Reine Angew.Math. 309(1979) pp 176-182.
12. Ribenboim P., Polynomials whose values are powers. J.Reine Agnew. Math. 168/169 (1974) pp 34-40.
13. Schinzel A., On a theorem of Bauer and some of its applications II, Acta Arith. 22 (1972) pp 221-231.
14. Schinzel A., On the relation between two conjectures on polynomials, Acta Arith. 38 (to appear).
15. Schinzel A., Families of curves having each an integer point, Acta.Arith. (1980) pp 285-322.
16. Schinzel A., An application of Hilbert's irreducibility theorem to Diophantine equations, Acta Arith.41 (to appear).
17. Shapiro H.S., The range of an integer-valued polynomial, Amer. Math.Monthly 64(1957) pp 424-425.
18. Skolem T., Untersuchungen über die möglichen Verteilungen ganzzahliger Lösungen gewisser Gleichungen, Kristiania Vid. Selskab. Skrifter I, 1921 No.17.
19. Skolem T., Über die Lösbarkeit gewisser linearer Gleichungen im Bereiche der ganzwertigen Polynome, Kong,Norske Vid. Selskab Forh. 9 (1937) No.34.
20. Skolem T., Einige Sätze über Polynome, Avh.Norske Vid.Akad. Oslo I 1940 No. 4.
21. Skolem T., Diophantische Gleichungen (Chelsea reprint, New York 1950).
22. Stephens N.M., Congruence properties of congruent numbers, Bull. London Math.Soc. 7 (1975) pp 182-185.
23. Watson G.L., Determination of a binary quadratic form by its values at integer points, Mathematika 26 (1979) pp 72-75.

GALOIS MODULE STRUCTURE OF RINGS OF INTEGERS

M.J. Taylor
Queen Mary College
London

Let E/K be a Galois extension of number fields, let $\mathcal{O}_E$ be the ring of integers of E and let $\Gamma = \mathrm{Gal}(E/K)$. We consider $\mathcal{O}_E$ as a (right) module over the integral group ring $\mathbb{Z}\Gamma$. The fundamental problem in the theory of Galois module structure of rings of integers is to determine whether or not $\mathcal{O}_E$ is a free $\mathbb{Z}\Gamma$-module (i.e. whether or not $\mathcal{O}_E$ possesses a $\mathbb{Z}$-basis of the form $\{a_i^\gamma\}\gamma \in \Gamma$, $i = 1 \ldots [K : \mathbb{Q}]$). If $\mathcal{O}_E$ is $\mathbb{Z}\Gamma$ free, then we refer to the $\{a_i\}$ as a normal integral basis.

In Section 1 we describe the history of the subject up to the appearance of the article [F1]. Then in the second section we describe the 'general methods' introduced by Fröhlich in [F1]. Lastly in section 3 we outline the main ideas involved in the proof of Fröhlich's conjecture.

1. HISTORY

The first known result in the subject is due to Hilbert (cf. [H]), who showed:

(1.1) *If* $K = \mathbb{Q}$, Γ *is abelian and the prime divisors of the order of* Γ, $|\Gamma|$, *are non-ramified in* $E/\mathbb{Q}$, *then* $\mathcal{O}_E$ *is a free (rank one)* $\mathbb{Z}\Gamma$*-module.*

The condition that the 'prime divisors of $|\Gamma|$ be non-ramified in $E/\mathbb{Q}$' was subsequently slackened to 'the extension $E/\mathbb{Q}$ be tame, i.e. at most tamely ramified'. The credit for this generalisation is usually given to Speiser - although there appears to be no evidence to support this claim.

More recently (cf. [T1]) this result has been extended by the author to arbitrary basefield, i.e.

(1.2) *If the extension* E/K *is tame and if* Γ *is abelian, then* $\mathcal{O}_E$ *is a free* $\mathbb{Z}\Gamma$*-module (of rank* $[K:\mathbb{Q}]$*).*

Returning to chronological order, the next important result after the Hilbert-Speiser Theorem is due to E. Noether who considered the local structure of $\mathcal{O}_E$ over $\mathcal{O}_K\Gamma$. In fact, she too only considered extensions where the divisors of $|\Gamma|$ are non-ramified (cf. [N]). However, it is usual to call the following Noether's Theorem.

(1.3) $\mathcal{O}_E$ *is locally free over* $\mathcal{O}_K\Gamma$ *at a prime* p *of* K *if, and only if,* p *is at most tamely ramified in* E .

For an outline of the proof of this result see page 21 of [CF].

In the sequel we shall always assume the extension E/K to be tame.

The first person to obtain good global results on the existence of normal integral bases for non-abelian Galois groups was J. Martinet. In [M1] he showed

(1.4) *If* $K = \mathbb{Q}$ *and* Γ *is dihedral of order* 2ℓ *(*ℓ *an odd prime), then* $\mathcal{O}_E$ *is* $\mathbb{Z}\Gamma$*-free.*

However, when he applied his techniques to Galois groups which were quaternion groups of order 8 , he discovered that there exist extensions where $\mathcal{O}_E$ is not $\mathbb{Z}\Gamma$-free (cf. [M3]). It was Fröhlich's interpretation of Martinet's apparently negative result which is really the richness of the whole subject. In order to describe his interpretation it is necessary to introduce some notation.

Let Γ be an arbitrary finite group, and let R_Γ be the Grothendieck group of virtual characters of Γ . For $\chi \in R_\Gamma$, we denote the 'extended' Artin L-function associated to χ by $\Lambda(s,\chi)$ (as defined in [M2]). $\Lambda(s,\chi)$ satisfies a functional equation

$$\Lambda(s,\chi) = W(\chi)\Lambda(1-s,\bar{\chi}) ,$$

where $\bar{\chi}$ is the complex conjugate of χ . The constant $W(\chi)$ is known as the Artin root number of χ . From the general theory of Artin root numbers (cf. [Te] for instance) it is known that for real valued characters χ , $W(\chi) = \pm 1$.

Now let $K = \mathbb{Q}$, let Γ be a quaternion group of order 8 and let χ be the unique non-abelian irreducible character of Γ . Then χ is a symplectic and hence real valued. In [F2] Fröhlich showed that

(1.5) $\mathcal{O}_E$ *is* $\mathbb{Z}\Gamma$*-free if, and only if,* $W(\chi) = 1$.

2. GENERAL METHODS

So far we have only considered special types of Galois group. We now introduce a number of results which will enable us to tackle the general problem.

Let $K_0(\mathbb{Z}\Gamma)$ be the Grothendieck group of locally free $\mathbb{Z}\Gamma$-modules. The rank homomorphism yields a surjection $K_0(\mathbb{Z}\Gamma) \to \mathbb{Z}$.

Let M be a locally free $\mathbb{Z}\Gamma$-module of rank m. We denote by (M) the stable isomorphism class of M in $K_0(\mathbb{Z}\Gamma)$ minus the stable isomorphism class of the direct sum of m copies of $\mathbb{Z}\Gamma$. Thus $(M) \in Cl(\mathbb{Z}\Gamma)$, and we refer to (M) as the class of M.

Let $\mathcal{M}$ be a maximal order of $\mathbb{Q}\Gamma$ which contains $\mathbb{Z}\Gamma$. Then extension of scalars yields a surjection $Cl(\mathbb{Z}\Gamma) \to Cl(\mathcal{M})$; we denote the kernel of this homomorphism by $D(\mathbb{Z}\Gamma)$. Jacobinski has shown that the sub-group $D(\mathbb{Z}\Gamma)$ is independent of the choice of maximal order $\mathcal{M}$.

The reason for our introducing $D(\mathbb{Z}\Gamma)$ is that Fröhlich, following a conjecture of Martinet, showed (cf. Theorem 11 of [F1])

$$(\mathcal{O}_E) \in D(\mathbb{Z}\Gamma) \ . \tag{2.1}$$

The main aim of this section is to firstly give a description of $D(\mathbb{Z}\Gamma)$, and then secondly, using this description, to give a representative of the class $(\mathcal{O}_E)$.

<u>Description of $D(\mathbb{Z}\Gamma)$</u> Let F be a number field which is 'large enough' and which is Galois over $\mathbb{Q}$. In particular F is to contain the $|\Gamma|^{th}$ roots of unity and all other number fields which we introduce. For any number field $M \subset F$, we write $\Omega_M = Gal(F/M)$. For a prime number ℓ we put

$$\mathcal{O}_{F_\ell} = \mathcal{O}_F \otimes_{\mathbb{Z}} \mathbb{Z}_\ell \qquad\qquad U_\ell = \mathcal{O}^*_{F_\ell}$$

and more generally, if S is a finite set of primes, we write $U_S = \prod_{\ell \in S} U_\ell$. Then U_S is an $\Omega_{\mathbb{Q}}$-module in the natural way, and, if $x \in F$ is an S-unit, we view x as an element of U_S via the diagonal embedding.

$Hom_{\Omega_{\mathbb{Q}}}(R_\Gamma, U_S)$ is the group of homomorphisms from R_Γ to U_S which commute with $\Omega_{\mathbb{Q}}$-action. (We view $\Omega_{\mathbb{Q}}$ as acting on R_Γ value-wise).

$Hom^+_{\Omega_{\mathbb{Q}}}(R_\Gamma, \mathcal{O}^*_F)$ is the group of homomorphisms from R_Γ to $\mathcal{O}^*_F$ which commute with $\Omega_{\mathbb{Q}}$ and which are totally positive on all symplectic characters. Note that any $\Omega_{\mathbb{Q}}$-homomorphism necessarily takes real values on symplectic characters since such characters are, of course, real valued.

We now wish to introduce a third group of homomorphisms on R_Γ. Let $z \in \mathbb{Z}_\ell\Gamma^*$. We want to describe a homomorphism $\mathrm{Det}(z) : R_\Gamma \to U_\ell$. Let χ be a character of Γ which is afforded by a representation $T : \Gamma \to GL_n(\mathcal{O}_F)$. We extend T to a homomorphism of algebras $T : \mathbb{Z}_\ell\Gamma \to M_n(\mathcal{O}_{F_\ell})$ and we define

$$\mathrm{Det}(z)(\chi) = \det(T(z)) \quad .$$

Because the character χ determines T up to conjugacy, the left hand side is well-defined. Now we extend $\mathrm{Det}(z)$ to the whole of R_Γ by $\mathbb{Z}$-linearity. In fact from Appendix I of [F1] we know that $\mathrm{Det}(z)$ is an $\Omega_{\mathbb{Q}}$-homomorphism. We write $\mathrm{Det}(\mathbb{Z}_\ell\Gamma^*)$ for the group of all such homomorphisms, and, more generally, we put

$$\mathrm{Det}(\mathbb{Z}_S\Gamma^*) = \prod_{\ell \in S} \mathrm{Det}(\mathbb{Z}_\ell\Gamma^*) \quad .$$

By II.2 in [F1] we have an isomorphism

$$D(\mathbb{Z}\Gamma) \cong \frac{\mathrm{Hom}_{\Omega_{\mathbb{Q}}}(R_\Gamma, U_S)}{\mathrm{Det}(\mathbb{Z}_S\Gamma^*)\mathrm{Hom}_{\Omega_{\mathbb{Q}}^+}(R_\Gamma, \mathcal{O}_F^*)} \tag{2.2}$$

where we take S to be the set of prime divisors of $|\Gamma|$.

Remark 1 In (2.2) we have written the group operation in $D(\mathbb{Z}\Gamma)$ multiplicatively instead of additively.

Remark 2 The two essential ideas to understand in this description are firstly that the class of a module is represented by a homomorphism, and secondly, if we wish to find out whether the class of the module is trivial or not, then we have to develop methods for distinguishing whether a given homomorphism lies in the denominator or not.

Example (due to Swan and Ullom) Let r be an integer prime to $|\Gamma|$ and let (r,Σ) be the two sided $\mathbb{Z}\Gamma$-ideal $r\mathbb{Z}\Gamma + \mathbb{Z}\sum_{\gamma\in\Gamma}\gamma$. Swan (cf. [S]) showed that (r,Σ) is a locally free $\mathbb{Z}\Gamma$ module. Ullom (cf. (2.4) of [U]) showed that its class lies in $D(\mathbb{Z}\Gamma)$ and that this class is represented under (2.2) by the homomorphism

$$\chi \mapsto r^{(\chi,\varepsilon)}$$

for $\chi \in R_\Gamma$. Here ε is the identity character of Γ and $(\ ,\ )$ is the standard inner product of character theory.

<u>Class of $\mathfrak{O}_E$</u> The remainder of this section is devoted to describing a representative of the class $(\mathfrak{O}_E)$ under (2.2).

I. We define the homomorphism $W' : R_\Gamma \to \pm 1$ by stipulating that

$$W'(\chi) = \begin{cases} W(\chi) & \text{if } \chi \text{ is irreducible and symplectic,} \\ 1 & \text{if } \chi \text{ is irreducible and non-symplectic.} \end{cases}$$

From Theorem 9 of [F1] we know that W' is an $\Omega_{\mathbb{Q}}$-homomorphism. Following an idea of Philippe Cassou-Noguès, we define $t(W)$ to be the class represented by W' under (2.2).

II. For $\chi \in R_\Gamma$, we denote the Galois Gauss sum of χ by $\tau(\chi)$. The reader is referred to [M2] for the definition of $\tau(\chi)$. However, it is worth pointing out that $\tau(\chi)$ is equal to $W(\bar{\chi})N\mathfrak{f}(\chi)^{1/2}$ multiplied by a certain fourth root of unity which is determined by the behaviour of the infinite primes. (Here $N\mathfrak{f}(\chi)^{1/2}$ is the positive square root of the absolute norm of the Artin conductor of χ.)

III. By Noether's Theorem (1.3), we can choose $a \in \mathfrak{O}_E$ so that the index $(\mathfrak{O}_E : a.\mathfrak{O}_K\Gamma)$ is prime to $|\Gamma|$. We define

$$A = \prod_\sigma \Big(\sum_{\gamma\in\Gamma} a^{\gamma\sigma}.\gamma^{-1} \Big)$$

where the product is taken (in any order) over $\sigma \in \Omega_{\mathbb{Q}}$ which are (arbitary) extensions to F of a complete set of embeddings of K into F.

We define $u : R_\Gamma \to F^*$ to be the homomorphism $\tau^{-1}.W'.\mathrm{Det}(A)$. In Section 9 of [F1] Fröhlich showed that

(a) u is an $\Omega_{\mathbb{Q}}$-homomorphism.

(b) u is S-unit valued.

(c) Viewing u as an element of $\mathrm{Hom}_{\Omega_{\mathbb{Q}}}(R_\Gamma, U_s)$, u represents the class $(\mathfrak{O}_E)$ under (2.2)

<u>Remark</u> That (a), (b) and (c) hold clearly represents a very deep relationship between the Galois Gauss sum and the Galois module structure of $\mathfrak{O}_E$. Indeed, in [F-T], the (tame) Galois Gauss sum is actually characterised completely in terms of module invariants.

3. FRÖHLICH'S CONJECTURE

One immediate application of Fröhlich's general methods was to permit the calculation of $(\mathfrak{O}_E)$ for many new types of Galois group Γ. Following calculations for quaternion groups of order $4\ell^r$

(ℓ an odd prime) in [F3], for groups of square free order (and more recently cube free order) in [C], and for ℓ-groups in [T2], Fröhlich conjectured the following:

<u>Conjecture</u> $(\mathbb{D}_E) = t(W)$

So that in particular

(a) $(\mathbb{D}_E)^2 = 1$

(b) *the only obstructions to the vanishing of the class of* $\mathbb{D}_E$ *are the signs of the Artin root numbers of the irreducible symplectic characters of* Γ .

It is worth pointing out that (1.2), (1.4), (1.5) together with certain self-duality results (cf. [T3] and [F4]) would follow very easily from the above conjecture.

The remainder of this article is concerned with describing the main ideas contained in the proof of the following result (for details see [T4]).

(3.1) *If the prime divisors of* $|\Gamma|$ *are non-ramified in* $E/\mathbb{Q}$ *, then* $(\mathbb{D}_E) = t(W)$ *(i.e. Fröhlich's conjecture is true for such extensions).*

<u>Remark</u> While it is clearly desirable to avoid all restrictions on the extension $K/\mathbb{Q}$, it is worth emphasising that the above result is genuinely relative in that $K/\mathbb{Q}$ can be wildly ramified.

The proof of this result naturally breaks up into three steps.

<u>Step 1</u> Using an idea of Deligne (cf. §5 of [D]) we adjust the usual Galois Gauss sum by multiplying by a certain root of unity valued, $\Omega_{\mathbb{Q}}^{+}$-homomorphism. We denote this adjusted Gauss sum by τ^* . Thus we see that $(\mathbb{D}_E)t(W)^{-1}$ is represented by the homomorphism $\tau^{*-1}\mathrm{Det}(A)$.

<u>Step 2</u> Next one shows that there is a number field M containing the normal closure of $E/\mathbb{Q}$, which is non-ramified over $\mathbb{Q}$ at S and which has the property that

$$\tau^* \in \mathrm{Det}(\mathbb{D}_{M_S}\Gamma^*) \quad . \tag{A}$$

The proof of this result involves developing a method which enables us to decide whether a given Ω_M homomorphism from R_Γ to

U_S lies in $\mathrm{Det}(\mathfrak{O}_{M_S}\Gamma^*)$ or not. This is really the fundamental problem in the theory of classgroups over integral group rings.

Step 3 By definition we know $A \in \mathfrak{O}_M\Gamma$. Moreover, since S is non-ramified, it can (quite easily) be shown that $A \in \mathfrak{O}_{M_S}\Gamma^*$ under the diagonal embedding $\mathfrak{O}_M\Gamma \hookrightarrow \mathfrak{O}_{M_S}\Gamma$. Consequently we see that the class $(\mathfrak{O}_E)t(W)^{-1}$ is represented by a homomorphism (v, say) which lies in $\mathrm{Det}(\mathfrak{O}_{M_S}\Gamma^*)$, but which is an $\Omega_{\mathbb{Q}}$ homomorphism. We consider this situation in some detail.

Let $\chi \in R_\Gamma$, let $\omega \in \Omega_{\mathbb{Q}}$ and choose $z \in \mathfrak{O}_{M_S}\Gamma^*$ so that $v = \mathrm{Det}(z)$. Then, because v is an $\Omega_{\mathbb{Q}}$-homomorphism $v(\chi^\omega) = v(\chi)^\omega$, i.e. $\mathrm{Det}(z)(\chi^\omega) = (\mathrm{Det}(z)(\chi))^\omega = \mathrm{Det}((z^\omega)(\chi^\omega))$ and so $\mathrm{Det}(z) = \mathrm{Det}(z^\omega)$ for each $\omega \in \Omega_{\mathbb{Q}}$.

Equivalently, if we view $\mathrm{Det}(\mathfrak{O}_{M_S}\Gamma^*)$ as an $\Omega_{\mathbb{Q}}$-module by stipulating that for $x \in \mathfrak{O}_{M_S}\Gamma^*$, $\omega \in \Omega_{\mathbb{Q}}$

$$\mathrm{Det}(\chi).\omega = \mathrm{Det}(x^\omega) .$$

Then clearly the above work translates as saying $\mathrm{Det}(z)$ is an $\Omega_{\mathbb{Q}}$-fixed point. However, by [T5], because S is non-ramified in $M/\mathbb{Q}$

$$\mathrm{Det}(\mathfrak{O}_{M_S}\Gamma^*)^{\Omega_{\mathbb{Q}}} = \mathrm{Det}(\mathfrak{O}_{M_S}\Gamma^{*\Omega_{\mathbb{Q}}}) = \mathrm{Det}(\mathbb{Z}_S\Gamma^*) . \tag{B}$$

Consequently the representative homomorphism v of the class $(\mathfrak{O}_E)t(W)^{-1}$ lies in the denominator of (2.2), as we require.

Remark It is interesting to note that the main idea in both the proof of (A) and (B) is the method of integral logarithms which was developed in [T5].

REFERENCES

[CF] J.W.S. Cassels and A. Fröhlich, Algebraic number theory, Academic Press, New York and London, 1967.

[C] Ph. Cassou-Noguès, Quelques théorèmes de base normale d'entiers, Ann.Inst. Fourier, Grenoble, 28, 3 (1978), 1-33.

[D] P. Deligne, Les constantes des équations fonctionnelles des fonctions L, Modular forms in one variable II, 1973, Lecture Notes in Mathematics 349, 501-597.

[F1] A. Fröhlich, Arithmetic and Galois module structure for tame extensions, J. reine angew.Math. 286/7 (1976), 380-440.

[F2] A. Fröhlich, Artin root numbers and normal integral bases for quaternion fields, Invent.Math. 17 (1972), 143-166.

[F3] A. Fröhlich, Module invariants and root numbers for quaternion fields of degree $4\ell^r$, Proc.Camb.Phil.Soc. 76 (1974), 393-399.

[F4] A. Fröhlich, to appear in the Springer Ergebnisse series.

[F-T] A. Fröhlich and M.J. Taylor, The arithmetic theory of local Galois Gauss sums for tame characters, to appear in the Phil. Trans. Royal Soc.

[H] D. Hilbert, Die Theorie der algebraischen Zahlkörper, Satz 132, Jahr.ber.d.d.Math.Ver (4) (1897), 175-546, or Ges.Abh. I, New York, 63-367.

[M1] J. Martinet, Sur l'arithmétique des extensions galoisiennes à groupe de Galois diédral d'ordre 2p, Ann.Inst. Fourier 19 (1969), 1-80.

[M2] J. Martinet, Character theory and Artin L-functions, Algebraic number fields (ed. A. Fröhlich), Academic Press, London (1977).

[M3] J. Martinet, Modules sur l'algèbre du groupe quaternonien, Ann.Sci. Ecole Norm. Sup. 4 (1971), 229-308.

[N] E. Noether, Normalbasis bei Körpern ohne höhere Verzweigung, J. reine angew.Math. 167 (1932), 147-152.

[S] R. Swan, Periodic resolutions for finite groups, Ann. of Math. 72 (1960), 267-291.

[Te] J. Tate, Local constants, Algebraic number fields (ed. A. Fröhlich), Academic Press, London (1977).

[T1] M.J. Taylor, Galois module structure of relative abelian extensions, J. reine angew.Math. 303/4 (1978), 97-101.

[T2] M.J. Taylor, Adams operations, local root numbers and Galois module structure of rings of integers, Proc.L.M.S. (3), 39 (1979), 147-175.

[T3] M.J. Taylor, On the self-duality of a ring of integers as a Galois module, Invent.Math. 46 (1978), 173-177.

[T4] M.J. Taylor, On Fröhlich's conjecture for rings of integers of tame extensions, to appear.

[T5] M.J. Taylor, A logarithmic approach to classgroups of integral group rings, to appear in the J.Alg.

[U] S.V. Ullom, Non-trivial lower bounds for classgroups, Illinois J.Math. 20 (1976), 361-367.

ON THE FRACTIONAL PARTS OF αn^3, βn^2 AND γn

R.C. BAKER

1. Introduction We denote by $\| \dots \|$ the distance to the nearest integer. Let ε be an arbitrary positive number. In the present note we prove a theorem announced in [1].

THEOREM *Let* α, β and γ *be real numbers. Let* η_1, η_2, η_3 *be numbers with* $0 < \eta_i < 1$,

$$\eta_1 \eta_2 \eta_3 \geq N^{-(1/4) + \varepsilon} .$$

Then for $N > c_1(\varepsilon)$ there is a natural number $n \leq N$ having

$$\| \alpha n^3 \| < \eta_1 , \quad \| \beta n^2 \| < \eta_2 \quad \text{and} \quad \| \gamma n \| < \eta_3 .$$

We mention some related results obtained since the appearance of the survey paper [1]. Let k be a natural number, $k \geq 2$, and let $K = 2^{k-1}$. Let $\alpha_1, \dots, \alpha_h$ be real numbers where $h \leq K/2$. Then for $N > c_2(k, \varepsilon)$ we have

$$\min_{1 \leq n \leq N} \max_{1 \leq i \leq h} \| \alpha_i n^k \| < N^{-(1/hK) + \varepsilon} .$$

The upper bound on h is larger than that of [2]. This inequality is due to myself and G. Harman (to appear).

Let $Q_1, \dots, Q_h$ be quadratic forms in s variables where $s \geq c_3(h, \varepsilon)$. Then for $N \geq 1$ there are integers $x_1, \dots, x_s$ not all zero, having $|x_i| \leq N$,

$$\max_{1 \leq i \leq h} \| Q_i(x_1, \dots, x_s) \| < N^{-(2/h) + \varepsilon} .$$

The exponent announced in [1] was $(-1/h) + \varepsilon$.

The new exponent, due to myself and G. Harman, is sharp (this is explained in [1]).

Finally, my student G. Harman has proved

$$\min_{1 \le n \le N} \| \alpha n^2 \| < N^{-(1/2) + \varepsilon}$$

(Heilbronn's theorem) by elementary means. That is, he uses no integrals or infinite series.

In what follows let $e(x) = e^{2\pi i x}$.

2. Preliminary lemmas. Let k and K be as above.

LEMMA 1. *Suppose that* $f(x) = \alpha_k x^k + \alpha_{k-1} x^{k-1} + \dots + \alpha_1 x$.

Suppose that $N > c_5(k, \varepsilon)$. *Now if*

$$\left| \sum_{x=1}^{N} e(f(x)) \right| \ge B, \tag{1}$$

where

$$B \ge N^{1 - (1/K) + (\varepsilon/2)}, \tag{2}$$

then there exists a natural number $r \le B^{-K} N^{K+\varepsilon}$, with

$$\| \alpha_k r \| \le B^{-K} N^{K-k+\varepsilon} \quad \text{and} \quad \| \alpha_{k-1} r \| \le B^{-K} N^{K-k+1+\varepsilon}. \tag{3}$$

Proof This is a special case of Lemma 11A of [6].

LEMMA 2. *Let* $f(x)$ *and* N *be as in Lemma* 1 *and suppose that* (1) *holds. Suppose that there is a natural number* m *having*

$$m \le N^{1-\varepsilon}, \quad \| m\alpha_j \| \le N^{1-j-\varepsilon} \ (j = 2, \dots, k); \tag{4}$$

suppose further that

$$m^{1-(1/k)} \le B N^{-\varepsilon}. \tag{5}$$

$$t \leq B^{-k} N^{k+\varepsilon} \tag{6}$$

and

$$\| t\alpha_j \| < B^{-k} N^{k-j+\varepsilon} \qquad (j = 1, \ldots, k). \tag{7}$$

Proof. This is Lemma 4 of [3].

Perhaps a few words of explanation would help. It is classical that given a large exponential sum as in (1), one obtains a 'good' rational approximation to the leading coefficient of f. Good rational approximation was extended to the leading pair of coefficients by Schmidt [6]. The idea in Lemma 2 is that given a good rational approximation to all but the lowest coefficient, one can use the large exponential sum to improve the rational approximation and at the same time extend it to the lowest coefficient. We now apply this to particular cubic polynomials.

3. Proof of the theorem. We write $|S|$ for the cardinality of a finite set S. The theorem is the case $|S| = 3$ of

PROPOSITION. *Let S be a nonempty subset of* $\{1,2,3\}$. *Let* $N > c_6(S,\varepsilon)$ *and let* $M_j > 1$ *be defined for all j in S with*

$$\prod_{j \in S} M_j \leq N^{(1/4)-\varepsilon} \quad . \tag{8}$$

For each j in S, let θ_j *be real. Then there is a natural number* $n \leq N$ *with*

$$\| n^j \theta_j \| < M_j^{-1} \quad \textit{for all } j \text{ in } S. \tag{9}$$

Proof. By induction on $|S|$. For $|S| = 1$ the result is well known [6]. In our inductive step let $2 \leq |S| \leq 3$. Suppose that the simultaneous inequalities (9) have no solution $n \leq N$. By standard arguments [4] it follows that there is a nonempty subset T of S such that

$$\sum_{m_j (j \in T)} \left| \sum_{x=1}^{N} e\left(\sum_{j \in T} m_j \theta_j x^j \right) \right| > c_7(k,\varepsilon) N ,$$

where $c_7 > 0$ and the outer sum is over sets of integers $m_j (j \in T)$ having $0 < |m_j| < M_j N^{\varepsilon/15}$ for all $j \in T$.

We see that there is one such set of m_j $(j \in T)$ having

$$\left| \sum_{x=1}^{N} e\left(\sum_{j \in T} m_j \theta_j x^j \right) \right| \geq N^{1-\varepsilon/4} \left(\prod_{j \in T} M_j \right)^{-1} .$$

Define B by

$$B = N^{1-\varepsilon/4} \left(\prod_{j \in T} M_j \right)^{-1} ,$$

then (1) holds. Write $k = \max\{j : j \in T\}$ and $K = 2^{k-1}$.

We distinguish two cases. Suppose first that $T \neq \{1\}$. We apply Lemma 1. The inequality (2) follows from (8), so there exists a natural number r with

$$r \leq B^{-K} N^{K+\varepsilon} \leq \left(\prod_{j \in T} M_j \right)^{K} N^{2\varepsilon} \leq N^{1-\varepsilon} ,$$

and

$$\| r m_i \theta_i \| < B^{-K} N^{K-1+\varepsilon} \leq \left(\prod_{j \in T} M_j \right)^{K} N^{-i+2\varepsilon}$$

for all $i > 1$ in T. Notice that

$$r^{1-(1/k)} \leq r^{3/4} \leq \left(\prod_{j \in T} M_j \right)^{3} N^{2\varepsilon} \leq B N^{-\varepsilon}$$

by (8) ; similarly

$$\| r m_i \theta_i \| \leq N^{-i+1-\varepsilon}$$

for all $i > 1$ in T. By Lemma 2, then, there is a natural number t ,

$$t \leq B^{-k} N^{k+\varepsilon} \leq \left(\prod_{j\in T} M_j \right)^3 N^{2\varepsilon} \tag{10}$$

having

$$\| t m_i \theta_i \| < B^{-k} N^{k-i+\varepsilon} \leq \left(\prod_{j\in T} M_j \right)^3 N^{-i+2\varepsilon} \tag{11}$$

for all i in T.

Suppose now that $T = \{1\}$. Then the inequalities (10), (11) hold for some natural number t. This is proved by arguing as at the end of Lemma 4 of [4]. Thus (10), (11) hold in both cases.

Write $s = t \prod_{j\in T} |m_j|$; then, from (10) and (11),

$$s \leq \left(\prod_{j\in T} M_j \right)^4 N^{3\varepsilon} \tag{12}$$

and

$$\| s \theta_i \| < \left(\prod_{j\in T} M_j \right)^4 M_i^{-1} N^{-i+3\varepsilon} \tag{13}$$

for all i in T. By induction (or trivially if $T = \{1,2,3\}$ there is a natural number v with

$$v \leq \left(\prod_{j\notin T} M_j \right)^4 N^{\varepsilon} \tag{14}$$

and

$$\| v^i s^i \theta_i \| < M_i^{-1} \quad \text{for all } i \notin T . \tag{15}$$

Let $n = sv$; then, from (12), (13) and (14),

$$n \leq \left(\prod_{j\in S} M_j \right)^4 N^{4\varepsilon} \leq N , \tag{16}$$

while if $i \in T$,

$$\begin{aligned} \| n^i \theta_i \| &\leq N^{i-1} v \| s \theta_i \| \\ &< N^{i-1} \left(\prod_{j\notin T} M_j \right)^4 \left(\prod_{j\in T} M_j \right)^4 M_i^{-1} N^{-i+4\varepsilon} \\ &\leq M_i^{-1} . \end{aligned} \tag{17}$$

Combining (15) , (16) and (17) we obtain a solution $n \leq N$ of (9). This is a contradiction, and the inductive step is complete.

References.

1. R.C. Baker, 'Recent results on fractional parts of polynomials'. Number Theory, Carbondale 1979, 10-18. Lecture Notes in Mathematics no. 751 (Springer, Berlin).
2. R.C. Baker, 'Fractional parts of several polynomials III'. Quart. J. Math. Oxford (2), 31 (1980), 19-36.
3. R.C. Baker, 'On the distribution modulo 1 of the sequence $\alpha n^3 + \beta n^2 + \gamma n$' , to appear, Acta Arith.
4. R.C. Baker and J. Gajraj , "Some non-linear Diophantine approximatior Acta Arith. 31 (1976), 325-341.
5. R.C. Baker and G. Harman, 'Small fractional parts of quadratic and additive forms'. To appear.
6. W. Schmidt, Small fractional parts of polynomials. Regional conference series no. 32, American Math. Soc. , Providence 1977.

Royal Holloway College,

Egham,

Surrey

IRREGULARITIES OF POINT DISTRIBUTION IN UNIT CUBES

W. W. L. CHEN

Let $U_0 = [0,1)$ and $U_1 = (0,1]$. Suppose we have a distribution $P(k,N)$ of N points in U_0^{k+1}, where k is a positive integer. For any $\underline{x} = (x_1,\ldots,x_{k+1})$ in U_1^{k+1}, let $Z[P(k,N);\underline{x}]$ denote the number of points $\underline{y} = (y_1,\ldots,y_{k+1})$ of $P(k,N)$ which lie in the box $0 \le y_i < x_i$ $(i = 1,\ldots,k+1)$, and write

$$D[P(k,N);\underline{x}] = Z[P(k,N);\underline{x}] - Nx_1\ldots x_{k+1}.$$

We are interested in measuring the irregularity of the distribution $P(k,N)$ by considering, for $0 < W < \infty$,

$$\|D[P(k,N)]\|_W = \left(\int_{U_1^{k+1}} |D[P(k,N);\underline{x}]|^W d\underline{x}\right)^{1/W};$$

we shall also consider

$$\|D[P(k,N)]\|_\infty = \sup_{\underline{x}\in U_1^{k+1}} |D[P(k,N);\underline{x}]|.$$

Roth [7] proved in 1954 that there exists a positive constant $c_1(k)$, depending only on k, such that for every $P(k,N)$,

$$(1) \qquad \|D[P(k,N)]\|_2 > c_1(k)(\log N)^{\frac{1}{2}k}.$$

It follows easily from (1) that there exists a positive constant $c_2(k)$, depending only on k, such that for every $P(k,N)$,

(2) $$\|D[P(k,N)]\|_\infty > c_2(k)(\log N)^{\frac{1}{2}k}.$$

For the special case k = 1, a sharp lower bound (see later) was obtained in 1972 by Schmidt [11]. He showed that there exists a positive absolute constant c_3 such that for every $P(1,N)$,

(3) $$\|D[P(1,N)]\|_\infty > c_3 \log N.$$

Recently, an alternative proof of (3) was obtained by Halász [3].

To obtain his estimate (1), Roth constructed an auxiliary function $F(\underline{\underline{x}})$ such that, writing $D(\underline{\underline{x}})$ for $D[P(k,N);\underline{\underline{x}}]$,

(4) $$\int_{U_1^{k+1}} F(\underline{\underline{x}})D(\underline{\underline{x}})\,d\underline{\underline{x}} > c_4(k)(\log N)^k,$$

and

(5) $$\int_{U_1^{k+1}} F^2(\underline{\underline{x}})\,d\underline{\underline{x}} < c_5(k)(\log N)^k.$$

These, together with Schwarz's inequality, give (1). A few years ago, Schmidt [12] showed that this auxiliary function $F(\underline{\underline{x}})$ also satisfies

(6) $$\|F\|_r < c_6(k,r)(\log N)^{\frac{1}{2}k} \quad (r > 0).$$

He did this by showing that

$$\int_{U_1^{k+1}} F^{2m}(\underline{\underline{x}})\,d\underline{\underline{x}} < c_7(k,m)(\log N)^{mk} \quad (m = 1,2,\ldots).$$

(4) and (6), together with Hölder's inequality, give

Theorem 1. For every $W > 1$, there exists a positive number $c_8(k,W)$, depending only on k and W, such that for every $\mathcal{P}(k,N)$,

$$\|D[\mathcal{P}(k,N)]\|_W > c_8(k,W)(\log N)^{\frac{1}{2}k}.$$

Schmidt [12] also showed that for some positive number $c_9(k)$, depending only on k, and for large N,

$$\|D[\mathcal{P}(k,N)]\|_1 > c_9(k)\frac{\log\log N}{\log\log\log N}.$$

Recently, Halász [3] has improved this to

(7) $$\|D[\mathcal{P}(k,N)]\|_1 > c_{10}(k)(\log N)^{\frac{1}{2}}.$$

That (3) is essentially best possible was established by Lerch [6] in 1904, and later by van der Corput [1] using a different method. In 1960, Halton [4] showed that for a suitable number $c_{11}(k)$, depending only on k, there exists, corresponding to every natural number $N \geq 2$, a distribution $\mathcal{P}(k,N)$ such that

(8) $$\|D[\mathcal{P}(k,N)]\|_\infty < c_{11}(k)(\log N)^k.$$

However, for $k \geq 2$, there remains a gap between (2) and (8).

On the other hand, Roth's lower bound (1) has been shown to be sharp, apart from the value of the constants. This was established in the cases $k = 1$ and $k = 2$ by Davenport [2] and Roth [9] respectively and more recently for general k by Roth [10]. Meanwhile, alternative proofs for the case $k = 1$ were given by Vilenkin [13], Halton and Zaremba [5] and Roth [8].

Recently, the author was able to show that Theorem 1 is also sharp, apart from the value of the constants.

Theorem 2. Let $W > 0$. For a suitable number $c_{12}(k,W)$, depending only on k and W, there exists, corresponding to every natural number $N \geq 2$, a distribution $P(k,N)$ such that

$$\|D[P(k,N)]\|_W < c_{12}(k,W)(\log N)^{\frac{1}{2}k}.$$

A detailed proof is too long to be included here, and will be published in Mathematika.

References

1. J.G. van der Corput. Verteilungsfunktionen, II, Proc. Kon. Ned. Akad. v. Wetensch., 38 (1935), 1058-1066.
2. H. Davenport. Note on irregularities of distribution, Mathematika, 3 (1956), 131-135.
3. G. Halász. On Roth's method in the theory of irregularities of point distributions, to appear in Proc. Conf. Analytic Number Theory at Durham (1979).
4. J.H. Halton. On the efficiency of certain quasirandom sequences of points in evaluating multi-dimensional integrals, Num. Math., 2 (1960), 84-90.
5. J.H. Halton and S.K. Zaremba. The extreme and L^2 discrepancies of some plane sets, Monatsh. für Math., 73 (1969), 316-328.
6. M. Lerch. Question 1547, L'Intermediaire Math., 11 (1904), 144-145.

7. K.F. Roth. On irregularities of distribution, Mathematika, 1 (1954), 73-79.

8. K.F. Roth. On irregularities of distribution, II, Communications on Pure and Applied Math., 29 (1976), 749-754.

9. K.F. Roth. On irregularities of distribution, III, Acta Arith., 35 (1979), 373-384.

10. K.F. Roth. On irregularities of distribution, IV, to appear in Acta Arith.

11. W.M. Schmidt. Irregularities of distribution, VII, Acta Arith., 21 (1972), 45-50.

12. W.M. Schmidt. Irregularities of distribution, X, Number theory and algebra, pp. 311-329, Academic Press, New York (1977).

13. I.V. Vilenkin. Plane nets of integration (Russian), Ž. Vyčisl. Mat. i Mat. Fiz., 7 (1967), 189-196; English translation in U.S.S.R. Comp. Math. and Math. Phys., 7(1) (1967), 258-267.

Imperial College,
London, England.

THE HASSE PRINCIPLE FOR PAIRS OF QUADRATIC FORMS

D.F. Coray*
Université de Genève, Section de mathématiques
2-4, rue du Lièvre
CH-1211, Genève 24, Switzerland

1. MOTIVATIONS

The arithmetic study of rational surfaces was initiated by B. Segre [13], who considered the case of cubic surfaces in some detail, over a field which was mostly $\mathbb{Q}$ or $\mathbb{R}$. This study was continued by Manin [9] and Iskovskih [7], who obtained a complete birational classification over an arbitrary perfect field k. Without going into the details of this classification, one may mention the subdivision into two main types:

(a) Del Pezzo surfaces, which include in particular all smooth quadric and cubic surfaces in projective space $\mathbf{P}^3_k$, and all smooth intersections of two quadrics in $\mathbf{P}^4_k$.

(b) Conic bundle surfaces, which can be described as fibrations over a rational curve with general fibre a conic. A standard example is the surface defined in affine space $\mathbf{A}^3_k$ by the equation:

$$y^2 + dz^2 = P(x) \tag{1.1}$$

where $d \in k^*$ and $P(x)$ is a polynomial with coefficients in k. The fibration is given by $(x,y,z) \mapsto x \in \mathbf{A}^1_k$; above each point $x_0 \in \mathbf{A}^1_k$, there lies a conic, with equation $y^2 + dz^2 = P(x_0)$.

Del Pezzo surfaces have been extensively studied with regard to unirationality ([10], chap. 4), a problem which is wide open for the surfaces of type (b): it is not even known whether (1.1) can have a rational solution without having infinitely many. On the other hand, there are some results on rational equivalence for conic bundle surfaces ([3];[1], lecture 7), whose analogues have not been fully investigated for the surfaces of type (a). Concerning the Hasse principle, the following example is due to Iskovskih [6]:

*This is a report on joint work with J-L. Colliot-Thélène and J-J. Sansuc [4].

<u>Example</u> For $c \in \mathbb{Z}$, let $V \subset \mathbb{A}^3_{\mathbb{Q}}$ be the surface with equation

$$y^2 + z^2 = (c - x^2)(x^2 - c + 1) \tag{1.2}$$

This is a special case of equation (1.1). For almost all $x_o \in \mathbb{A}^1_{\mathbb{Q}}$, the fibre above x_o is irreducible. The only exceptional points are $x_o = \pm\sqrt{c}$ and $x_o = \pm\sqrt{c-1}$, above which the fibre is a union of two lines: $(y+iz)(y-iz) = 0$.

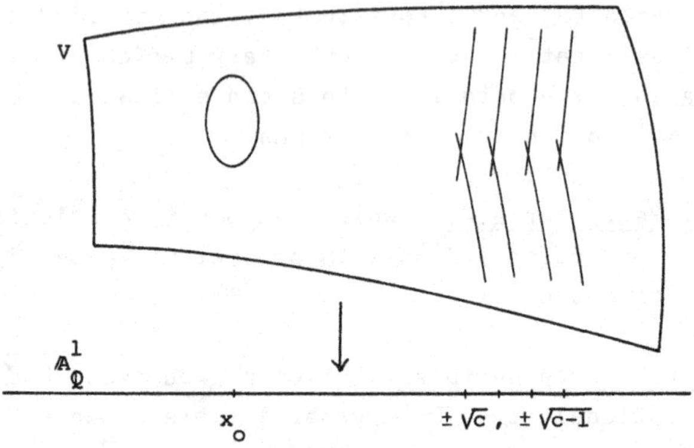

As Iskovskih showed in [6], the Hasse principle fails for this variety whenever c is positive and congruent to 3 modulo 4. In other words, (1.2) has no solution in rational numbers (x,y,z), although it can be solved p-adically for every prime p and also over $\mathbb{R}$. Actually the proof is fairly simple: the well-known characterization of sums of two squares in $\mathbb{Z}$ implies readily that (1.2) can be solved over $\mathbb{Q}$ if and only if the following system can:

$$\left.\begin{array}{l} u_1^2 + v_1^2 = c - x^2 \\ u_2^2 + v_2^2 = x^2 - c + 1 \end{array}\right\} \tag{1.3}$$

For $c \equiv 3 \ (4)$, this system has no 2-adic solution.

Iskovskih was interested in (1.2) because he knew how to get from it a (singular) intersection of two quadrics in $\mathbb{P}^4_{\mathbb{Q}}$ which violates the Hasse principle. Our interest in (1.2) stems rather from its relationship to (1.3): this defines a variety V_1, which is an intersection of two quadrics in $\mathbb{A}^5_{\mathbb{Q}}$. If $c \equiv 3 \ (4)$, then V_1 has no point with coordinates in $\mathbb{Q}_2$, which is an easy way of satisfying the Hasse principle! As a matter of fact, our main result (§2) implies that the Hasse principle holds for (1.3) for every value of c . As a consequence we get:

Proposition *Let* $c \in \mathbb{Z}$ *and* $V \subset \mathbb{A}^3_{\mathbb{Q}}$ ***be the conic bundle surface defined by (1.2). Then*** V ***has no rational point if and only if:*** $c < 0$ ***or*** $c \equiv 3 \ (4)$ ***or*** $c = 4^n(8m+7)$. ***In all other cases, the set*** $V(\mathbb{Q})$ ***of*** $\mathbb{Q}$***-rational points of*** V ***is infinite:*** V ***is even*** $\mathbb{Q}$***-unirational.***

The beauty of this result is that V , in general, does not satisfy the Hasse principle, as we saw above. And yet we can say precisely for what values of c the equation (1.2) has a solution! In fact, this is only a special case of a general procedure, explored systematically by Manin ([10], chap.6), for bringing to light possible obstructions to the Hasse principle over a number field k . For instance, let us consider the variety V defined by the equation:

$$y^2 + dz^2 = P_1(x)P_2(x) \ \ldots \ P_n(x) \qquad (1.4)$$

where $-d \in k^*$ is not a square and the polynomials $P_i(x) \in k[x]$ are irreducible over $k(\sqrt{-d})$ and coprime in pairs. For simplicity let us assume also that the degree of each P_i is even. Suppose now that (1.4) has solutions everywhere locally, which we write: $V(k_v) \neq \emptyset$ for every place v . Then one can produce a ***finite*** set of ***varieties of descent*** V_i such that the following alternative holds:

Either: (a) for all i , there is a place v for which $V_i(k_v) = \emptyset$. This condition is equivalent to the ***Manin obstruction*** attached to the Brauer group $\mathrm{Br}\, V$, and it implies that $V(k) = \emptyset$.

Or: (b) there exists an i such that $V_i(k_v) \neq \emptyset$ for every place v . In this case, Manin's obstruction is ***empty*** for V , and also for V_i . And if $V_i(k) \neq \emptyset$ then $V(k) \neq \emptyset$.

It is not known whether Manin's obstruction is the only obstruction to the Hasse principle for rational surfaces. As the above alternative implies, it is worth while to investigate the Hasse principle for the very special varieties $V_i \subset \mathbb{A}^{2n+1}_k$. They can be described by a system of relations:

$$\{0 \neq u_i^2 + dv_i^2 = \alpha_i P_i(x)\}_{i=1,\ldots,n} \tag{1.5}$$

with $\prod_{i=1}^{n} \alpha_i = 1$.

Since Manin's obstruction is guaranteed to be empty for these varieties, it is natural to ask whether the Hasse principle holds for them, in which case one has an explicit procedure for deciding whether or not the conic bundle surface (1.4) has a k-rational point. The results described in the forthcoming section furnish a positive answer to that question for the case $n = 2$, $\deg P_1 = \deg P_2 = 2$. A positive answer may also be expected in the general case: indeed, for $k = \mathbb{Q}$, Colliot-Thélène and Sansuc have shown that the Hasse principle for (1.5) can be derived from Schinzel's conjecture H (see [5]). But this hypothesis is very much stronger than the twin prime conjecture. The results discussed below suggest that the Hasse principle question is less inaccessible. On the other hand, more sceptical mathematicians may find it easier to produce a counter-example to the Hasse principle for a system like (1.5) than to disprove Schinzel's conjecture H by a direct attack!

2. THE CLEAN HASSE PRINCIPLE

The main result of [4] can be stated as follows:

THEOREM. *Let* k *be a number field,* ϕ , ϕ_1 , ϕ_2 *three non-degenerate binary quadratic forms with coefficients in* k . *Let* $V \subset \mathbb{P}_k^5$ *be the 3-dimensional variety defined by the following pair of equations*:

$$\left.\begin{aligned} \phi(u_1,v_1) &= \phi_1(x,y) \\ \phi(u_2,v_2) &= \phi_2(x,y) \end{aligned}\right\} \tag{2.1}$$

Assume, moreover, that ϕ_1 *and* ϕ_2 *are not both isotropic. Then the 'Clean Hasse Principle' holds for* V : *if* V *contains a smooth point defined over* k_v *, for every completion* k_v *of* k *, then every proper model of* V *contains a point with coordinates in* k . *In fact,* V *is even* k-unirational *(hence* V(k) *is infinite).*

Scholium. In order to verify the conclusion, it suffices to find one model of V on which there is a smooth k-point. In the present case, V itself has this property. For details concerning the birational invariance of the Clean Hasse Principle, see [4], §3. This notion clears away a number of pathological situations, like affine plane curves with rational points only at infinity, or projective curves whose only rational points are a few accidental singularities. A typical flaw in the usual definition of the Hasse principle is illustrated by the following example, which is due to W. Ellison:

$$\left.\begin{aligned} u_1^2 - 3v_1^2 &= 23x^2 + y^2 \\ u_2^2 - 3v_2^2 &= -(23x^2 + y^2) \end{aligned}\right\} \qquad (2.2)$$

This is a special case of (2.1), in which ϕ_2 is proportional to ϕ_1. It is a counter-example to the ordinary Hasse principle over $\mathbb{Q}$ (add the two equations; 3 is not a sum of two squares). None the less, the Clean Hasse Principle does hold, as there are only finitely many solutions with coordinates in $\mathbb{Q}_2$ or $\mathbb{Q}_3$ and they are all singular. If ϕ_1 and ϕ_2 are coprime, then the Clean Hasse Principle reduces to the ordinary one.

Remark. If ϕ_1 and ϕ_2 are both isotropic, the theorem is false. This can be seen on the following example:

$$\left.\begin{aligned} u_1^2 - 5v_1^2 &= 2xy \\ u_2^2 - 5v_2^2 &= 2(x+20y)(x+25y) \end{aligned}\right\} \qquad (2.3)$$

for which Manin's obstruction is non-empty, as a local analysis reveals. Furthermore, a fine study of the invariant $H^1(k,\mathrm{Pic}\bar{V})$ shows the Hasse principle of the theorem to be new, in the sense that no birational transformation can carry the variety V into one for which the Hasse principle is well-known to hold (quadric, Severi-Brauer variety, etc.).

3. SOME WORDS ABOUT THE PROOF

(a) The first idea is to replace the pair of equations (2.1) by one quadratic form over $k(t)$, where t is an indeterminate. The following result of Brumer therefore plays a crucial role in the argument:

THEOREM [2]. *Let Φ_1, Φ_2 be two quadratic forms with coefficients in k. The system $\Phi_1 = \Phi_2 = 0$ can be solved non-trivially over k if and only if the form $\Phi_1 + t\Phi_2$ is isotropic over $k(t)$.*

Thus it is equivalent to solve (2.1) over k or the equation

$$\phi(u_1,v_1) + t\phi(u_2,v_2) = \phi_1(x,y) + t\phi_2(x,y) \qquad (3.1)$$

over $k(t)$. There is no strong Hasse principle for quadratic forms over $k(t)$. So it is not immediately clear what has been gained from this reformulation.

(b) Without loss of generality we may assume that ϕ is of the form $\phi(u,v) = u^2 + dv^2$ with $d \in k^*$. This is a ***multiplicative*** form (also called a ***Pfister form***). Now the left-hand side of (3.1) turns out to be $\phi \oplus t\phi$, which is also multiplicative, by general theory (cf. [8], chap. 10).

In [12], Pfister proved a kind of Hasse principle over $k(t)$ for equations of the type $\Phi = f$, where Φ is a multiplicative k-form and f a polynomial with coefficients in k. It is therefore tempting to substitute suitable elements of $k[t]$ for x and y, so that the right-hand side of (3.1) becomes a polynomial f, and to ask whether a convenient type of Hasse principle applies to the equation $\phi \oplus t\phi = f$. If f is irreducible, Pfister's method leads to the consideration of the equation $\phi \oplus \tau\phi = 0$, where τ is the class of t in the number field $k[t]/(f)$. This equation can be handled using the local-to-global principle for number fields.

Example. Written homogeneously, (1.3) becomes:

$$\left.\begin{aligned} u_1^2 + v_1^2 &= cy^2 - x^2 \\ u_2^2 + v_2^2 &= x^2 - (c-1)y^2 \end{aligned}\right\} \qquad (3.2)$$

It is of no use choosing x and y in $\mathbf{Q}$: this choice would lead us back to the original problem. The next simplest possibility is:

$$x = x_1 + tx_2 \quad , \quad y = y_1 + ty_2 \qquad (3.3)$$

where the x_i and the y_i are suitable integers, to be determined later. The problem is then reduced to that of solving:

$$(u_1^2 + v_1^2) + \tau(u_2^2 + v_2^2) = 0 \qquad (3.4)$$

in $\mathbf{Q}[t]/(f)$, where τ is a root of the cubic polynomial f. Let us take a look at that polynomial:

$$f = (x_2^2 - (c-1)y_2^2)t^3 + \ldots + (cy_1^2 - x_1^2) \qquad (3.5)$$

Notice that if the leading and the constant coefficients of f are units, then τ is a unit of $\mathbf{Q}(\tau)$ and (3.4) can be solved everywhere locally, except possibly above the prime 2 and at the places at infinity. Note also that the places at infinity are harmless if all the coefficients of f happen to be positive, for then τ is negative for every real embedding of $\mathbb{Q}(\tau)$. The problem is then reduced to a 2-adic computation, in which the product formula is useful.

The main difficulty we are thus faced with amounts to solving the two simultaneous equations:

$$x_2^2 - (c-1)y_2^2 = 1 \tag{3.6}$$

$$cy_1^2 - x_1^2 = 1 \tag{3.7}$$

But it should strike any one that these two equations can be solved independently, since the variables are disjoint. In other words, ***Brumer's theorem, together with the transformation (3.3), have the effect of separating the variables***! This is a very important feature of the general proof.

(3.6) is nothing else than a Pell equation, and it has non-trivial solutions whenever $c-1$ is positive and not a square. Of course, (3.7) does not always admit a solution (it is a mock Pell equation, with a -1 instead of a $+1$). But the argument can be slightly modified, so as to replace (3.7) by

$$cy_1^2 - x_1^2 = p \tag{3.8}$$

where p is a suitable sum of two squares. For showing that all the coefficients of f can be made positive, one has to use the fact that (3.6) has infinitely many solutions. It goes without saying that such a proof, based on the properties of the Pell equation, does not generalize easily to an arbitrary number field or to the case of arbitrary forms ϕ_1, ϕ_2. The forthcoming points list some of the main tools that are needed for this generalization.

(c) Although there is no strong Hasse principle over $k(t)$, there are two weak Hasse principles:

Proposition *Let* k *be a number field. The natural homomorphism of Witt rings*

$$W(k(t)) \to \prod_v W(k_v(t)) ,$$

where v *runs through the set of places of* k *, is injective.*

Hence, if two $k(t)$-forms are equivalent over $k_v(t)$ for every place v, they are equivalent over $k(t)$. But there is another class of valuations in $k(t)$, namely those corresponding to prime ideals (π) in the principal ideal domain $k[t]$. The following proposition, due to Harder and Milnor, is more general than Pfister's results mentioned in (b):

Proposition [11] *Let* k *be any field. If two* $k(t)$*-forms are equivalent over all completions* $k(t)_\pi$ *, then they are equivalent over* $k(t)$.

(d) For Pfister forms, a weak Hasse principle is just as good as a strong one. This is because a multiplicative K-form Φ represents an element $f \in K^*$ if and only if Φ is equivalent to $f\Phi$ over K. For suppose Φ represents f everywhere locally. Then Φ is equivalent to $f\Phi$ everywhere locally, hence also globally if the weak Hasse principle holds. It follows that Φ represents f globally.

(e) Pell equations are replaced by a powerful theorem, which basically goes back to Hecke. Such a result deserves to be used much more widely than has been the case so far in the literature. See [4], §2, for a more comprehensive version than is given here.

THEOREM (Hecke). *Let* k *be a number field,* A *its ring of integers,* ψ *a binary quadratic form with coefficients in* A. *Suppose* ψ *is* anisotropic *and primitive. Let* x^0, y^0 *be given elements of* A *and* $m \subset A$ *a non-zero ideal such that* $\psi(x^0,y^0)$ *is prime to* m. *Then there exist elements* x, y *in* A *and a prime ideal* $p \subset A$ *such that*

$$x \equiv x^0 \bmod m \quad , \quad y \equiv y^0 \bmod m \tag{3.9}$$

and

$$(\psi(x,y)) = p \tag{3.10}$$

In fact x, y *and* p *can be chosen in infinitely many ways and further proximity conditions can be imposed: we may require* x *and* y *to be arbitrarily close to specified elements* x^v, y^v *for all archimedean valuations* v *except one,* v_0 *say. And, assuming* v_0 *is real, the pair* (x,y) *may also be required to lie in a given angular region of* $k_{v_0} \times k_{v_0}$, *via the embedding* $k \hookrightarrow k_{v_0}$.

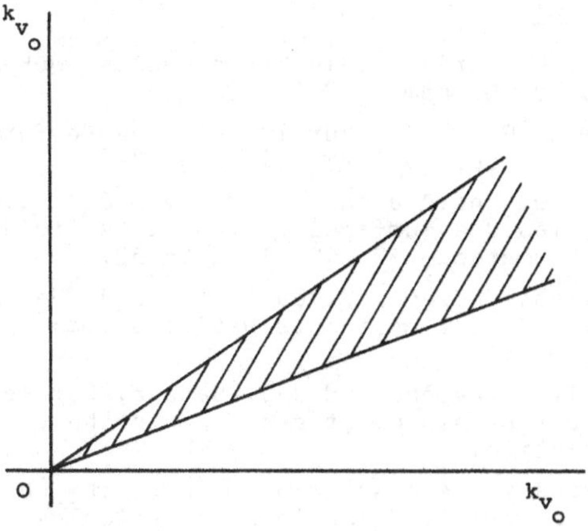

(3.10), with $\psi = \phi_1$, should be compared with equation (3.8). This is also where the assumption that ϕ_1 or ϕ_2 is anisotropic comes into the proof. The choice of an angular region can be made so as to ensure, whenever necessary, that the coefficients of the polynomial f are all positive.

(f) In (3.8), p was a sum of two squares, and not necessarily a prime. Now, in the situation of the example, (3.10) yields a prime p, which - a priori - is not a sum of two squares. The argument is therefore wound up by showing that, as a matter of fact, it is! More generally one shows that the prime ideal p that appears in (3.10) splits completely in the extension $k(\sqrt{-d})/k$ (recall, from (b), that $\phi(u,v) = u^2 + dv^2$). This is made possible by a series of consistent choices (in particular the ideal m in Hecke's theorem) and by a final application of the product formula.

(g) Once we know that V contains a smooth point P with coordinates in k, it is easy to show that $V(k)$ is infinite. Indeed, let W be a k-rational hyperplane section of V passing through P. Since V has only finitely many singular points, a theorem of Bertini guarantees that W can be chosen non-singular. Then W is a del Pezzo surface of degree 4, and it is known that such a surface is k-unirational if $W(k) \neq \emptyset$. Hence $W(k)$ is infinite, a fortiori $V(k)$. The k-unirationality of V is obtained by the same argument, in which we replace k by $k(t)$ and W by the generic element W_t of a pencil of hyperplane sections through P.

REFERENCES

1. S. Bloch, Lectures on algebraic cycles, Duke University Mathematics Series IV, Durham, N.C., 1980.

2. A. Brumer, Remarques sur les couples de formes quadratiques; C.R. Acad.Sci. Paris, 286A (1978), 679-681.

3. J-L. Colliot-Thélène and D. Coray, L'équivalence rationnelle sur les points fermés des surfaces rationnelles fibrées en coniques, Compositio Mathematica, 39 (1979), 301-332.

4. J-L. Colliot-Thélène, D. Coray and J-J. Sansuc, Descente et principe de Hasse pour certaines variétés rationnelles; J. Crelle, 320 (1980), 150-191.

5. J-L. Colliot-Thélène and J-J. Sansuc, Sur le principe de Hasse et l'approximation faible, et sur une hypothèse de Schinzel; to appear in Acta Arithmetica.

6. V.A. Iskovskih, A counter-example to the Hasse principle for a system of two quadratic forms in five variables, Mat. Zametki, 10 (1971), 253-257 (transl. Math. Notes, 10 (1971), 575-577).

7. V.A. Iskovskih, Minimal models of rational surfaces over arbitrary fields (in Russian), Izv.Akad. Nauk S.S.S.R. 43 (1979), 19-43.

8. T.Y. Lam, The algebraic theory of quadratic forms, Benjamin, Reading, 1973.

9. Ju.I. Manin, Rational surfaces over perfect fields, Publ.Math. I.H.E.S. 30 (1966), 55-113 (transl. Amer.Math.Soc.Transl. 84 (1969), 137-186).

10. Ju.I. Manin, Cubic forms, Nauka, Moscow, 1972 (transl. North-Holland, Amsterdam, 1974).

11. J. Milnor, Algebraic K-theory and quadratic forms, Inventiones Math. 9 (1970), 318-344.

12. A. Pfister, Sums of squares in the function field $\mathbf{R}(x,y)$; in: Computers in Number Theory, ed. by Atkin & Birch, Academic Press, London, 1971, pp.77-81.

13. B. Segre, On the rational solutions of homogeneous cubic equations in four variables, Math. Notae Univ. Rosario 11 (1951), 1-68.

ALGORITHME D'APPROXIMATION DIOPHANTIENNE

Eugène Dubois
Département de Mathématiques
Université de Caen
Esplanade de la Paix
14032 CAEN CEDEX

RÉSUMÉ

Nous introduisons une notion de meilleure approximation de zéro par une forme linéaire $p_0 + p_1w_1 + \dots + p_nw_n$, relativement à une fonction F de $(p_0,p_1,\dots,p_n)$. Dans (II) nous avons étudié le cas $n = 2$, F étant une forme quadratique. Le cas $n = 1$ est complétement résolu par l'algorithme des fractions continues. Nous donnons ici une condition suffisante sur F pour que les meilleures approximations soient de 'bonnes' approximations et nous donnons dans le cas $n = 2$ et $F(p_0,p_1,p_2) = \mathrm{Max}(p_1^2,p_2^2)$ un algorithme pour construire la suite des meilleures approximations (définition habituelle).

I. DÉFINITION ET PROPRIÉTÉS D'APPROXIMATION

Soient $1,w_1,\dots,w_n$ des nombres réels supérieurs à 1, $\mathbb{Q}$-linéairement indépendants, $P = (p_0,p_1,\dots,p_n)$ un (n+1)-uple d'entiers. Considérons la fonction L définie sur $\mathbb{Z}^{n+1}$ ou sur $\mathbb{R}^{n+1}$ par:

$$L(P) = p_0 + p_1w_1 + \dots + p_nw_n$$

et une fonction distance, F, définie sur $\mathbb{R}^{n+1}$ (c'est-à-dire que pour tout X, Y dans $\mathbb{R}^{n+1}$ et tout nombre réel t on a $F(X) \geq 0$, $F(X+Y) \leq F(X) + F(Y)$, $F(tX) = |t|f(X)$) et telle que

$$\forall \varepsilon > 0,\ \exists C_0,\ \forall C > C_0,\ 0 < \#\{P \in \mathbb{Z}^{n+1}, 0 < L(P) < \varepsilon, F(P) < C\} < \infty \qquad (1)$$

Cette condition est, en particulier, vérifiée lorsque F est la racine d'une forme quadratique positive de rang n et lorsque F est une fonction distance ne dépendant que de n paramètres.

Définition 1 P dans $\mathbb{Z}^{n+1}$ est dit meilleure approximation de zéro par L, relativement à F (F - M.A. de 1, $w_1,\dots,w_n$) si

$$0 < L(P) \leq 1$$

$$\forall P' \in \mathbb{Z}^{n+1},\ P' \neq P,\ 0 < L(P') < L(P) \Rightarrow F(P') > F(P)$$

Puisque $1, w_1, \ldots, w_n$ sont $\mathbb{Q}$-linéairement indépendants, pour tout $\varepsilon > 0$ l'ensemble $\{P \in \mathbb{Z}^{n+1} \;/\; 0 < L(P) < \varepsilon\}$ n'est jamais vide. Soit Q un point de cet ensemble. D'après (1) $\{P \in \mathbb{Z}^{n+1}$, $0 < L(P) < \varepsilon$, $F(P) \leq F(Q)\}$ est fini. Les F.M.A. forment donc une suite $(P_k)_{k \geq 0}$. $L(P_k)$ décroit vers zéro et $F(P_k)$ croit vers l'infini.

Graphiquement, en représentant $X_k = (L(P_k), F(P_k))$ dans un repère (L,F), les régions (I) et (II) (fig 1) ne contiennent aucun point image d'un point P' de $\mathbb{Z}^{n+1}$

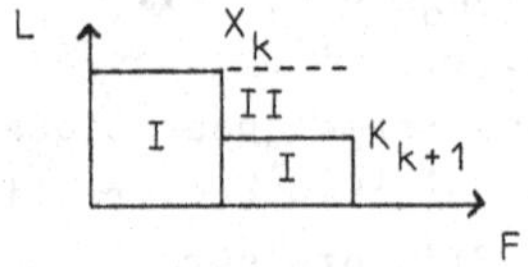

fig. 1

<u>Remarque 1</u> Les fonctions λL (en considerant $\lambda, \lambda w_1, \ldots, \lambda w_n$) et F^{μ} ou νF (λ, μ, ν strictement positifs) définissent les mêmes F.M.A.

<u>Remarque 2</u> La définition usuelle correspond au cas

$$F(P) = H(p_0, p_1, \ldots, p_n) = \mathrm{Max}(|p_1|, \ldots, |p_n|)$$

Nous ne connaissons pas de méthode de calcul pour $n > 1$. Pour $n = 1$, il s'agit de l'algorithme des fractions continues. Récenment, T.W. Cusick (I) a donné une méthode pour $n = 2$ dans le cas où $1, w_1, w_2$ est une base d'un corps cubique à conjugués complexes, à condition de connaître l'unité fondamentale du corps. Nous donnerons au §2 une méthode générale pour le cas $n = 2$. Le problème pour $n > 2$ reste à ma connaissance un problème ouvert.

<u>Remarque 3</u> Si $1, w_1, \ldots, w_n$ sont $\mathbb{Q}$-linéairement dépendants, non tous rationnels on peut encore montrer l'existence d'une suite en supposant que la fonction (L,F) est injective sur $\{P \in \mathbb{Z}^{n+1}$, $L(P) > 0\}$. Dans ce cas l'existence d'un algorithme permettant de construire les F.M.A. devrait permettre de déterminer les relations de dépendance. Mais H.R.P. Fergusson et R.W. Forcade (III) viennent de définir un algorithme résolvant cette question. D'autre part, pour la construction de rationnels vérifiant le bon degré d'approximation nous n'avions pas à considérer ce cas.

<u>Remarque 4</u> J.C. Lagarias (IV) a étudié une généralisation de la notion de meilleure approximation sous forme simultanée et

relativement à différente norme.

Si il existe des constantes c_1 et c_2 ne dépendant que de L et de F telles que pour tour P dans $\mathbf{Z}^{n+1}$ vérifiant $0 < L(P) < 1$ on ait

$$c_1 H(P) \leq F(P) \leq c_2 H(P) \tag{2}$$

Dans ce cas la condition (1) est vérifiée et on a:

THÉORÈME 1. ***Si*** F ***est une fonction distance vérifiant*** (2), ***il existe une constante*** $c_3 = c_3(L,F)$ ***effectivement calculable telle que les F-M.A.,*** P_k ***, vérifient***

$$L(P_k)F(P_{k+1})^n \leq c_3 \tag{3}$$

$$L(P_k)H(P_k)^n \leq c_3 c_1^{-n} \tag{4}$$

On sait d'après le principe des tiroirs de Dirichlet que l'inégalité

$$L(P)H(P)^n < \lambda \tag{5}$$

a une infinité de solution $P \in \mathbf{Z}^{n+1}$ pour $\lambda = 1$. On dit alors que P est une λ-bonne approximation. D'autre part, d'après les travaux de W.M. Schmidt (V), si $w_1, \ldots, w_n$ sont des nombres algébriques et si on remplace n par $n + \varepsilon$ $(\varepsilon > 0)$ l'inégalité (5) ainsi obtenue n'a qu'un nombre fini de solutions.

Pour démontrer le théorème 1, il suffit de considérer dans $\mathbb{R}^{n+1}$ le convexe défini par:

$$|L(X)| \leq L(P_k)\,, \quad F(X) < c$$

Ce convexe est un cylindre limité par deux plans ayant pour base un convexe en dimension n. Son volume est proportionnel à $V = c^n L(P_k)$. Si V est assez grand, il contient, d'après le théorème de Minkowski-Blichfeldt, un point entier. Alors P_{k+1} vérifient (3) et puisque $F(P_k) < F(P_{k+1})$ on a (4) d'après (2).

II. ALGORITHME

Dans le cas $n = 1$ et en notant $H(p,q) = |q|$, les H-M.A. de $\alpha = w_1$ sont les réduites du développement en fraction continue. Dans (II) nous avons montré que pour toute forme quadratique positive $F^{(1)}$ de rang 1 sur $\mathbb{R}^2$, la suite des F^1-M.A. est la suite des réduites du développement en fraction continue. Le théorème 1 illustre alors la propriété $q|q\alpha-p| < 1$ si p/q est une réduite de α . L'algorithme que nous défini (II) dans le cas $n = 2$ et F forme quadratique repose sur l'existence d'une approximation auxiliaire Q_k dans $\mathbf{Z}^3$ telle que P_k , Q_k , P_{k-1} forment une base de $\mathbf{Z}^3$. Pour $n = 1$, ceci correspond à la propriété $p_n q_{n+1} - p_{n+1} q_n = \pm 1$ si p_n/q_n et p_{n+1}/q_{n+1} sont deux réduites successives.

II.1 Existence d'une base d'approximation $(n \geq 2)$

Soient 1 , w_1 , w_2 des nombres réels $\mathbb{Q}$-linéairement indépendants, $L(P) = p_0 + p_1 w_1 + p_2 w_2$, $H(P) = \mathrm{Max}(p_1, p_2)$. Les H.M.A. sont les meilleures approximations au sens habituel.

THÉORÈME 2. *Soit* $(P_k)_{k \geq 0}$ *la suite des H-M.A. Pour tout* $k \geq 1$ *, il existe une approximation auxiliaire* Q_k *formant avec* P_k *et* P_{k-1} *une base de* $\mathbb{Z}^3$.

On a $P_0 = (1,0,0)$ et $P_1 = (p_0, p_1, p_2)$ avec p_1 , p_2 égaux à 0 , 1 ou -1 et non tous les deux nuls. Le théorème est donc vrai pour $k = 1$. Supposons le vrai pour tout $k' \leq k$ et considerons un point Q'_{k+1} tel que $D_k = |\det(P_{k+1}, Q'_{k+1}, P_k)|$ soit minimum non nul. On veut montrer $D_k = 1$. Sinon soit ℓ un diviseur premier de D_{k+1} et étudions les points P de la forme

$$P = \frac{1}{\ell} (\ell_1 P_{k+1} + \ell_2 Q'_{k+1} + \ell_3 P_k) \qquad \ell_i \in \mathbb{Z}$$

En exprimant P_{k+1} , Q'_{k+1} , P_k dans la base P_k , Q_k , P_{k-1} ;

$$P_{k+1} = cP_k + bQ_k + aP_{k-1}$$

$$Q'_{k+1} = c'P_1 + b'Q_k + a'P_{k-1}$$

les composantes de ℓP sur la base P_k , Q_k , P_{k-1} sont alors

$$\ell_1 c + \ell_2 c' + \ell_3$$

$$\ell_1 b + \ell_2 b'$$

$$\ell_1 a + \ell_2 a'$$

P est à composantes entières si et seulement si ces trois entiers sont divisibles par ℓ . Mais le déterminant, D_k , de ce système est multiple de ℓ . Il existe donc des entiers ℓ_1, ℓ_2, ℓ_3 définis modulo ℓ tels que P soit entier. Choisissons ℓ_i dans $(-\ell/2 , \ell/2)$.

Si $\ell_2 = 0$ on a $P = \frac{1}{\ell} (\ell_1 P_{k+1} + \ell_3 P_k)$ et donc:

$$|L(P)| \leq \frac{|\ell_1| + |\ell_3|}{\ell} L(P_k) \leq L(P_k)$$

$$H(P) \leq \frac{|\ell_1| + |\ell_3|}{\ell} H(P_{k+1}) \leq H(P_{k+1})$$

L'une des deux inégalités étant stricte suivant que ℓ_3 est nul ou non, on a $H(P) < H(P_{k+1})$ et donc une contradiction avec la définition 1 pour P_{k+1} . Cet argument est aussi valable pour toute fonction distance F .

Si $\ell_2 \neq 0$, on a

$$|\det(P_{k+1}, P, P_k)| = |\frac{\ell_2}{\ell}| D_k \leq \frac{1}{2} D_k$$

et on obtient une contradiction avec le choix de D_k minimum non nul.

Corollaire Pour toute fonction distance F vérifiant (1) et (2) soit $(P_k)_{k\geq 0}$ la suite des F.M.A. Si il existe Q_1 tel que P_o , Q_1 , P_1 soit une base de $\mathbb{Z}^3$ alors pour tout $k \geq 0$ il existe Q_{k+1} tel que P_k , Q_{k+1} , P_{k+1} soit une base de $\mathbb{Z}^3$.

Ce corollaire est encore vrai en dimension quelconque. Par une démonstration identique on montre qu'il existe $Q_k^{(i)}$ tels que $P_{k-1} , Q_k^{(1)} , \ldots , Q_k^{(n-1)} , P_k$ soit une base de $\mathbb{Z}^{n+1}$. L'hypothèse que le résultat soit vrai au départ se vérifie, pour une fonction F donnée en calculant P_o et P_1 . D'une manière générale on peut montrer que si F vérifie (2) avec $2c_1 \geq c_2$ ou si F ne dépend que de p_1 , p_2 (ou de $p_1 , p_2 , \ldots , p_n$ en dimension n) on peut compléter P_o , P_1 pour obtenir une base de $\mathbb{Z}^3$ (ou $\mathbb{Z}^{n+1}$).

Remarque 5 Voronoi (VI) en généralisant l'algorithme des fractions continues, a défini des minimas relatifs qui correspondent dans le cas d'une base $(1 , \alpha , \beta)$ d'un corps cubique K non totalement réel aux F_c-meilleures approximations avec $F_c(p_o , p_1 , p_1) = N(p_o + p_1\alpha + p_2\beta) / (p_o + p_1\alpha + p_2\beta)$ où N représente la norme de K sur $\mathbb{Q}$. L'algorithme de Voronoi ou le F_c algorithme (IIa) donne dans ce cas un moyen de calculer l'unité fondamentale de K .

II.2 Methode de calcul $(n = 2)$

Dans (IIa) nous avons donné une méthode de calcul des F-M.A. dans le cas $n = 2$, F forme quadratique. Posons $F_1(p_0,p_1,p_2) = \sqrt{p_1^2 + p_2^2}$. Nous avons alors:

$$H(P) < F_1(P) < H(P)\sqrt{2} \tag{6}$$

Nous recherchons P_{k+1} sous la forme $iP_{k-1} + jQ_k + \ell P_k$.

P_{k+1} étant déterminé, nous obtenons Q_{k+1} sous la forme $i'P_{k+1} + j'Q_k + \ell' P_k$ en recherchant i', j' tels que $|ij'-i'j| = 1$ puis ℓ' par le calcul d'une partie entière.

Notons $\alpha_1 = L(Q_k)/L(P_k)$, $\alpha_2 = L(P_{k-1})/L(P_k)$ et écrivons:

$$P = P(i,j) = i(P_{k-1} - \alpha_2 P_k) + j(Q_k - \alpha_1 P_k) + \varepsilon P_k$$

$$= iX + jY + \varepsilon P_k$$

avec $\varepsilon = \varepsilon(i,j)$ dans $(-1,+1)$.

Parmi ces points P on recherche le minimum de $H(iX + jY + \varepsilon P_k)$. En réduisant une forme quadratique comme dans (II), on détermine i_1, j_1 tels que $F_1(i_1X + j_1Y)$ soit minimum. On calcule $h = H(i_1X + j_1Y + \varepsilon_1 P_k)$ puis on étudie tous les couples (i,j) tels que $H(P(i,j)) \leq h$. Tous ces points vérifient d'après (6)

$$F_1(iX + jY + \varepsilon P_k) < h\sqrt{2}$$

et donc il suffit de considérer les couples (i,j) vérifiant

$$F_1(iX + jY) < B_f = h\sqrt{2} + F_1(P_k) \tag{7}$$

On obtient comme dans (II) des majorations de i et j. Il est alors facile de dècrire cette famille, de choisir le plus petit $H(P(i,j))$ et donc d'obtenir P_{k+1}.

Notons $X = (x_1,x_2,x_3)$, $Y = (y_1,y_2,y_3)$, $A = x_2^2 + x_3^2 = F_1(X)^2$, $B = x_2y_2 + x_3y_3$, $C = y_2^2 + y_3^2 = F_1(Y)^2$, $D = AC - B^2$ on obtient en considérant dans $\mathbf{R}^2$ des produits scalaires de $i\binom{x_2}{x_3} + j\binom{y_2}{y_3}$ avec des vecteurs unitaires respectivement perpendiculaires à (x_2,x_3), (y_2,y_3) :

$$|i| \le B_f \sqrt{C/D}$$
$$|j| \le B_f \sqrt{A/D} \qquad (8)$$

Nous n'avons pas de borne absolue pour ces majorations. Il se peut que D soit petit par rapport à A, B, C. Dans la pratique, nous avons en moyenne 2 à 3 couples (i,j) à considerer à chaque étape pour obtenir P_{k+1} et Q_{k+1}.

L'organigramme du calcul est le suivant:

1) Lecture des données.

2) Calcul de P_1, Q_1, P_0 par l'étude de $jw_1 + kw_2$ pour j, k égaux à 0 ou ±1. On range alors P_1, Q_1, P_0 dans les tableaux P, Q, R.

3) Calcul des constantes α_1, α_2, A, B, C, D en fonction de P, Q, R. Détermination du minimum de $Ai^2 + 2Bij + Cj^2 = F_1(P(i,j))^2$ en considérant les premières réduites du développement en fraction continue de $-B/A$. Calcul de B_f.

4) Boucle sur les (i,j) vérifiant (8). Le sous programme HMIN calcule les deux points $P(i,j)$ et choisit $H(P(i,j))$ minimum. Détermination de P_{k+1}.

5) Calcul de Q_{k+1} en déterminant i',j' tels que $ij' - ij' = \pm 1$. Rangement de P_{k+1}, Q_{k+1}, P_k dans P, Q, R et retour à l'étape 3 jusqu'à un test d'arrêt.

Applications

La méthode développée ici est valable pour toute fonction distance G vérifiant (2). On choisit une forme quadratique F_2 telle que

$$c_1\sqrt{F_2(P)} < G(P) < c_2\sqrt{F_2(P)}$$

On applique les étapes 2 et 3 pour F_2. Dans (7) et (8) et donc dans l'exemple du programme donné en II.b) il suffit de remplacer B_f par $\frac{1}{c_1} G(u_1X + j_1Y + \varepsilon_1P_k) + \sqrt{F_2(P_k)}$.

Ceci nous a permis de trouver un contre exemple à la conjecture de Szekeres. Soit $\theta = \sqrt[3]{4}$. L'algorithme de G. Szekeres donne une suite d'approximations $p_0 \frac{1}{\theta^2} + p_1 \frac{1}{\theta} + p_2$. La conjecture est que cette suite contient toutes les meilleures approximations au sens habituel. Or en considérant $H_1(p_0,p_1,p_2) = \mathrm{Max}(|p_0|,|p_1|)$ nous avons trouve que $(-10,19,-8)$ et $(3075,4422,-4006)$ sont des

H_1.M.A. de $(1, \theta, \theta^2)$ donc des H_1.M.A. de $(1/\theta^2, 1/\theta, 1)$ mais ne sont pas fournies par l'algorithme de G. Szekeres.

II.3 Cas des approximations simultanées

On dit que l'entier q est une meilleurs approximation simultanée (M.A.S.) de w_1, w_2 relativement à la norme N (sur $\mathbb{R}^2$) si en notant $\delta_q = \operatorname{Min}_{p_1,p_2} N(qw_1-p_1, qw_2-p_2)$ et $Q = (q, p_1, p_2)$ le point correspondant on a :

$$\text{pour tout } q' \quad 0 < q' < q \text{ on a } \delta_{q'} > \delta_q .$$

Si $(q_k)_{k\geq 0}$ ou $(Q_k)_k$ est la suite des N.M.A.S. on montre comme au théorème 2 qu'il existe R_k tel que Q_{k-1}, R_k, Q_k soit une base de $\mathbb{Z}^3$. Pour la méthode de calcul des M.A.S. on utilise les mêmes idées qu'au paragraphe précédent. Lorsque N est la norme euclidienne la méthode est équivalente à celle utilisée par Furtwangler Math. Annalen Bd 99 (1928).

REFERENCE

(I) T.W. Cusick, Best diophantine approximation for linear ternary form (à paraitre).

(II) E. Dubois, a) Approximations diophantiennes simultanées de nombres algébriques, Calcul des meilleures approximations, Thèse, Paris, 1980. b) Calculation of F-best approximation of zero by a ternary form, Computation of units (à paraitre).

(III) H.R.P. Ferguson and R.W. Forcade, Generalization of Euclidian algorithm for real numbers to all dimensions higher than two, Bull. Amer.Math.Soc. 1 (1979), no. 6, 912-914.

(IV) J.C. Lagarias, Best simultaneous diophantine approximations II, Behaviour of consecutive best approximations (preprint).

(V) W.M. Schmidt, Linear forms with algebraic coefficients, J. Number Theory 3 (1971), 253-277.

(VI) Voronoi, A generalization of the algorithm of continued fractions, Thèse, Warsaw (1896) (En Russe).
Delone-Fadeev, The theory of irrationalities of the third degree, Trans.Math.Mono., 10 (1964).

On the group $PSL_2(\mathbb{Z}[i])$

by J. Elstrodt, F. Grunewald, J. Mennicke

<u>0.</u> The present article is an extended version of a survey lecture given by one of the authors. It consists of two parts. In the first part, we discuss the connection of the above-mentioned group with elliptic curves defined over $\mathbb{Q}(i)$. This connection is almost entirely in a conjectural state. The main conjectures are analogues of the Weil conjecture for elliptic curves over $\mathbb{Q}$. There are links to topology and to hyperbolic geometry. The second part is concerned with certain Poincaré series which arise naturally in the study of discontinuous groups on hyperbolic 3-space. These Poincaré series are non-Euclidean analogues of Jacobi theta functions. They are closely related to other classical functions such as Bessel functions. We study the Mellin transforms of these functions and show that they are meromorphic functions in the complex plane. One of these functions possesses a very simple functional equation. The singularities are, as usual, tied up with arithmetical questions. Invoking Siegel's main theorem on quadratic forms, we can produce information about certain sums over class numbers. Using the above mentioned generalised theta function, we can produce an explicit version of Selberg's trace formula for $\Gamma = PSL_2(\mathbb{Z}[i])$

Practically no proofs will be given. Parts of the details have appeared in [3],[5]. The remaining details will appear in three subsequent papers [4],[1],[7].

We thank G. Harder for useful discussions on the first part of the following results.

<u>1.1.</u> We shall fix some notations.

$$H = H^3 = \{(z,r) \mid z \in \mathbb{C}, r \in \mathbb{R}^+\} \cong SL_2(\mathbb{C}) / SU_2 .$$

H is a symmetric domain in the sense of E. Cartan.
The hyperbolic metric is given by

$$(1) \qquad ds^2 = \frac{dx^2 + dy^2 + dr^2}{r^2} \quad , \quad z = x + iy \quad .$$

$PSL_2(\mathbb{C})$ is a group of isometries, by the action

$$\begin{pmatrix} \alpha & \beta \\ \gamma & \delta \end{pmatrix} (z,r) = \left(\frac{(\alpha z + \beta)(\overline{\gamma z} + \bar{\delta}) + \alpha\bar{\gamma}r^2}{N} , \frac{r}{N}\right) .$$

$$N = |\gamma z + \delta|^2 + r^2|\gamma|^2 \ , \quad \begin{pmatrix} \alpha & \beta \\ \gamma & \delta \end{pmatrix} \in PSL_2(\mathbb{C}) \quad .$$

$$\Gamma = PSL_2(\mathbb{Z}[i]) \ .$$

$\mathfrak{a} \subseteq \mathbb{Z}[i]$ is an ideal, $N(\mathfrak{a})$ its norm.

$$\Gamma_o(\mathfrak{a}) = \{ \begin{pmatrix} \alpha & \beta \\ \gamma & \delta \end{pmatrix} \in \Gamma \mid \gamma \equiv 0 \bmod \mathfrak{a} \} \quad .$$

$Q(\mathfrak{a})$ is the normal closure in Γ of the set of unipotent elements contained in $\Gamma_o(\mathfrak{a})$.
$U(\mathfrak{a})$ is the subspace generated by the image of $Q(\mathfrak{a})$ in the rationalized commutator quotientgroup $\Gamma_o(\mathfrak{a})^{ab} \otimes \mathbb{Q}$.

$$V(\mathfrak{a}) = \Gamma_o(\mathfrak{a})^{ab} \otimes \mathbb{Q}/U(\mathfrak{a}) \quad , \quad r(\mathfrak{a}) = \dim_{\mathbb{Q}}(V(\mathfrak{a})).$$

Conjugation by the element $\begin{pmatrix} 1 & 0 \\ 0 & i \end{pmatrix}$ induces involutions on $V(\mathfrak{a})$ and $U(\mathfrak{a})$. $V^{\pm}(\mathfrak{a})$, $U^{\pm}(\mathfrak{a})$ are their ± 1 eigenspaces. $r^{\pm}(\mathfrak{a})$ is the dimension over $\mathbb{Q}$ of $V^{\pm}(\mathfrak{a})$.
If $\mathfrak{a} \subseteq \mathfrak{b}$ are two ideals let

$$\text{Tra} : \Gamma_o(\mathfrak{b})^{ab} \otimes \mathbb{Q} \to \Gamma_o(\mathfrak{a})^{ab} \otimes \mathbb{Q}$$

be the map induced by the transfer. We have
$\text{Tra}(U(\mathfrak{b})) \subseteq \text{Tra}(U(\mathfrak{a}))$. Tra also respects the spaces $V^{\pm}$.
We write $V^{\pm}_{old}(\mathfrak{a})$ for the subspace generated by all elements $\underline{v} \in V^{\pm}(\mathfrak{a})$ so that a conjugate of $\underline{v}$ by an appropriate element of $GL_2(\mathbb{Q}(i))$ is of the form $\text{Tra}(\underline{u})$ with $\underline{u} \in V^{\pm}(\mathfrak{b})$ for some $\mathfrak{b} \supseteq \mathfrak{a}$. Put $V^{\pm}_{new}(\mathfrak{a}) = V^{\pm}(\mathfrak{a})/V^{\pm}_{old}(\mathfrak{a})$.

1.2. One of our first objectives of interest was the finite dimensional rational vectorspace $V(\mathfrak{a})$, in particular its dimension $r(\mathfrak{a})$. Since one knows an explicit finite presentation of Γ one may compute $r(\mathfrak{a})$ for each particular ideal $\mathfrak{a}$ by solving a finite linear system of equations. If $\mathfrak{a} = \mathfrak{p}$ is a prime ideal of degree 1 the following table contains some numerical information on $r(\mathfrak{p})$.

$p = N(\mathfrak{p})$	137	233	257	277	433	509	569
$(r^+(\mathfrak{p}), r^-(\mathfrak{p}))$	(0,1)	(1,0)	(1,0)	(1,0)	(0,2)	(1,0)	(0,1)

733	757	853	941	953	977
(0,1)	(1,1)	(1,1)	(0,3)	(2,0)	(0,1)

For all other prime ideals $\mathfrak{p}$ of degree 1 with $N(\mathfrak{p}) < 1000$ we have found $r(\mathfrak{p}) = 0$.
In the $PSL_2(\mathbb{Z})$ situation it is not difficult to compute the analogue of $r(\mathfrak{a})$ explicitely. In fact Hecke [10] and followers have given closed formulae for these dimensions. This analogue also occurs as the dimension of an Eichler cohomology group, see [15] appendix.
In our case the number $r(\mathfrak{p})$, $\mathfrak{p}$ a prime ideal , reappears in many different forms. It is the first Betti number of a certain compact, closed 3-manifold [3],[5]. For a prime ideal $\mathfrak{p}$ of degree 1 it is also the multiplicity of the irreducible representation of dimension $p = N(\mathfrak{p})$ of $PSL_2(\mathbb{Z}/p\mathbb{Z})$ in $N_{\mathfrak{p}}^{ab} \otimes \mathbb{C}$, where $N_{\mathfrak{p}}$ is the full congruence subgroup of modulus $\mathfrak{p}$ [3]. Note that $V(\mathbb{Z}[i]) = \langle 0 \rangle$. Hence $V^{\pm}(\mathfrak{p})$ coincides with $V^{\pm}_{new}(\mathfrak{p})$ for every prime ideal $\mathfrak{p}$.

1.3. Here we wish to discuss the construction of the action of the Heckealgebra on $V^+(\mathfrak{a})$ and $U^+(\mathfrak{a})$. Let $\mathfrak{q} = (q)$ be a prime ideal of $\mathbb{Z}[i]$, generated by $q \in \mathbb{Z}[i]$.

Put $\delta = \delta(q) = \begin{pmatrix} 1 & 0 \\ 0 & q \end{pmatrix}$. Consider the diagram

$$\begin{array}{ccc} \Gamma_o(a)^{ab} & & \Gamma_o(a)^{ab} \\ \downarrow \text{Tra} & & \uparrow \text{in} \\ (\Gamma_o(a) \cap \delta\Gamma_o(a)\delta^{-1})^{ab} & \xrightarrow{\tilde{\delta}} & (\delta^{-1}\Gamma_o(a)\delta \cap \Gamma_o(a))^{ab} \end{array}$$

Tra is the transfer map, $\tilde{\delta}$ the map induced by conjugation with δ^{-1} ; in is the homomorphism induced by the inclusion. We define then

$$T(q) = \text{in} \circ \tilde{\delta} \circ \text{Tra} \quad .$$

It can be verified that $T(q)(U^{\pm}(a)) \subseteq U^{\pm}(a)$. Furthermore $T(q)$ induces a map in the quotient $V(a)$ respecting $V^{\pm}(a)$ and $V^{\pm}_{old}(a)$. Hence $T(q)$ induces endomorphisms of $U^{+}(a)$, $V^{+}(a)$, $V^{+}_{new}(a)$. All these do not depend on the choice of the generator q for q . They will all be called $T(q)$ in the sequel. Put $H(a)$ for the algebra generated by all $T(q)$ in the ring of endomorphisms of $V^{+}_{new}(a)$. It is also possible to define the Heckealgebra on $V^{-}(a)$ and $U^{-}(a)$. In this situation one has to make an appropriate choice of the generators q , see[4]. For a construction similar to the above for $PSL_2(\mathbb{Z})$ see [15] appendix. For reasons explained in the next section we can now prove:

<u>Theorem 1.3.1.</u> $H(a)$ is a commutative, semisimple algebra. It is diagonalizable over a totally real extension of $\mathbb{Q}$.

The letter V will in the following always stand for a one-dimensional eigenspace for $H(a)$ in $V^{+}_{new}(a)$. If V is generated by a vector $\underline{v}$, then

$$T(q)\underline{v} = a_q\underline{v} \quad .$$

The a_q are rational integers by construction. Our setup makes it clear that given a V , the a_q can be effectively computed. For $a = (8 + 13i)$ the table in 1.2 gives the existence of an eigenspace V .

We give here a few examples of the corresponding eigenvalues a_q.

$q = (q)$	1+i	3	1+2i	1-2i	7	11	-3+2i	1+4i
a_q	-2	-4	-2	-3	1	-10	-3	-3

	1-4i	19	23
	-4	35	-5

More examples are discussed in [3], [4]. To V we associate a Dirichlet series by the Euler product

$$(2) \qquad L(V,s) = \prod_{\substack{q|a \\ q^2 \nmid a}} (1-a_q N(q)^{-s})^{-1} \cdot \prod_{q \nmid a} (1-a_q N(q)^{-s}+N(q)^{1-2s})^{-1}$$

By elementary methods it can now be proved that

<u>Theorem 1.3.2.</u> $L(V,s)$ converges if the real part of s is big.

In fact what one proves here is $|a_q| \leq cN(q)(N(q)+1)$, [3]. By analogy with the results for $PSL_2(\mathbb{Z})$ one expects that $|a_q| \leq 2\sqrt{N(q)}$.

Using more technique we can establish some further facts about the Dirichlet series $L(V,s)$. To do this one uses the Weil, Jacquet, Langlands theory for $GL_2(\mathbb{C})$. We shall now explain the information one gets from this approach. Let χ be a Heckecharacter of $\mathbb{Q}(i)$. For our purposes we interprete χ as a homomorphism of the group of fractional ideals of $\mathbb{Q}(i)$ into the multiplicative group of $\mathbb{C}$. We shall here consider only χ which are of type A_0 in the terminology of [17]. If $L(s) = \sum_{a} a_a N(a)^{-s}$ is a Dirichlet series write $L_\chi(s)$ for the series $\sum_{a} a_a \chi(a) \cdot N(a)^{-s}$. Furthermore define

$$\Lambda(V,s) = (2\pi)^{-2s}(N(a))^{s/2} \cdot 4^s \cdot (\Gamma(s))^2 \cdot L(V,s) \quad .$$

Theorem 1.3.3.

(i) $L(V,s)$ can be continued to a holomorphic function on the whole of $\mathbb{C}$. It is then bounded on vertical strips.

(ii) $\Lambda(V,s)$ satisfies the functional equation

$\Lambda(V,s) = \pm\Lambda(V,2-s)$

with a certain choice of sign.

(iii) Every $L_\chi(V,s)$ can also be holomorphically continued to the whole of $\mathbb{C}$. The continuation satisfies a functional equation similar to the one in (ii).

In the next section we shall explain how to derive this result from the theory of Weil, Jacquet, Langlands. This will also clarify the exact meaning of (iii).

Consider now a one dimensional eigenspace U for all Heckeoperators $T(q)$ in $U^+(a)$. Say U is generated by $\underline{u}$ and $T(q)\underline{u} = a_q\underline{u}$ with $a_q \in \mathbb{Z}$. Define a Dirichlet series $L(U,s)$ by the formula (2). One can then prove that

$$L(U,s) = L(\chi_1,s)\cdot L(\chi_2,s) ,$$

where χ_1,χ_2 are appropriately chosen nontrivial Heckecharacters of $\mathbb{Q}(i)$. The series $L(\chi,s)$ is defined as

$$L(\chi,s) = \prod_{q \nmid \mathfrak{f}} (1-\chi(q)N(q)^{-s})^{-1}$$

where $\mathfrak{f}$ is the conductor of the Heckecharacter χ. $L(U,s)$ satisfies then (i), (ii) of Theorem 1.3.3. but not (iii).

1.4. Here we explain briefly the relation of $V^+(a)$, $U^+(a)$ and the various $L(V,s)$ to the theory of automorphic forms of Weil, Jacquet, Langlands. We give a translation into the language of [18]. For this it is convenient to introduce the group

$$\tilde{\Gamma}_0(a) = \{\begin{pmatrix}\alpha & \beta\\ \gamma & \delta\end{pmatrix} \in PGL_2(\mathbb{Z}[i]) \mid p \equiv 0 \bmod a\}.$$

Note that

$$\tilde{\Gamma}_o(\mathfrak{a})^{ab} \otimes \mathbb{Q} = (\Gamma_o(\mathfrak{a})^{ab} \otimes \mathbb{Q})^+ ,$$

where the + denotes the +1 eigenspaces for the involution induced by conjugation with $\begin{pmatrix} 1 & 0 \\ 0 & i \end{pmatrix}$. We write $H^1(\tilde{\Gamma}_o(\mathfrak{a}),\mathbb{C})$ for the first Eilenberg, Maclane cohomology group of $\tilde{\Gamma}_o(\mathfrak{a})$ with coefficients in $\mathbb{C}$. $\tilde{\Gamma}_o(\mathfrak{a})$ also acts as a discontinuous group of isometries on 3-dimensional hyperbolic space H. $H^1(H/\Gamma_o(\mathfrak{a}), \mathbb{C})$ stands for the cohomology group computed with $\tilde{\Gamma}_o(\mathfrak{a})$ - invariant differential 1-forms on H. From the contractability of H and from de Rham's theorem we deduce an identification

$$H^1(\tilde{\Gamma}_o(\mathfrak{a}),\mathbb{C}) = H^1(H/\Gamma_o(\mathfrak{a}),\mathbb{C}) .$$

Put $\mathcal{H}^1(H/\Gamma_o(\mathfrak{a}),\mathbb{C})$ for the space of harmonic, $\tilde{\Gamma}_o(\mathfrak{a})$-invariant 1-forms on H. A main result of [8] says that the obvious map

$$\mathcal{H}^1(H/\Gamma_o(\mathfrak{a}),\mathbb{C}) \to H^1(H/\Gamma_o(\mathfrak{a}),\mathbb{C})$$

is an isomorphism. In [8] there is also given a decomposition

$$H^1(H/\Gamma_o(\mathfrak{a}),\mathbb{C}) = H^1_{cusp}(H/\Gamma_o(\mathfrak{a}),\mathbb{C}) \oplus H^1_{inf}(H/\Gamma_o(\mathfrak{a}),\mathbb{C}) .$$

In our special case the cuspidal part is the subspace consisting of classes of compactly supported differential forms. The inf component is generated by certain Eisenstein classes. The duality between homology and cohomology gives then isomorphisms

$$(V^+(\mathfrak{a}) \otimes \mathbb{C})^* \cong H^1_{cusp}(H/\Gamma_o(\mathfrak{a}),\mathbb{C})$$

$$(U^+(\mathfrak{a}) \otimes \mathbb{C})^* \cong H^1_{inf}(H/\Gamma_o(\mathfrak{a}),\mathbb{C}) .$$

For details see [8],[13].

Let $A(a,M^3)$ be the space of (h,a)-automorphic functions as defined in [18]. Here M^3 stands for the unique irreducible 3-dimensional representation of $U(2)$, the maximal compact subgroup of $GL_2(\mathbb{C})$. $A_o(a,M^3) \subseteq A(a,M^3)$ is the subspace consisting of functions which are cuspidal in every cusp of $\tilde{\Gamma}_o(a)$. For definitions see [18],[12]. The discussion in [18] and the above shows that there are isomorphisms

$$\rho : \qquad (\tilde{\Gamma}_o(a)^{ab} \otimes \mathbb{C})^* \to A(a,M^3)$$

$$\rho_o : \qquad (V^+(a) \otimes \mathbb{C})^* \to A_o(a,M^3) \quad .$$

If $A_o^{new}(a,M^3)$ stands for the space of new forms in the sense of [12], then we have also an isomorphism

$$\rho_o^{new} : \qquad (V^+_{new}(a) \otimes \mathbb{C})^* \to A_o^{new}(a,M^3) \ .$$

Using the results of [12] and the isomorphism ρ_o^{new} one gets then Theorem 1.3.1. Weil associates to every automorphic function $\Phi \in A(a,M^3)$ a Dirichlet series $Z(\Phi,s)$. From [18],[12] it follows that if $\underline{v} \in ((\Gamma_o(a) \otimes \mathbb{C})^+)^*$ is an eigenvector for all Heckeoperators then

$$L(<\underline{v}>,s) = Z(\rho(\underline{v}),-1+s) \quad .$$

From this one gets using [18] (i) and (ii) of Theorem 1.3.3. If moreover $\underline{v} \in (V^+(a) \otimes \mathbb{C})^*$ one has to use the converse theorem in [11] to obtain (iii).

Here we would like to add some remarks on the converse theorem in our situation. Suppose one has a Dirichlet series

$$L(s) = \prod_{\substack{q|a \\ q^2 \nmid a}} (1-a_q N(q)^{-s})^{-1} \prod_{q \nmid a} (1-a_q N(q)^{-s} + N(q)^{1-2s})$$

convergent in some halfplane. Assume also $a_q \in \mathbb{Z}$. Suppose one knows (i), (ii) of Theorem 1.3.3. Assume further that (iii) is satisfied for all Heckecharacters χ with conductor prime to a. Weils converse theorem [18],[16] implies then that

$L(s) = L(V,s)$ for an eigenspace $V \subseteq (\Gamma_0(\mathfrak{a}) \otimes \mathbb{Q})^+$. If one knows (iii) for all Heckecharacters χ then Langlands converse theorem implies that $L(s) = L(V,s)$ for an eigenspace $V \subseteq V^+(\mathfrak{a})$.

If for example $L(s)$ is the product of two Heckecharacters over $\mathbb{Q}(i)$ or a Heckecharacter of a field extension then one is in a position to employ one of the converse theorems.

1.5. In this part we discuss the possible connection of the eigenspaces $V \subseteq V^+_{new}(\mathfrak{a})$ to elliptic curves defined over $\mathbb{Q}(i)$. An elliptic curve E defined over $\mathbb{Q}(i)$ is represented by a cubic equation

$$E : y^2 + a_1xy + a_3y = x^3 + a_2x^2 + a_4x + a_6$$

where the $a_j \in \mathbb{Z}[i]$. Let $\mathfrak{f}(E) \subseteq \mathbb{Z}[i]$ be the conductor ideal of E. By $\zeta(E,s)$ we denote the Hasse, Weil ζ-function of E. It is defined as an Eulerproduct

$$\zeta(E,s) = \prod_{\substack{q \mid \mathfrak{f}(E) \\ q^2 \nmid \mathfrak{f}(E)}} (1-b_q N(q)^{-s})^{-1} \cdot \prod_{q \nmid \mathfrak{f}(E)} (1-b_q N(q)^{-s}+N(q)^{1-2s})^{-1}$$

with appropriate $b_q \in \mathbb{Z}$. Note that the typical Eulerfactor is the same as in (2). It is a general belief that $\zeta(E,s)$ has the same analytic properties as stated in Theorem 1.3.3. (i) and (ii), see e. g. [13].

One has to distinguish 3 cases:

(i) E has complex multiplication by an order in $\mathbb{Q}(i)$. Then $\zeta(E,s) = L(\chi_1,s) \cdot L(\chi_2,s)$ for two Heckecharacters χ_1, χ_2 of $\mathbb{Q}(i)$. By application of the converse theorem one finds that $\zeta(E,s) = L(U,s)$ for an eigenspace $U \subseteq U^+(\mathfrak{f}(E))$. It is not possible for a V with this property to occur in $V^+(\mathfrak{f}(E))$, since $L_{\chi_1^{-1}}(U,s)$ cannot be holomorphically continued to the whole of $\mathbb{C}$.

(ii) E has complex multiplication by an order in $Q(\sqrt{-d})$ with $Q(\sqrt{-d}) \neq Q(i)$.
Then $\zeta(E,s) = L(\chi,s)$, where χ is a Hecke-character of $Q(\sqrt{-d},i)$. One finds then that $\zeta(E,s) = L(V,s)$ for an eigenspace $V \subseteq V^+(\mathfrak{f}(E))$.

(iii) E has only trivial complex multiplications.
Nothing is known on this case, but Weil [16],[18] expects that $\zeta(E,s) = L(V,s)$ for a $V \subseteq V^+(\mathfrak{f}(E))$.

We want to concentrate now on the case where $\mathfrak{f}(E) = \mathfrak{p}$ is a prime ideal of $Z[i]$. The following table contains all E such that $\mathfrak{f}(E)$ is a prime ideal of degree 1 with $N(\mathfrak{p}) \leq 1000$ and so that E has a representing equation with "small" coefficients a_j.

	a_1	a_2	a_3	a_4	a_6	$N(\mathfrak{f}(E))$
E_1	0	1+i	i	i	0	233
E_2	1	-i	i	-i	0	257
E_3	i	i	1	5-i	-1-3i	257
E_4	i	i	1	5+4i	-12-4i	257
E_5	1	-i	i	85-6i	34+274i	257
E_6	i	-1+i	1+i	1-2i	0	277
E_7	1+i	-1	i	1-i	0	509
E_8	1+i	-1+i	1	0	-2	757
E_9	1+i	-1-i	1	-1	0	853

The curves E_2, E_3, E_4, E_5 are isogenous. Of course there is an obvious coincidence of this table with the eigenspaces in $V^+(\mathfrak{a})$ in the table of 1.2. We have also checked that the first few hundred Eulerfactors of $\zeta(E,s)$ and of the corresponding

$L(V,s)$ coincide. For details see [4].

We have carried this computation further and also considered the fields $\mathbb{Q}(\sqrt{-2})$, $\mathbb{Q}(\sqrt{-3})$. From our tables [4] there could be a correspondence between the sets

$\{V \subseteq V^+(p)$ is a 1-dimensional eigenspace for $H(p)$, and p is a prime ideal in $\mathbb{Z}[i]\}$

and

$\{E$ is an elliptic curve defined over $\mathbb{Q}(i)$ with $\mathfrak{f}(E)$ a prime ideal$\}$

so that the Dirichlet series of corresponding objects are the same. Note that curves falling under (i) cannot have prime conductor!

For the eigenspaces $V \subseteq V^-(p)$ listed in 1.2 we have also found elliptic curves corresponding to them in the above manner. But here the conductor of the corresponding curve is not p but $(1+i)^4 \cdot p$. For example the curve

$$y^2 + (1+i)xy = x^3 - (1+i)x - 1$$

satisfies $\mathfrak{f}(E) = (1+i)^4 p$ where $N(p) = 137$. Furthermore its first few hundred b_q coincide with the eigenvalues of the (appropriately defined) $T(q)$ on the eigenspace mentioned in 1.2.

1.6. Here we would like to make some general comments on the correspondence between the eigenspaces $V \subseteq V^+_{\text{new}}(a)$ and certain abelian varieties defined over $\mathbb{Q}(i)$. All of the following is inspired by the Eichler, Shimura theory and the Weil conjecture over $\mathbb{Q}$. See [2] for a survey of these results.

The first question arising is

Problem 1

Let E be an elliptic curve defined over $Q(i)$ with conductor $\mathfrak{f}(E)$. Is there a $V \subseteq V^+_{new}(\mathfrak{f}(E))$ with $(E,s) = L(V,s)$?

From the discussion in 1.5. it is clear that one has to assume that the endomorphism ring of E does not contain an order of $\mathbb{Z}[i]$. In this case $L(E,s) = L(\chi_1,s)\cdot L(\chi_2,s)$ for two Heckecharacters of $Q(i)$, and $L(E,s) = L(U,s)$ for an eigenspace $U \subseteq U^+(\mathfrak{a})$. Otherwise our tables suggest that the answer to problem 1 is 'yes'.

Problem 2

Given an eigenspace $V \subseteq V^+_{new}(\mathfrak{a})$ is there an elliptic curve E defined over $\mathbb{Q}(i)$ so that $L(V,s) = \zeta(E,s)$?

The answer here is 'no'. We shall explain this phenomenon in the case of an explicit example. If A is an abelian variety defined over $Q(i)$ we write $\zeta(A,s)$ for the Hasse,Weil-ζ-function of A computed over $Q(i)$, see [14] for a definition. $End_K(A)$ stands for the ring of endomorphisms of A defined over the extension K of $Q(i)$. We shall describe the ζ-function of the Jacobians of the curves $y^2 = x^5 + x$.

To do this we give two Heckecharacters χ_1, χ_2 of $Q(\sqrt{-2})$. Put $\mathcal{O}_{-2} = \mathbb{Z} + \mathbb{Z}\sqrt{-2}$. The group $(\mathcal{O}_{-2}/4\sqrt{-2}\cdot\mathcal{O}_{-2})^*$ has order 16.

Let G be the subgroup generated by the images of 3, $1+\sqrt{-2}$, it has index 2. For an ideal (x) with $(\sqrt{-2}) \nmid (x)$ put

$$\chi_1((x)) = x \qquad \text{if } x \in G \quad .$$

This describes a Heckecharacter of conductor $4\cdot\sqrt{-2}\cdot\mathcal{O}_{-2}$. Let ε be the nontrivial character of $(\mathcal{O}_{-2}/2\cdot\mathcal{O}_{-2})^*$ and define $\chi_2 = \chi_1\cdot\varepsilon$. Put

$$L_1(s) = L(\chi_1,s)\cdot L(\chi_2,s)$$

then $\zeta(\mathrm{Jac}(y^2 = x^5 + x),s) = (L_1(s))^2$. This phenomenon is described in more detail in [6]. On the other hand there is an eigenspace $V \subseteq V^+((1+i)^{12})$ so that $L(V,s) = L_1(s)$. Note that $L_1(s)$ has the correct decomposition as an Eulerproduct over $\mathbb{Q}(i)$. This result is obtained by application of the converse theorem as explained in 1.4.

On the other hand one can show that there is no elliptic curve E defined over $\mathbb{Q}(i)$ with $L_1(s) = \zeta(E,s)$. There is also the possibility to consider a two dimensional abelian variety A defined over $\mathbb{Q}$ so that $\mathbb{Z}[i] \subseteq \mathrm{End}_{\mathbb{Q}(i)}(A)$ and $\mathrm{End}_{\mathbb{C}}(A)$ is a quaternion algebra. Then it also happens that $L(A,s) = (L(s))^2$ where $L(s)$ has the correct type of Eulerproduct and should perhaps come from an eigenspace $V \subseteq V^+(\mathfrak{a})$. We owe this construction to P. Deligne. These and other examples suggest that one has to modify problem 2.

Problem 2'

Given an eigenspace $V \subseteq V^+_{new}(\mathfrak{a})$ is there an abelian variety A defined over $\mathbb{Q}(i)$ so that $L(V,s)^2 = \zeta(E,s)$ and such that A satisfies $\mathbb{Z}[i] \subseteq \mathrm{End}_{\mathbb{Q}(i)}(A)$ and $\mathrm{rk}_{\mathbb{Z}}(\mathrm{End}_{\mathbb{C}}(A)) = 4$?

Note that all counterexamples to problem 2 mentioned so far seem to be only possible if $\mathfrak{a}$ is not a prime ideal. Of course there is then

Problem 3

Given an eigenspace $V \subseteq V^+_{new}(\mathfrak{a})$ is there a possibility of constructing a corresponding A geometrically?

There is the suggestion in [16] to try to use the periods of a differential form ω representing a generator of V. One can also bring the periods of the differential 2-form $*\omega$ into play. But these seem to be too few data. Another approach could be to represent A as an emersed surface in $H/\Gamma_o(\mathfrak{a})$.

The example of elliptic curves with complex multiplication over Q(i) seems to suggest this. In section 1.8. we report on some further hints in this direction.

1.7. Here we want to report on a congruence relation for Heckeoperators stressing the correspondence discussed in 1.5. and 1.6. We discuss this phenomenon in the case of a prime ideal p with $N(p) = 257$.

In general there is a homomorphism

$$\varphi : \Gamma_o(p)^{ab} \to (\mathcal{O}/p)^*/\langle 1,-1\rangle$$

defined by

$$\varphi : \begin{pmatrix} \alpha & \beta \\ \gamma & \delta \end{pmatrix} \mapsto \bar{\alpha} .$$

Let $\mathrm{Ker}\,\varphi$ be its kernel and let $\mathrm{Tor}(p)$ be the torsion subgroup of $\Gamma_o(p)^{ab}$. Put

$$\Lambda(p) = \Gamma_o(p)^{ab} / (\mathrm{Ker}\,\varphi + \mathrm{Tor}(p)) .$$

A computation shows that

$$(T(q) - N(q) - 1)\underline{v} \in \mathrm{Ker}\,\varphi$$

for every $\underline{v} \in V^+(p)$. Again a computation shows that $A(p)$ is cyclic of order 4 for a prime ideal p with $N(p) = 257$. This gives

$$4 \mid (a_q - N(q) - 1)$$

if a_q is the corresponding eigenvalue of $T(q)$. The curves E_2, E_3, E_4, E_5 appearing in 1.6. also satisfy the same congruence with a_q replaced by their b_q. This arises from the fact that E_3 is the Klein 4 group as group of Q(i)-rational points.

In each case we found a curve with nontrivial torsiongroup, we were able to prove the corresponding congruence for the Heckeoperator eigenvalues by the above methods.

1.8. The quotient space $\Gamma_o(p)\backslash H$ has two cusps. We have

$$\begin{pmatrix} i & 0 \\ 0 & -i \end{pmatrix} \in \Gamma_o(p) .$$

This implies that the cusps are bounded by a sphere, not by a torus as in the general case. The quotient $\Gamma_o(p)\backslash H$ has a unique compactification which is a closed oriented compact 3-manifold. We have studied this phenomenon in [5].

We use the notation

$$M_p = \overline{\Gamma_o(p)\backslash H} .$$

We have

$$r = p^1(M_p) .$$

By Poincaré duality, we have

$$r = p^2(M_p) .$$

Hence, in a r=1 case, $H_2(M_p,\mathbb{Z}) = \langle F\rangle$ is a free cyclic group, generated by a 2-cycle F. It is natural to look out for a generating 2-cycle which has additional structure.

There is an abstract version M'_p of M_p which is defined in a combinatorial setup. M'_p is a union of cells of dimension 0,1,2,3. This presentation is quite explicit, the details are given in [3]. Helling has proved that M'_p and M_p are homeomorphic:

$$M'_p = M_p .$$

In fact, there is a translation which carries the combinatorial structure form M'_p to M_p. There is a proof of Helling's theorem in [19].

It is not difficult to show for the first few r = 1 cases, for $Np = 137, 233, 277$, that there is an abstract surface $F'_2 \subset M'_p$ which is of genus 2 and which generates H_2. Using the above homeomorphism, one can carry F'_2 to H^3. The result is a surface $F \subset H^3$ which has some additional structure. We have carried out the study in one particular case. Here is the result:

<u>Proposition 1.8.1.</u> Consider $p = (11 - 4i)$, $Np = 137$, and $\Gamma_o(p)$. There is a surface $F \subset H^3$ with the following properties:

(i) $F = F_1 \cup F_2$
is the union of two geodesic surfaces $F_1, F_2 \subset H^3$.

(ii) $\Gamma_o(p)\backslash F$ is a closed surface of genus 2.

(iii) $\Gamma_o(p)\backslash F$ generates H_2, i. e.
$\langle\Gamma_o(p)\backslash F\rangle = H_2(M_p, \mathbb{Z})$.

(iv) The intersection of F_1, F_2 with a fundamental domain of $\Gamma_o(p)$ are Riemann surfaces of genus 0 with 5 holes each.

(v) The surface $\Gamma_o(p)\backslash F$ has 5 edges, corresponding to the above holes, where the adjacent 2-cells intersect under constant angles $\alpha = \frac{\pi}{2}$.

(vi) There is a group $G < \Gamma_o(p)$ such that
$G\backslash F = \Gamma_o(p)\backslash F$,
i. e. certain elements of G identify certain edges on F such that the result is a closed surface.
G has the abstract presentation

$$G = \langle T_1, T_2, T_3, T_4, T_5, T_6 \; ; \; T_1T_2T_3T_4T_1^{-1}T_2^{-1}T_3^{-1}T_4^{-1} \cdot T_5T_6 = 1 \, , \; T_5^2 = T_6^2 = 1\rangle$$

It seems that in our case, H_2 does not have a generator which is a smooth closed geodesic surface. Rather, the above object F seems to be natural. We do not yet know how to relate it to the elliptic curve corresponding to p mentioned in 1.5.

2.1 Let $P = (z,r)$, $Q = (z',r') \in H^3$. The metric (1.1 (1)) defines the hyperbolic distance d:

$$\operatorname{Cos} d(P,Q) = \delta(P,Q)$$

$$\delta(P,Q) := \frac{|z-z'|^2 + r^2 + r'^2}{2rr'} .$$

Let $\Gamma < PSL_2(\mathbb{C})$ be a discrete subgroup. Unless mentioned otherwise, we shall take $\Gamma = PSL_2(\mathbb{Z}[i])$. For $g \in \Gamma$

$$g = \begin{pmatrix} \alpha & \beta \\ \gamma & \delta \end{pmatrix},$$

the expression $\delta(P,gQ)$ is an Hermitian form of the matrix coefficients $\alpha,\beta,\gamma,\delta$:

$$\delta(P,gQ) = \frac{1}{2rr'}\{|-z(\gamma z'+\delta)+ \alpha z' + \beta|^2 + |\alpha-\gamma z|^2 r'^2$$

$$+ |\gamma z'+\delta|^2 r^2 + |\gamma|^2 r^2 r'^2\}.$$

For $P = Q = (0,1)$, this reduces to

$$\delta_0 = \frac{1}{2}\{|\alpha|^2 + |\beta|^2 + |\gamma|^2 + |\delta|^2\}.$$

It is natural to introduce the Poincaré series

$$\vartheta^*(P,Q,t) = \sum_{g\in\Gamma} e^{-t\delta(P,gQ)} .$$

For $P = 0 = (0,1)$, and $\Gamma = PSL_2(\mathbb{Z}[i])$, this reduces to

$$\vartheta^*(t) = \frac{1}{2} \sum_{\substack{\alpha,\beta,\gamma,\delta \in \mathbb{Z}[i] \\ \alpha\delta-\beta\gamma = 1}} e^{-\frac{t}{2}(|\alpha|^2 + |\beta|^2 + |\gamma|^2 + |\delta|^2)}.$$

Notice that $\vartheta^*(P,Q,t)$ has as its Mellin transform

(3) $$H(P,Q,s) = \sum_{g\in\Gamma} \frac{1}{\delta(P,gQ)^{s+1}} .$$

It is not difficult to see that the series (3) converges uniformly on compact sets for $\mathrm{Re}\, s > 1$, and has a singularity for $s = 1$, if $\Gamma \backslash H$ has finite volume.

2.2 We shall now list a few elementary properties of $\vartheta^*(P,Q,t)$.

Proposition 2.2.1

$\vartheta^*(P,Q,t)$ converges uniformly on compact sets for all $t > o$. Moreover, we have

$$\vartheta^*(P,Q,t) = O(e^{-\kappa t}) \quad \text{for} \quad t \to \infty, \quad \text{for some} \quad \kappa > 0.$$

Proposition 2.2.2 We have

$$\int_{\Gamma \backslash H} \vartheta^{*2}(t,P,Q)\,dv(P) \qquad < \infty .$$

Here $dv(P)$ denotes the hyperbolic volume element

$$dv(P) = \frac{dx\,dy\,dr}{r^3} .$$

Proposition 2.2.3 $\vartheta^*(P,Q,t)$ satisfies a hyperbolic differential equation:

$$(4) \qquad L_P\, \vartheta^* = t^2\, \vartheta^*_{tt} + 3t\, \vartheta^*_t - t^2\, \vartheta^* .$$

Here $L_P = L$ denotes the Laplace operator with respect to the metric (1):

$$L = r^2\left(\frac{\partial^2}{\partial x^2} + \frac{\partial^2}{\partial y^2} + \frac{\partial^2}{\partial r^2}\right) - r\frac{\partial}{\partial r}$$

The last term on the right hand side of (4), which makes the right hand side into a differential operator of Bessel type, results from the summation over Γ in the definition of ϑ^*.

Proposition 2.2.4 The function $t^2\vartheta^*(P,Q,t)$ has a positive lim inf and a finite lim sup as $t \to o$.

This proposition holds for arbitrary discrete groups $\Gamma < PSL_2(\mathbb{C})$ such that the hyperbolic volume of $\Gamma \backslash H$ is finite.

Proposition 2.2.5

$$\int_{\Gamma\setminus H} \vartheta^*(P,Q,t)\,dv(Q) = \frac{4\pi}{t} K_1(t) .$$

$K_1(t)$ is the usual modified Bessel function.

2.3 In order to produce more information, we have to make recourse to spectral theory. Let

$$L_2(\Gamma \setminus H^3)$$

denote the space of functions which are invariant under Γ and square integrable over $\Gamma \setminus H^3$. As usual, this space is made into a Hilbert space by defining the inner product

$$\langle f,g\rangle = \int_{\Gamma\setminus H} f\,\bar{g}\,dv(P) \quad \text{for} \quad f,g \in L_2(\Gamma\setminus H).$$

The Laplace operator L is symmetric with respect to this inner product.

$$\langle Lf,g\rangle = \langle f,Lg\rangle \qquad \text{for} \quad f,g \in L_2(\Gamma\setminus H)$$

and f,g in the domain of L. There is a unique extension of L which is a selfadjoint operator on $L_2(\Gamma\setminus H)$.

The relevant Eisenstein series can be defined as follows:

$$E(P,s) = \sum_{(\gamma,\delta) = (1)} \left(\frac{r}{|\gamma z+\delta|^2 + r^2|\gamma|^2}\right)^{1+s} .$$

The series converges for $\mathrm{Re}\ s > 1$.
Using the Hecke representation, one can easily see that $E(P,s)$ can be continued into the complex s-plane. It is a meromorphic function without pc on the imaginary axis and satisfies a functional equation. One has

(5) $$\int_0^T E(P,i\mu)\,\mu d\mu \in L_2(\Gamma\setminus H) .$$

The space $L_2(\Gamma\backslash H)$ has the following decomposition:

$$(6) \qquad L_2(\Gamma\backslash H) = \mathbb{C} \oplus L_2^{cusp} \oplus L_2^{eis}$$

The space L_2^{eis} is the smallest closed subspace containing all funtions (5). The space L_2^{cusp} is the closed space spanned by all eigenfunctions of L:

$$L\, e_n = \lambda_n e_n \quad , \qquad e_n \in L_2(\Gamma\backslash H)$$

$$\lambda_n = -1 - \mu_n^2 \quad , \qquad \mu_n \in \mathbb{R}$$

It turns out that all e_n can be chosen such that they are an orthogonal system of cusp forms. For $f \in L_2(\Gamma\backslash H)$, the decomposition (6) implies a decomposition

$$(7) \qquad f(P) = a_0 + \sum_n a_n e_n(P) + \int_{-\infty}^{+\infty} a(\mu)\; E\;(P,i\mu)d\mu\;.$$

Here the coefficients a_n and $a(\mu)$ are the Fourier coefficients of f. The decomposition (7) converges in the Hilbert norm sense.

If f is in the domain of L, it can be shown that (7) holds also pointwise, and even uniformly on compact sets.

2.4. Using the setup discussed in 2.3, one can establish the following decomposition

Theorem 2.4.1

$$(8) \qquad \vartheta^*(t,P,Q) = \frac{4\pi}{t\;\mathrm{vol}(\Gamma\backslash H)} K_1(t) + \frac{4\pi}{t} \sum_n K_{i\mu_n}(t)\; e_n(P)\; \overline{e_n(Q)}$$

$$+ \frac{1}{8t} \int_{-\infty}^{+\infty} K_{i\mu}(t)\; E(P,i\mu)E(Q,-\,i\mu)d\mu.$$

The theorem amounts to a computation of the Fourier coefficients. Although none of the eigenfunctions e_n is known explicitly, the computation can be carried out using the method of Selberg transform. The computation of the Fourier coefficient for the Eisenstein part is a routine matter.

Theorem 2.4.2

(9) $$\lim_{t\to o} t^2\, \vartheta^*(P,Q,t) = \frac{4\pi}{\mathrm{vol}(\Gamma\backslash H)}$$

Theorem 2.4.3

$$t(\vartheta^* - \frac{4\pi}{\mathrm{vol}(\Gamma\backslash H)}\frac{1}{t^2}) = O(1) \quad \text{for} \quad t \to o$$

Note that the right hand side of (9) is independent of P and Q.

2.5 In 2.1, we introduced the Mellin transform H(P,Q,s) of ϑ^*. The decomposition theorem discussed in 2.3 yields the following result:

Theorem 2.5.1 For

$$H(P,Q,s) = \sum_{g\in\Gamma} \frac{1}{\delta(P,gQ)^{1+s}} \quad , \text{ we have}$$

$$H(P,Q,s) = \frac{\pi}{\mathrm{vol}(\Gamma\backslash H)} \frac{2^s}{\Gamma(s+1)} \Gamma(\frac{s-1}{2})\, \Gamma(\frac{s+1}{2})$$

$$+ \pi \frac{2^s}{\Gamma(s+1)} \sum_n \Gamma(\frac{s-i\mu_n}{2})\, \Gamma(\frac{s+i\mu_n}{2})\, e_n(P)\, \overline{e_n(Q)}$$

$$+ \frac{1}{32} \frac{2^s}{\Gamma(s+1)} \int_{-\infty}^{+\infty} \Gamma(\frac{s-i\mu}{2})\, \Gamma(\frac{s+i\mu}{2})\, E(P,i\mu)\, E(Q,-i\mu)d\mu.$$

This result implies at once:

Theorem 2.5.2 The function H(P,Q,s) has an analytic continuation into the whole complex s-plane. It is a meromorphic function. There is a pole of order 1 at $s = 1$. There are poles at $s = i\mu_n$, $\lambda_n = -1 - \mu_n^2$. The remaining poles are obtained by translating the imaginary axis to the left by an even integer.

2.6 It is natural to consider solutions of the Laplace equation which depend on the distance to a fixed point only:

$$\delta = \delta(P,Q), \quad Q \text{ fixed}$$
$$\varphi = \varphi(\delta)$$
$$L\varphi = \lambda\varphi. \tag{10}$$

The solution of (10) is a hypergeometric function, which in our case happens to be elementary:

$$\varphi_s(\delta) = \frac{1}{\sqrt{\delta^2-1}}\,(\delta + \sqrt{\delta^2-1}\,)^{-s}$$

$$\lambda = s^2-1 .$$

We can use Poincaré summation over this function, obtaining

$$F(P,Q,s) = \sum_{g\in\Gamma} \varphi_s(P,gQ) . \tag{11}$$

Note that F has a singularity for P = Q. The series (11) converges for Re s > 1. We call it a Maaß-Selberg series.
The decomposition theorem yields

Theorem 2.6.1 The Maaß-Selberg series has the expansion

$$\begin{aligned} F(P,Q,s) = {} & \frac{4\pi}{\mathrm{vol}\ \Gamma\backslash H}\,\frac{1}{s^2-1} + \sum_n \frac{4\pi}{s^2+\mu_n^2}\, e_n(P)\,\overline{e_n(Q)} \\ & + \frac{1}{8}\int_{-\infty}^{+\infty} \frac{E(P,i\mu)E(Q,-i\mu)}{s^2+\mu^2}\, d\mu \,. \end{aligned} \tag{12}$$

If s is on the imaginary axis, the path of integration must be shifted such that the two poles lie either both to the left or both to the right.

The expansion (12) converges in the Hilbert norm sense.

Using this result, and the result of 2.5, one can prove

Theorem 2.6.2 The Maaß-Selberg function $F(P,Q,s)$ has an analytic continuation into the whole complex s-plane. It is a meromorphic function with poles at $s = \pm 1$ and $i\mu_n$, $\lambda_n = -1-\mu_n^2$.
$F(P,Q,s)$ satisfies the functional equation

$$F(P,Q,s) = F(P,Q,-s).$$

We can identify the Maaß-Selberg function with another object.

Theorem 2.6.3 The Maaß-Selberg function $F(P,Q,s)$ is the kernel of an integral operator which is, up to a constant factor, the inverse of the operator $-L-\lambda$. Thus $F(P,Q,s)$ is what is known as the resolvent kernel.

2.7 We shall now discuss an application of our results to number theory. Write ϑ^* in the form

$$\vartheta^*(t) = \sum_{n=2}^{\infty} c(n)\, e^{-\frac{nt}{2}}$$

$$c(n) = \# \left\{ \begin{pmatrix} \alpha \\ \beta \\ \gamma \\ \delta \end{pmatrix} \in \mathbb{Z}[i]^4, \quad \begin{matrix} \alpha\delta-\beta\gamma = 1 \\ |\alpha|^2+|\beta|^2+|\gamma|^2+|\delta|^2 = n \end{matrix} \right\}.$$

Using elementary techniques, one can tie up the multiplicities $c(n)$ with objects from the integral theory of quadratic forms.

Lemma: Consider the quadratic forms

$$f = x^2 + y^2 + u^2 + v^2$$

$$g = (n+2)\,\xi^2 + (n-2)\,\eta^2,$$

and the representations of g by f:

$$X = (u,v), \quad u,v \in \mathbb{Z}^4,$$

(13) $$^tXX = \begin{pmatrix} n+2 & 0 \\ 0 & n-2 \end{pmatrix}$$

satisfying the side conditions

$$(14)\qquad \begin{aligned} &u = (u_i),\quad v = (v_i),\\ &u_i \equiv v_{5-i} \bmod 2,\quad i = 1,2,3,4. \end{aligned}$$

$c(n)$ equals the number of representations (13) satisfying (14).

On the other hand, the form f has only one class in its genus. Thus Siegel's main theorem can be used to compute $c(n)$. One has to work out the local densities and invoke Dirichlets formulae for class numbers. The result is

Theorem 2.7.1

$$c(n) = \begin{cases} 0 & \text{for } n \equiv 0 \bmod 8 \\ 32H(4n^2-16) & \text{for } n \equiv 1 \bmod 2 \end{cases}$$

For $n \equiv 0 \bmod 2$, put

$$n^2-4 = 2^{2e+2}\, m^2\, N,\quad N \text{ square-free, } m \text{ odd.}$$

Then

$$c(n) = \begin{cases} 48\,H\,(n^2-4) & \text{for } n \equiv 4 \bmod 8, \\ 0 & \text{for } n \equiv \pm 2 \bmod 8,\ N \equiv -1 \bmod 8 \\ 192\,H\left(\dfrac{n^2-4}{2^{2e+2}}\right) & \text{for } n \equiv \pm 2 \bmod 8,\ N \equiv 3 \bmod 8 \\ 96\,H\left(\dfrac{n^2-4}{2^{2e}}\right) & \text{for } n \equiv \pm 2 \bmod 8,\ N \equiv 1,2 \bmod 4. \end{cases}$$

Here $H(M)$ counts the number of equivalence classes of positiv definite binary quadratic forms $(a,b,c) = ax^2 + bxy + cy^2$ of discriminant $-M = b^2 - 4ac$

Notice that non-primitive forms are included. The form (a,a,a) is counted with weight $\frac{1}{3}$.

Now invoking the Tauberian theorem of Karamata, we deduce from Theorem 2.4.2:

Theorem 2.7.2

$$\sum_{n=2}^{N} c(n) \sim \frac{\pi}{\mathrm{vol}\,\Gamma\setminus H} N^2 .$$

This seems to be a new result on a certain type of sums over class numbers. One can also use Theorem 2.4.3 to produce a weak remainder term.

2.8. In this last section we shall discuss another application of our results.

On the diagonal $P = Q$, the function $\vartheta^*(P,Q,t)$ is not integrable over the fundamental domain of Γ. However, using the theory of theta functions, it is quite easy to isolate the terms which produce divergence of the integral.

Proposition 2.8.1:
The function

$$\vartheta^*_c(P, P, t) = \begin{cases} \vartheta^*(P, P, t) - \frac{4\pi}{t} e^{-t} r^2 & \text{for } r \geq 1 \\ \vartheta^*(P, P, t) & \text{otherwise} \end{cases} \tag{15}$$

is integrable over the standard fundamental domain of Γ. The standard fundamental domain of $\Gamma = PSL_2(\mathbb{Z}[i])$ is the three-dimensional analogue of the standard fundamental domain for $PSL_2(\mathbb{Z})$.

Now, on the one hand, we can integrate the expansion (8) of Theorem 2.4.1. On the other hand, we can evaluate the integrals over the various conjugacy classes of Γ. The result is a trace formula.

Theorem 2.8.2: For $\Gamma = PSL_2(\mathbb{Z}[i])$, we have the following trace formula

$$\begin{aligned} &\frac{2\pi}{t} e^{-t} - \frac{\pi}{t} K_0(t) + \frac{2\pi}{t} K_1(t) + \frac{4\pi}{t} \sum_n K_{i\mu_n}(t) \\ &+ \frac{\pi}{t} \sum_m \left(K_{\rho_m}(t) - \rho_m H_{-1,\rho_m}(t) \right) \\ &= \mathrm{vol}(\Gamma\setminus H)\, e^{-t} + \frac{\pi}{2t} e^{-t} \left(\gamma + 2 \log \frac{2\Gamma(\frac{3}{4})}{\Gamma(\frac{1}{4})} \right) \\ &+ \frac{\pi}{8t} e^{-t} \left\{ E_1(\tfrac{t}{32}) - E_1(\tfrac{t}{4}) + E_1(\tfrac{t}{2}) - E_1(t) \right\} + \frac{2\pi}{t} e^{t} E_1(2t) \\ &+ \end{aligned} \tag{16}$$

$$+ \frac{5\pi}{8t} e^{-t} \log 2 + \frac{4\pi}{9t} e^{-t} \log \varepsilon$$

$$+ \sum_{\substack{\{Y \text{ primitive}\}_\Gamma \\ |N(Y)| \neq \varepsilon}} \sum_{n=1}^{\infty} \frac{8\pi \log |N(Y)|}{t\,|N(Y)^n - N(Y)^{-n}|^2} e^{-\frac{t}{2}(|N(Y)|^{2n} + |N(Y)|^{-2n})}$$

$$+ \sum_{n=1}^{\infty} \frac{8\pi \log \varepsilon}{3t(\varepsilon^n - \varepsilon^{-n})^2} e^{-\frac{t}{2}(\varepsilon^{2n}+\varepsilon^{-2n})} + \sum_{n=1}^{\infty} \frac{16\pi \log \varepsilon}{3t(\varepsilon^{2n}+\varepsilon^{-2n} - \zeta^{2n} - \zeta^{-2n})} e^{-\frac{t}{2}(\varepsilon^{2n}+\varepsilon^{-2n}) \cdot \varepsilon^{-2n}} .$$

Here we use the following notation. ρ_m runs through the zeros of the function $\zeta_k(s)$, $k = \mathbb{Q}(i)$, in the critical strip $0 < \operatorname{Re} s < 1$. The zeros are counted with their multiplicities. The function $H_{-1,\rho}(t)$ is an associated Bessel function, in the notation of [20], p. 112. The number γ is the Euler constant. The function $E_1(t)$ is the exponential integral function, see [21], p. 342. A hyperbolic or loxodromic element $Y \in \Gamma$ is called primitive, if it is not the power of another element of Γ. The summation over Y runs over a set of representatives of Γ-conjugacy classes of hyperbolic and loxodromic primitive elements. In $G = PSL_2(\mathbb{C})$, such an element $Y \in \Gamma$ is conjugate to $\begin{pmatrix} \xi & 0 \\ 0 & \xi^{-1} \end{pmatrix}$, $|\xi| > 1$, $\operatorname{Re} \xi > 0$. Put $N(Y) = \xi$. The number $\varepsilon = 2 + \sqrt{3}$ is the basic unit in $\mathbb{Q}(\sqrt{3})$. $\zeta = \frac{1}{2} + \frac{i}{2}\sqrt{3}$ is a third root of unity.

<u>Remark:</u> Comparing the formula (16) with Selberg's formula in the case of Fuchsian groups [22], p. 74 and p. 78, or with Tanigawa's formula [23], or with Venkov's formula [24], we see that all these formulae are equivalent to our explicit formula. They differ from our formula by an appropriate integral transform, which in our case happens to be a Lebedev transform, see [21], p. 398.
Our method undoubtedly applies in more general situations, but we have not yet studied generalisations.

References

[1] Elstrodt, J.; Grunewald, F.; Mennicke, J.:
Spectral theory of the Laplacian on 3-dimensional hyperbolic space and number theoretic applications.
To appear

[2] Gelbart, S.:
Elliptic curves and automorphic representations.
Advances in Mathematics 21, 235-292 (1976).

[3] Grunewald, F,; Helling, H.; Mennicke, J.:
$SL_2(O)$ over complex quadratic numberfields I.
Algebra i Logica 17 512 - 580 (1978).
(= Algebra and Logic 17, 332 - 382 (1978).)

[4] Grunewald, F.; Mennicke, J.:
$SL_2(O)$ and elliptic curves.
To appear

[5] Grunewald, F.; Mennicke, J.:
Some 3-manifolds arising from $PSL_2(\mathbb{Z}[i])$.
Archiv für Mathematik, 35, 275-291 (1980).

[6] Grunewald, F.; Mennicke, J.:
The ζ - function of the curves $y^2 = x^5 + Dx$.
To appear.

[7] Gushoff, A.C.; Mennicke, J.; Grunewald, F.
Komplex-quadratische Zahlkörper kleiner Diskriminante und Pflasterungen des dreidimensionalen hyperbolischen Raumes.
To appear.

[8] Harder, G.:
On the cohomology of $SL_2(\mathcal{O})$.
In: Lie groups and their representations.
Edited by I.M. Gelfand. London, Hilger, 1975.

[9] Harder, G.:
Period integrals of cohomology classes which are represented by Eisenstein series.
Preprint 1980

[10] Hecke, E.:
Über ein Fundamentalproblem aus der Theorie der elliptischen Modulfunktionen.
Abh. Math. Sem. Univ. Hamburg **6**, 235-257 (1928) (= Mathematische Werke, 525-547).

[11] Jacquet, H.; Langlands, R.P.:
Automorphic forms on GL(2).
Springer LNM 114 (1970).

[12] Miyake, T.:
On automorphic forms on GL_2 and Hecke operators.
Annals of Math., **94**, 174-189 (1971).

[13] Serre, J.P.:
Le problème des groupes de congruence pour SL_2.
Annals of Math. **92** (1970), 489-527.

[14] Serre, J.P.:
Facteurs locaux des fonctions zéta des variétés algébriques.
Séminaire Delange - Pisot - Poitou (1969/1970).

[15] Shimura, G.:
Introduction to the arithmetic theory of automorphic funtions.
Iwanami Shoten and Princeton Univ. Press (1971).

[16] Weil, A.:
Zeta - functions and Mellin transforms.
Proc. of the Bombay Coll. on Algebràic Geometry, pp. 409-426, Bombay 1968 (= Oeuvres Scientifiques, Vol.III, pp. 179-196).

[17] Weil, A.:
On a certain type of characters of the idèle-class group of an algebraic numberfield.
Proc. Intern. Symb. on Algebraic Number Theory, pp. 1-7, Tokyo-Nikko 1955 (= Oeuvres Scientifiques, Vol.II, pp. 255-261).

[18] Weil, A.:
Dirichlet series and automorphic forms.
Springer LNM 189

[19] F. Grunewald, J. Mennicke:
$SL_2(o)$ over complex quadratic number fields II, (forthcoming).

[20] Y. Luke:
Integrals of Bessel functions,
New York, 1962

[21] W. Magnus, F. Oberhettinger, R.P. Soni:
Formulas and Theorems for the Special Functions of Mathematical Physics, 3rd edition, New York, 1966

[22] A. Selberg:
Harmonic analysis and discontinuous groups in weakly symmetric Riemannian spaces with applications to Dirichlet series,
Journ. Indian Math. Soc. 20 (1956), p. 47-87

[23] Yoshio Tanigawa:
Selberg trace formula for Picard groups,
Proceed. Int. Symp. Algebraic number theory, Tokyo, 1977, p. 229-242

[24] A.B. Venkov:
Expansion in automorphic eigenfunctions of the Laplace-Beltrami operator in classical symmetrie spaces of rank one, and the Selberg trace formula.
Proc. Steklov Inst. Math. 125 (1973), p. 1-48.

J. Elstrodt
Mathematisches Institut der
Universität Münster
Roxeler Str. 64
4400 Münster

F. Grunewald
Sonderforschungsbereich "Theoretische Mathematik" der
Universität Bonn
Beringstr. 4
5300 Bonn

J. Mennicke
Fakultät für Mathematik der
Universität Bielefeld
Universitätsstr. 1
4800 Bielefeld 1

SUITES A FAIBLE DISCREPANCE EN DIMENSION s

H. Faure

1 *INTRODUCTION.*

1.1 Le tore à s dimensions $\mathbb{T}^s$ est identifié au cube unité $[0,1[^s$; un pavé P de $\mathbb{T}^s$ est le produit de s intervalles $[a_k, b_k[$ de $[0,1[$: $P = \prod_{k=1}^{s} [a_k, b_k[$; son volume est $|P| = \prod_{k=1}^{s} (b_k - a_k)$.

Etant donnée une suite $X = (X_n)_n$ de points de $\mathbb{T}^s$, on note A(P, N, X) le nombre de termes de la suite X, d'indices inférieurs à N, qui appartiennent à P ; on pose alors :

$E(P, N, X) = A(P, N, X) - |P|N$, puis

$D(N, X) = \sup_{P \in \mathcal{P}_s} |E(P, N, X)|$ et

$D^*(N, X) = \sup_{P \in \mathcal{P}_s^*} |E(P, N, X)|$,

où $\mathcal{P}_s$ est l'ensemble des pavés de $\mathbb{T}^s$ et $\mathcal{P}_s^*$ l'ensemble des pavés de la forme $\prod_{k=1}^{s} [0,b_k[$.

Les fonctions D et D* sont respectivement la discrépance et la discrépance à l'origine de la suite X ; elles sont reliées l'une à l'autre par les inégalités :

$D^* \leqslant D \leqslant 2^s D^*$.

Dans la suite, l'entier s est toujours supposé au moins égal à 2.

1.2 Le comportement asymptotique des fonctions D et D* quand N augmente indéfiniment a été étudié par de nombreux auteurs :

K.F. Roth a d'abord établi une minoration générale [4]:

Il existe une constante K_s telle que pour toute suite infinie on ait, pour une infinité de N :

$D^*(N) \geqslant K_s (\operatorname{Log} N)^{s/2}$.

Puis J.H. Halton a construit des suites infinies en dimension s, généralisant la suite de Van Der Corput, pour lesquelles il obtient les majorations [1]:

$$D^*(N) \leqslant A_s(\mathrm{Log}\ N)^s \text{ avec } \mathrm{Log}\ A_s = O\ (s\ \mathrm{Log}\ s).$$

Ensuite I.M. Sobol′ a obtenu de nouvelles suites infinies, basées sur le système de numération binaire, qui vérifient [5]:

$$D^*(N) \leqslant B_s(\mathrm{Log}\ N)^s \quad \text{avec} \quad \mathrm{Log}\ B_s = O(s\ \mathrm{Log}\ \mathrm{Log}\ s).$$

Tous ces résultats figurent dans un article très détaillé de H.Niederreiter [3] qui fait le point sur la question et qui contient une bibliographie complète jusqu'en 1977.

1.3 En travaillant avec les systèmes de numération en base quelconque, nous avons construit des suites pour lesquelles nous obtenons les majorations suivantes :

$$D^*(N) \leqslant C_s(\mathrm{Log}\ N)^s \quad \text{avec} \quad C_s = 0(1).$$

Ces suites ont les plus faibles discrépances actuellement connues (voir la table en annexe). Remarquons qu'on ne sait rien pour l'instant de l'ordre exact de $D^*(N)$ pour $s \geqslant 2$.

2 *DEFINITIONS ET RESULTATS.*

2.1 Soit r un entier au moins égal à s (pratiquement r sera toujours un nombre premier).

On appelle *pavé élémentaire en base* r un pavé de la forme $\prod_{k=1}^{s} \left[\frac{u_k}{r^{p_k}}, \frac{u_k+1}{r^{p_k}} \right[$, avec u_k et p_k entiers positifs ou nuls, et

$u_k < r^{p_k}$ pour tout k.

Soit m un entier positif ou nul et $X = (X_1,\ldots,X_{r^m})$ une suite de r^m points de $\mathbb{T}^s$; on dit que X est une *suite de type* $P^m_{r,s}$ si tout pavé élémentaire en base r de volume r^{-m} contient un terme et un seul de la suite X.

Soit $X = (X_n)_{n \geqslant 1}$ une suite infinie dans $\mathbb{T}^s$; on dit que X est une *suite de type* $P_{r,s}$ si, quelquesoient m et ℓ entier positifs ou nuls, la suite finie $X_m^{\ell} = (X_{\ell r^m + 1}, \ldots, X_{(\ell+1)r^m})$ est une suite de type $P_{r,s}^m$.

2.2 *Principaux résultats.*

Théorème 1.

Quels que soient les entiers $s \geq 2$ *et* r *premier au moins égal à* s, *il existe des suites de type* $P_{r,s}$.

De telles suites, notées $S_{R_s}^r$, seront définies au paragraphe 3 ci-dessous.

Théorème 2.

Quels que soient les entiers $s \geq 2$, r *premier au moins égal à* $s-1$ *et* $m \geq 0$, *il existe des suites de type* $P_{r,s}^m$.

Ces suites finies s'obtiennent aisement à partir des suites de type $P_{r,s}$.

Remarques :

Les suites de type $P_{2,2}$ sont les LP_0- suites de $\mathbb{T}^2$ introduites par I.M. Sobol' [5], suites également étudiées par S. Srinivasan [6].

Les suites de type $P_{2,2}^m$ et $P_{2,3}^m$ sont les P_0- réseaux de $\mathbb{T}^2$ et $\mathbb{T}^3$ considérés par Sobol'.

Théorème 3.

(i) *Pour toute suite* X *de type* $P_{2,2}$, *on a la majoration :*

$$D^*(N,X) \leq \frac{3}{16(\operatorname{Log} 2)^2}(\operatorname{Log} N)^2 + O(\operatorname{Log} N) \text{ pour tout } N \geq 1.$$

(ii) *Soient* $s \geq 2$ *et* r *impair (premier) au moins égal à* s ; *pour toute suite* X *de type* $P_{r,s}$ *on a la majoration :*

$$D^*(N,X) \leq \frac{1}{s!}\left(\frac{r-1}{2 \operatorname{Log} r}\right)^s (\operatorname{Log} N)^s + O((\operatorname{Log} N)^{s-1}) \text{ pour tout } N \geq 1.$$

Remarques :

Les suites de type $P_{2,2}$ sont les LP_0-suites de $\mathbb{T}^2$ pour lesquelles Sobol' obtient le majorant $\frac{1}{2(\operatorname{Log} 2)^2} = 1,04\ldots$, alors que

$\frac{3}{16(\text{Log } 2)^2} = 0,39\dots.$

Soit q_s le premier nombre premier au moins égal à s ; alors le majorant $C_s = \frac{1}{s!}\left(\frac{q_s - 1}{2 \,\text{Log}\, q_s}\right)^s$ tend vers 0 quand s tend vers l'infini ; les suites de type $P_{r,s}$ sont d'autant meilleures que s est grand alors que les majorants associés aux suites de Halton et aux LP_τ- suites de Sobol' tendent vers l'infini avec s.

En adaptant la méthode qui permet de remplacer (r-1) par $\frac{r-1}{2}$ dans les majorations, on peut diviser par 2^s la constante obtenue par Meijer [2] pour les suites de Halton $\phi_{g_1,\dots,g_s}$; on obtient ainsi :

$$D^*(N,\phi_{g_1,\dots,g_s}) \leqslant (\text{Log } N)^s \prod_{k=1}^{s} \left(\frac{g_k - 1}{2\text{Log } g_k}\right) + O((\text{Log } N)^{s-1}).$$

Les démonstrations des théorèmes ci dessus ainsi que d'autres résultats annexes seront publiées dans un article à paraître prochainement.

3 *DEFINITION DES SUITES* $S^r_{R_s}$.

3.1 Soit $C = (\binom{n}{p})$ la matrice des coefficients binomiaux (matrice triangulaire d'ordre infini).

Pour tout entier $t \geqslant 1$, on a $C^t = (t^{n-p}\binom{n}{p})$; en particulier $C^t = I \pmod t$, I étant la matrice unité.

3.2 Soit r un entier au moins égal à 2, et soit A_r l'ensemble des réels de la forme $k\,r^{-n}$ avec $n \geqslant 1$ et $0 \leqslant k < r^n$; si $x \in A_r$ s'écrit $x = \sum_{j=0}^{\infty} x_j \, r^{-j-1}$ (avec les x_j tous nuls sauf un nombre fini), on définit $y = Cx \pmod r$ par $y = \sum_{j=0}^{\infty} y_j \, r^{-j-1}$ avec $y_j = \sum_{i \geqslant j} \binom{i}{j} x_i \pmod r$.

3.3 Soit alors $n \geqslant 1$ et $n-1 = \sum_{j=0}^{\infty} a_j(n) r^j$ l'écriture de (n-1)en base r ; posons $x_n^1 = \sum_{j=0}^{\infty} a_j(n) r^{-j-1}$, puis $x_n^k = C^{k-1} x_n^1 \pmod r$ pour $1 \leqslant k \leqslant r$.

Si s est un entier compris entre 2 et r, soient $r_1,\dots,r_s$, s entiers compris entre 1 et r ; posons $R_s = (r_1,\dots, r_s)$; nous définissons alors la suite $S^r_{R_s}$ à termes dans $\mathbb{T}^s$ par :

$$S^r_{R_s}(n)=(x_n^{r_1},\ldots,x_n^{r_s}).$$

En particulier si s = r et $r_k = k$, on obtient la suite de terme général $S^r(n)=(x_n^1,\ldots,x_n^r)$, et dans le cas r = 2, n retrouve une LP_0-suite étudiée par Sobol' (6.4 [5]).

La démonstration du théorème 1 se ramène à montrer qu'un déterminant, généralisant le déterminant de Van Der Monde, n'est pas nul modulo r. Le théorème 2 se déduit facilement du théorème 1. Pour le théorème 3, on distingue les cas r = 2 et r impair ; ce dernier cas repose sur une méthode de réduction par recurrence portant la longueur de la séquence X_m^ℓ (cf. 2.1) et sur la dimension s.

ANNEXE

Comparaison des constantes A_s, B_s *et* C_s.

Pour les *suites de Halton* $\phi_{p_1,\ldots,p_s}$ où p_k est le k-ième nombre premier, on : $A_s = \prod_{k=1}^{s} \left(\frac{p_k - 1}{2 \operatorname{Log} p_k}\right)$ (voir la dernière remarque suivant le théorème 3).

Pour les *suites* LP_{τ_s} *de Sobol'*, on a $B_s = \frac{2^{\tau_s}}{s!(\operatorname{Log} 2)^s}$, où τ_s est un entier dont l'ordre de grandeur est compris entre $\frac{s \operatorname{Log} s}{\operatorname{Log} \operatorname{Log} s}$ et $s \operatorname{Log} s$ (4.5 [5]); les premières valeurs sont les suivantes :

s	2	3	4	5	6	7	8	9	10	11	12	13
τ_s	0	1	3	5	8	11	15	19	23	27	31	35

Pour les *suites de type* $P_{q_s,s}$, avec q_s premier nombre premier supérieur ou égal à s on a :

$$C_2 = \frac{3}{16(\operatorname{Log} 2)^2} \quad \text{et} \quad C_s = \frac{1}{s!} \left(\frac{q_s - 1}{2 \operatorname{Log} q_s}\right)^s \text{ pour } s \geqslant 3.$$

On en déduit le tableau suivant :

s	2	3	4	5	6	7	8	13
A_s	.65	.81	1.25	2.62	6.13	17.3	52.9	90580
B_s	1.04	1.00	1.44	1.66	3.20	5.28	15.2	647
C_s	.39	.12	.099	.024	.018	.0041	.0088	.000010

BIBLIOGRAPHIE

[1] HALTON (J.H.).- On the efficiency of certain quasi-random sequences of points in evaluating multi-dimensional integrals.
Numerische Mathematik, 2, p. 84 - 90, 1960.

[2] MEIJER (H.G.).- The discrepancy of a g-adic sequence, *Indag - Math.*, 30, P. 54 - 66, 1968.

[3] NIEDERREITER (H.).- Quasi Monte-Carlo methods and pseudo-random numbers, *Bull. of A.M.S.*, 84, n° 6 , p. 957 - 1041, 1978.

[4] ROTH (K.F.).- Irregularities of distribution, *Mathematika*, 1, part. 2, p. 73 - 79, 1954.

[5] SOBOL' (I.M.).- On the distribution of points in a cube and the approximate evaluation of integrals, *U.S.S.R. Computational Math. and Math. Physics*, 7, n° 4 , p. 86 - 112, 1967.
Voir la bibliographie de [3] pour les autres publications de I.M. SOBOL'.

[6] SRINIVASAN (S.).- On two dimensional Hammersley's sequences, *J. of. Number Theory*, 10, n° 4 , p. 421 - 429, 1978.

CANONICAL DIVISIBILITIES OF VALUES OF p-ADIC L-FUNCTIONS

By Georges GRAS

1. INTRODUCTION AND GENERALITIES

The classical theory of p-adic L-functions of abelian fields may be regarded as a purely analytic one (p-adically) ; if so, one obtains results of the kind that we recall later (Res. 1 to 5), which constitute a synthesis of congruence properties of Bernoulli numbers. However, this theory contains important arithmetic aspects, which are independant of the p-adic ones, and due for instance to the fact that the values at $s=0$ and 1 of these functions are connected with the order and the structure of certain groups attached to the arithmetic of abelian fields . We recall the main facts :

1. 1 Arithmetic aspects

Analytic formulas. We write $u \approx v$ when uv^{-1} is a p-adic unit.

If k is an imaginary abelian field, and if $\mathcal{H}(k)$ denotes the p-group of relative classes of k, we have $|\mathcal{H}(k)| \approx p^{m_0(k)} \prod_\alpha \frac{1}{2} B_1(\alpha^{-1})$, where α runs over all odd primitive characters of k, $p^{m_0(k)} \approx Q(k)\,|\mu(k)|$ ($Q(k)=1$ or 2 is the unit index of k, $\mu(k)$ is the group of roots of unity contained in k). Recall the relation which expresses Bernoulli numbers in terms of the L_p-functions :

$$-\frac{1}{2} B_1(\alpha^{-1})\left(1-\alpha^{-1}(p)\right) = \frac{1}{2} L_p(0,\psi), \quad \psi = \theta\alpha^{-1} \qquad (1.1)$$

where θ is the Teichmüller character mod $q=4$ or p according as $p=2$ or not.

If k is real, let $\mathcal{T}(k)$ be the $\mathbb{Z}_p$-torsion module of the Galois group of the maximal abelian p-ramified p-extension of k ; then (see [C], App. I, and [G5], th. 2. 1) we have $|\mathcal{T}(k)| \approx$

$p^{n_o(k)} \prod_{\psi \neq 1} \frac{1}{2} L_p(1, \psi)$, where ψ runs over all non trivial primitive characters of k, $p^{n_o(k)} = [k \cap \mathbb{Q}_\infty : \mathbb{Q}]$, where $\mathbb{Q}_\infty$ is the cyclotomic $\mathbb{Z}_p$-extension of $\mathbb{Q}$.

Conjectures on abelian fields. Beyond these relations, we have the "main conjecture" which expresses a remarkable link between L_p-functions and the structure of certain modules of Iwasawa's theory (for a statement see [C], 5) ; it is connected with the fact that partial products of Bernoulli numbers (resp. of L_p-functions at $s = 1$), corresponding to irreducible p-adic characters ϕ, express the order of suitable groups depending only on ϕ : $\mathfrak{H}(\phi)$ in the odd case, $\mathcal{T}(\phi)$ in the even one (for the definitions, see [G1], I, 2, and [G5], 3, 5 ; see also 2. 1 and 3. 1 in §§ 2, 3). More precisely :

Conjecture 1. 1. If ϕ is odd (and if, for any $\alpha | \phi$, $\psi = \theta\alpha^{-1}$ is not a character of $\mathbb{Q}_\infty$), we have $|\mathfrak{H}(\phi)| \approx p^{m_o(\phi)} \prod_{\alpha|\phi} \frac{1}{2} B_1(\alpha^{-1})$, where $m_o(\phi) = 0$ except in the case $p = 2$ and α is of p-power order, in which case $m_o(\phi) = 1$.

Conjecture 1. 2. If ϕ is even (and not a character of $\mathbb{Q}_\infty$), we have $|\mathcal{T}(\phi)| \approx \prod_{\psi|\phi} \frac{1}{2} L_p(1, \psi)$.

1. 2 Analytic results

To simplify, we consider only primitive characters. For such a character β, we call : v_β the valuation, of the ring $\mathbb{Z}_p(\beta)$ of values of β, such that $v_\beta(\mathbb{Q}_p(\beta)^\times) = \mathbb{Z}$, β_p (resp. β_o) the component of p-power order (resp. of order prime to p) of β, K_β the field fixed by the kernel of β, and f_β the conductor of β. If c is a rational prime to p, we put $c\theta^{-1}(c) = 1 + qp^{n(c)}u$, $(u, p) = 1$.

Let ψ be an even character , and let s be any element of $\mathbb{Z}_p$:

Result 1. 1. If $\psi \neq 1$, then $v_\psi\left(\frac{1}{2} L_p(s, \psi)\right) \geq 0$ except if ψ is a character of $\mathbb{Q}_\infty$, in which case $v_\psi\left(\frac{1}{2} L_p(s, \psi)\right) = -1$.

Result 1. 2. (see [R], 4, [G3], IV, cor. 2). If $\psi_o \neq 1$ then a N. S. C. to have $v_\psi\left(\frac{1}{2} L_p(s, \psi)\right) \geq 1$ is : $v_{\psi_o}\left(\frac{1}{2} L_p(s, \psi_o)\right) \geq 0$ or there exists a prime number $\ell \neq p$, $\ell | f_\psi$, such that $\psi_o(\ell) = 1$.

Result 1.3. (see [R], 5, [G3], cor. to prop. V4). If ψ is of p-power order then a N. S. C. to have $v_\psi\left(\frac{1}{2} L_p(s,\psi)\right) = 0$ is that $\sum_{\ell \neq p} p^{n(\ell)} = 1$ $(\ell | f_\psi, \ell \neq p)$.

Now, introduce the invariant $\lambda(\psi)$ (see [R], 5, or [G3], prop. V 2) :

Result 1.4. (see [G3], prop. V 2, V 3, V 4). We have $\lambda(\psi) = \lambda(\psi_o) + \sum_{\ell \neq p} p^{n(\ell)} \left(\ell | f_\psi, \ell \neq p, \psi_o(\ell) = 1\right)$; $\lambda(\psi_o) \geq 0$ if $\psi_o \neq 1$, $\lambda(1) = -1$. We have $\lambda(\psi) \geq 1$ if and only if $v_\psi\left(\frac{1}{2} L_p(s,\psi)\right) \geq 1$.

Result 1.5. (see [Gr]). If $\psi_o \neq 1$, and if $\psi_o(2) = 1$, then $v_\psi\left(\frac{1}{2} L_2(s,\psi)\right) \geq 1$.

Then it is rather easy to see that these results express, analyticaly, some phenomena of "genera theory". The purpose of this talk is, precisely, to show how the use of arithmetic arguments permits to us to find (or conjecture) stronger analytic properties of L_p-functions, mainly canonical divisibilities in $\mathbb{Z}_p(\psi)$, when ψ is of order divisible by p. We do not know how to obtain these divisibilities by analytic methods in full generality ; the problem of knowing if it is logically possible or not remains open (notice that the two approaches are very distinct : the "analytic" side is based on the study of explicit p-adic measures ("of Stickelberger"), with, in addition, a very deep property of "regularity" implied by Ferrero - Washington's result ; the "arithmetic" side is based on class field theory, on the complex analytic formula of class numbers, as well as the fact that Leopoldt's conjecture is true for abelian fields). Of course, in the area of the theory developed, at the present time, by Mazur and Wiles, for the "main conjecture" in the abelian case of composite conductor, such divisibilities will probably appear, in so far as this theory contains, in some sense, an "explicit" form of class field theory.

In all the sequel, ψ is an even primitive character of order divisible by p ; moreover we suppose that ψ is not a character of $\mathbb{Q}_\infty$ because in this case everything is known.

2. CANONICAL DIVISIBILITIES AT s = 0

Define $\alpha = \theta \psi^{-1}$ and let ϕ (resp. φ) be the p-adic character above α (resp. α_o). Let L be the field K_α and let K be its subfield of index p.

The norm map $N_{L/K}$ induces an exact sequence (where $C\ell$ denotes the class group in ordinary sense) :

$$1 \to \mathcal{H}(\phi) \to C\ell(L)^{\varphi} \to C\ell(K)^{\varphi} \to 1 \qquad (2.1)$$

Next, we have the formula of φ-classes fixed by Gal (L/K) ([G2], th. II 1) :

$$|C\ell_1(L/K)^{\varphi}| = |C\ell(K)^{\varphi}|\, u_{\varphi} p^{\varphi(1)(-r+\Sigma D(\ell))} \qquad (2.2)$$
$$\left(\ell | f_\alpha ,\ \alpha_o(\ell)=1\right)$$

where $u_\varphi = 1$ for $p \neq 2$, $u_\varphi = |\left(E(K) \cap N_{L/K} L^{\times} / E(K)^2\right)^{\varphi}|$ for $p=2$, E(K) is the group of units of K ; $D(\ell)$ is the degree over $\mathbb{Q}$ of the splitting field of ℓ in the p-subextension of $L/\mathbb{Q}$, and $r=0$ except in the case $\psi_o = 1$, in which case $p^r = 1$ or p is the index $\left(\mu(K) : \mu(K) \cap N_{L/K} L^{\times}\right)$.

As $|C\ell_1(L/K)^{\varphi}|$ is a divisor of $|C\ell(L)^{\varphi}|$, it follows, from 2.1 and 2.2, that $p^{\varphi(1)(-r+\Sigma D(\ell))}$ is a divisor of $|\mathcal{H}(\phi)|$ (in the case $p=2$, we neglect the index u_φ which does not have any canonical divisor) ; then we have the following general conjecture which is a direct consequence of conjecture 1.1 :

<u>Conjecture 2.1</u>. Let ψ be even and not a character of $\mathbb{Q}_\infty$; let α be the primitive character $\theta\psi^{-1}$. Then $v_\alpha\left(\frac{1}{2}B_1(\alpha^{-1})\right) \geq -\delta - r + \sum_{\ell} D(\ell)$ $\left(\ell | f_\alpha ,\ \alpha_o(\ell)=1\right)$, where $\delta = 0$ except in the case $p=2$ and $\psi_o = 1$, in which case $\delta = 1$.

By 1.2, this gives a lower bound for $v_\psi\left(\frac{1}{2}L_p(0,\psi)\right)$.

In some cases, conjecture 1.1 is proved, and conjecture 2.1 is then true ; for example :

<u>Theorem 2.1</u>. When the p-adic character above α_o is rational over $\mathbb{Q}$, then $v_\alpha\left(\frac{1}{2}B_1(\alpha^{-1})\right) \geq -\delta - r + \sum_{\ell} D(\ell)$ $\left(\ell | f_\alpha ,\ \alpha_o(\ell)=1\right)$.

3. CANONICAL DIVISIBILITIES AT s = 1

Now we call ϕ (resp. φ) the p-adic character above ψ (resp. ψ_o).

The principle is then the same as for the odd case ; it is based on the analogue, for the groups $\mathcal{T}$, of the formula of invariant φ-classes. The properties of the groups $\mathcal{T}$ have been studied in [G5] where we have shown the existence of a norm map $N_{L/K}$ and an extension map $j_{L/K}$ (in any extension L/K of real abelian fields) which have, formally, the same properties as the analogous maps for class groups ; especially, as with relative classes, $j_{L/K}$ is injective (this fact needs the exactness of Leopoldt's conjecture for abelian fields). If we apply this to L/K with $L = K_\psi$ and K subfield of index p in L, we obtain the exact sequence :

$$1 \to \mathcal{T}(\emptyset) \to \mathcal{T}(L)^\varphi \to \mathcal{T}(K)^\varphi \to 1 \qquad (3.1)$$

and the formula ([G5], th. 3.2 and 3.3) :

$$|\mathcal{T}_1(L/K)^\varphi| = |\mathcal{T}(K)^\varphi| \, p^{\varphi(1)(\rho - 1 + \Sigma D(\ell))} \qquad (3.2)$$
$$\left(\ell | f_\psi, \ \ell \neq p, \ \psi_o(\ell) = 1\right)$$

where $D(\ell)$ is the degree over $\mathbb{Q}$ of the splitting field of ℓ in the p-subextension of $L/\mathbb{Q}$, and $\rho = 0$ or 1 is defined by : $\rho = 1$ if $\psi_o \neq 1$, and, for $\psi_o = 1$, $\rho = \mathrm{Min}\left(n_o(k) + 1, \ldots, n(\ell) + f(\ell), \ldots\right) - n_o(k)$ $\left(\ell | f_\psi, \ \ell \neq p\right)$, where $p^{f(\ell)}$ is the residual degree of ℓ in $L/\mathbb{Q}$.

Then we remark that $|\mathcal{T}_1(L/K)^\varphi|$ is a divisor of $|\mathcal{T}(L)^\varphi|$, and, consequently, by 3.1 and 3.2, $p^{\varphi(1)(\rho - 1 + \Sigma D(\ell))}$ is a divisor of $|\mathcal{T}(\emptyset)|$; thus we have the following general conjecture which is a direct consequence of conjecture 1.2 :

<u>Conjecture 3.1</u>. Let ψ be even and not a character of $\mathbb{Q}_\infty$. Then

$$v_\psi\left(\tfrac{1}{2} L_p(1, \psi)\right) \geq \rho - 1 + \sum_{\ell \neq p} D(\ell) \quad \left(\ell | f_\psi, \ \ell \neq p, \ \psi_o(\ell) = 1\right).$$

A particular and interesting case where conjecture 1.2 is trivial is the case $\psi_o = 1$ (it is also true when φ is rational over $\mathbb{Q}$) :

<u>Theorem 3.1</u>. Let ψ be even of p-power order and not a character of $\mathbb{Q}_\infty$, then $v_\psi\left(\frac{1}{2} L_p(1, \psi)\right) \geq \rho - 1 + \sum_{\ell \neq p} D(\ell) \left(\ell | f_\psi, \ \ell \neq p\right)$.

4. REMARKS ON THE VALUATION OF $\frac{1}{2} L_p(s, \psi)$

It is easy to find examples where the canonical divisibilities give, or, on the contrary, do not give the exact valuation of

$\frac{1}{2} L_p(s, \psi)$ at $s = 0$ or 1:

Example 4. 1. Consider the case $p = 2$, with ψ of order 2 and $s = 1$: An elementary computation, involving the 2-adic analytic formula of class number, shows that for all integers m of the form $m = 4^t u^2 \pm 1$, $t \geq 1$, u odd, we have, for the character ψ such that $K_\psi = \mathbb{Q}(\sqrt{m})$: $v_\psi\left(\frac{1}{2} L_2(1, \psi)\right) = t - 1 + v_\psi(h)$, where h is the class number of $\mathbb{Q}(\sqrt{m})$. But for such a character, th. 3. 1 gives $v_\psi\left(\frac{1}{2} L_2(1, \psi)\right) \geq \rho - 1 + \sum_{\ell \neq 2} 1 \left(\ell \mid f_\psi, \ell \neq 2\right)$, and, in general, equality does not occur (especially as soon as t is large enough).

However there exists a particular and remarkable situation for which the divisibilities of conjectures 2. 1 and 3. 1 are verified, and in some sense, the best possible ; the proof which needs Ferrero-Washington's result is analytic, and this gives a partial answer for the problem raised in the introduction :

Example 4. 2. Replace ψ by the character $\psi\gamma_n$, $n \geq 0$, where γ_n is a character of order p^n of $\mathbb{Q}_\infty$; we know that for n large enough $v_{\psi\gamma_n}\left(\frac{1}{2} L_p(s, \psi\gamma_n)\right) = \lambda(\psi)$ is independant of n and $s \in \mathbb{Z}_p$ and that we have $\lambda(\psi) = \lambda(\psi_0) + \sum_{\ell \neq p} p^{n(\ell)} \left(\ell \mid f_\psi, \ell \neq p, \psi_0(\ell) = 1\right)$ (Res. 1. 4). We will compare this expression of $\lambda(\psi)$ with the values predicted by conjectures 2. 1 and 3. 1. Before, we remark that for n large enough, $D(\ell) = p^{n(\ell)}$ for $\ell \neq p$; further $\theta(\ell) = 1$ for all $\ell \mid f_\psi$, $\ell \neq p$.

(i) Divisibility at $s = 0$. Here we have $\alpha = \theta\, \psi^{-1} \gamma_n^{-1}$, $L = K_\alpha$ and K subfield of index p in L.

Case $\psi_0 \neq 1$. Then $\delta = r = 0$, and conjecture 2. 1 predicts $\lambda(\psi) \geq \sum_\ell D(\ell)$, thus $\lambda(\psi_0) + \sum_{\ell \neq p} D(\ell) \geq \sum_\ell D(\ell) \left(\ell \mid f_\alpha, \alpha_0(\ell) = 1\right)$; then we see the appearance of the following phenomena : if $\alpha_0(p) = 1$, we obtain $\lambda(\psi_0) \geq D(p) = 1$; show that this is not a contradiction and for this verify directly that $\lambda(\psi_0) \geq 1$: for $p \neq 2$, 1. 1 implies $L_p(0, \psi_0) = 0$, and $\lambda(\psi_0) \geq 1$ (see Res. 1. 4) ; for $p = 2$, it is the result of [Gr] recalled in Res. 1. 5, which can also be shown arithmetically : by 1. 1, we have $\frac{1}{2} L_2(0, \psi_0) = \frac{-1}{2} B_1(\theta\, \psi_0)$ (because $\alpha_0 = \psi_0^{-1}$ here) ; consider the extension L'/K', $L' = K_{\theta\psi_0}$, $K' = K_{\psi_0}$; by 2. 2 and the fact that 2 splits in K', we see that

$|Cl_1(L'/K')^{\varphi}|\ |Cl(K')^{\varphi}|^{-1}$ is an even integer, therefore the quotient $Cl(L')^{\varphi}/j_{L'/K'}\,Cl(K')^{\varphi}$ is not trivial ; then it is sufficient to apply th. 1 of [G4] which implies that the valuation of $\frac{1}{2}B_1(\theta\psi_0)$ is not zero ; thus $\lambda(\psi_0) \geq 1$.

Case $\psi_0 = 1$. If $p=2$, $r=1$; if $p \neq 2$, as ψ is not a character of $\mathbb{Q}_\infty$, there exists $\ell \neq p$ totally ramified in $L/\mathbb{Q}(\mu(K))$, and an easy computation shows that, for n large enough, $\mu(K)$ is not contained in the group of local norms at ℓ (in L/K), therefore, $r=1$. Then the lower bound given by conjecture 2. 1 is $-\delta-1+\sum_\ell D(\ell)$ $\left(\ell \mid f_\alpha,\ \alpha_0(\ell)=1\right)$: for $p\neq 2$, $\delta=0$; as $\alpha_0=\theta$ here, p is excluded from the summation, and the lower bound is $\lambda(\psi)$, because $\lambda(\psi_0)=-1$ here. If $p=2$, then $\delta=1$ and $\alpha_0=1$; therefore the term $D(p)=1$ appears in the summation and the lower bound is still $\lambda(\psi)$.

(ii) Divisibility at $s=1$. Here we have $L=K_{\psi\gamma_n}$, and K subfield of index p in L.

Case $\psi_0 \neq 1$. We have $\rho=1$, $\lambda(\psi_0)\geq 0$, and the expression for $\lambda(\psi)$ proves immediately conjecture 3. 1 for the characters $\psi\gamma_n$ (n large enough).

Case $\psi_0=1$. An elementary computation shows that ρ (corresponding to $\psi\gamma_n$) equals 0 for n large enough ; th. 3. 1 implies $\lambda(\psi) \geq -1+\sum_{\ell\neq p} p^{n(\ell)}\left(\ell \mid f_\psi,\ \ell\neq p\right)$, but $\lambda(\psi_0)=-1$, and the lower bound is still an equality :

<u>Theorem 4. 1.</u> When n is large enough, conjectures 2. 1 and 3. 1 corresponding to the characters $\psi\gamma_n$ are true.

In conclusion, we point out the following problem, justified by the study of the above example ; this will be studied in another paper :

<u>problem</u>. Let ψ be even and not a character of $\mathbb{Q}_\infty$. Is it true that $v_\psi\left(\frac{1}{2}L_p(s,\psi)\right)$ is bounded, independantly of $s\in\mathbb{Z}_p$, by $\rho-1+\sum_{\ell\neq p} D(\ell)\left(\ell \mid f_\psi,\ \ell\neq p,\ \psi_0(\ell)=1\right)$?

5. THE CASE OF p-ADIC ZETA FUNCTION

We will prove a canonical divisibility for its residue at $s=1$; it is not based on those (conjectured) for L_p-functions, but uses a direct proof. It is interesting to note that the divisibility obtained is exactely what would give the application of conjecture 3. 1.

If k is a real abelian field, $\zeta_p(s,k) = \prod_\psi L_p(s,\psi)$, where ψ runs over all primitive characters of k ; the residue at s = 1 is given by:

$$\mathrm{Res}_p(k) = \left(1 - \frac{1}{p}\right) 2^{[k:\mathbb{Q}]-1} \prod_{\psi \neq 1} \frac{1}{2} L_p(1,\psi) \qquad (5.1)$$

Now introduce the groups $\mathcal{T}(\chi)$, relative to the irreducible rational characters of k, for which we have the formula ([G5], th. 3. 1) : $|\mathcal{T}(\chi)| \approx p^{n_0(\chi)} \prod_{\psi|\chi} \frac{1}{2} L_p(1,\psi)$, where $n_0(\chi) = 0$ except in the case χ character of $\mathbb{Q}_\infty$, in which case $n_0(\chi) = 1$. We have $\prod_{\psi \neq 1} \frac{1}{2} L_p(1,\psi) \approx \prod_{\chi \neq 1} |\mathcal{T}(\chi)|\, p^{-n_0(\chi)}$, thus :

$$\prod_{\psi \neq 1} \frac{1}{2} L_p(1,\psi) \approx p^{-n_0(k)} \prod_{\chi \neq 1} |\mathcal{T}(\chi)| \qquad (5.2)$$

Decompose k in the form $k = k' k_0$ ($[k':\mathbb{Q}]$ p-power, $[k_0:\mathbb{Q}]$ prime to p) ; for all subfields L of k cyclic over k_0, we call K its subfield of index p, when $L \neq k_0$. There exists a semi-simple decomposition of the form $\mathcal{T}(L) = \oplus\, \mathcal{T}(L)^{\chi_0}$, χ_0 running over all rational characters of k_0 ; if $L \neq k_0^{\chi_0}$, and if $\mathcal{T}^*(L/K)$ denotes the kernel of $N_{L/K} : \mathcal{T}(L) \to \mathcal{T}(K)$, we have the exact sequence $1 \to \mathcal{T}^*(L/K)^{\chi_0} \to \mathcal{T}(L)^{\chi_0} \to \mathcal{T}(K)^{\chi_0} \to 1$, and it is easy to verify that $\mathcal{T}^*(L/K)^{\chi_0} = \mathcal{T}(\chi_0^L)$, where χ_0^L is the character corresponding to the field $(k' \cap L) K_{\chi_0}$. Then we have, for each $L \neq k_0$: $|\mathcal{T}(L)|\,|\mathcal{T}(K)|^{-1} = \prod_{\chi_0} |\mathcal{T}(\chi_0^L)|$; therefore $\prod_{\chi \neq 1} |\mathcal{T}(\chi)| = \prod_{\chi_0 \neq 1} |\mathcal{T}(\chi_0)| \prod_{L \neq k_0} |\mathcal{T}(L)|\,|\mathcal{T}(K)|^{-1}$. Now we use the global formula giving $|\mathcal{T}_1(L/K)|$ ([G5], Rem. 3. 5) : $|\mathcal{T}_1(L/K)| = |\mathcal{T}(K)|\, p^{\rho(L) - 1 + t(L)}$, where $\rho(L) = 0$ or 1 is computed in $k' \cap L/\mathbb{Q}$ with reference to the unit character, and where t(L) is the number of prime ideals of K which ramify in L/K and do not divide p. Then

$|\mathcal{T}(L)|\,|\mathcal{T}(K)|^{-1}$ is a multiple of $p^{\rho(L)-1+t(L)}$, and by 5.1 and 5.2, we obtain (where v is the usual valuation on $\mathbb{Z}_p$) :

Theorem 5.1. Let k be a real abelian field, and let k_o be its maximal subfield of degree prime to p. Then the residue at s = 1 of $\zeta_p(s,k)$ verifies : $v\left(\mathrm{Res}_p(k)\right) \geq -1 - n_o(k) + \sum_L \left(\rho(L) - 1 + t(L)\right) + ([k:\mathbb{Q}] - 1)\,v(2)$, where L runs over all the subfields of k cyclic over k_o and distinct from k_o, $n_o(k)$ is defined by $p^{n_o(k)} = [k \cap \mathbb{Q}_\infty : \mathbb{Q}]$, $\rho(L) = 0$ or 1, and t(L) is the number of prime ideals of K (subfield of index p in L) which ramify in L/K and do not divide p.

Remark. Res. 1.5 allows us to add, to the right member of the above inequality, the term $(d-1)\,v(2)$, where d is the degree of the splitting field of 2 in k_o.

REFERENCES

[C] J. COATES, p-adic L-functions and Iwasawa's theory, Proceedings of Durham Symposium (1975), Academic Press, 1977, p. 269-353.

[G1] G. GRAS, Etude d'invariants relatifs aux groupes des classes des corps abéliens, Astérisque, 41-42, 1977, p. 35-53.

[G2] G. GRAS, Nombre de φ-classes invariantes, application aux classes des corps abéliens, Bull. Soc. Math. France, 106, 1978, p. 337-364.

[G3] G. GRAS, Sur la construction des fonctions L p-adiques abéliennes, Sém. Delange-Pisot-Poitou, 20e année, n° 22, 1978/79.

[G4] G. GRAS, Sur l'annulation en 2 des classes relatives des corps abéliens, C. R. Math. Rep. Acad. Sci. Canada, 1, n° 2, 1979, p. 107-110.

[G5] G. GRAS, Module de torsion de la p-extension abélienne p-ramifiée maximale d'un corps abélien réel et fonctions L p-adiques (à paraitre).

[Gr] R. GREENBERG, On 2-adic L-functions and cyclotomic invariants, Math. Z., t. 159, 1978, p. 37-45.

[R] K. RIBET, p-adic L-functions attached to characters of p-power order, Sém. Delange-Pisot-Poitou, 19e année, 1977/78, n° 9.

Georges GRAS
Université de Franche-Comté - Faculté des Sciences
Mathématiques, E. R. A. au C. N. R. S. n° 070654
F - 25030 Besançon Cedex

MINIMAL ADDITIVE BASES AND RELATED PROBLEMS

George P. Grekos
Department of Mathematics
University of Crete
Iraklio, Crete, Greece

1. MINIMAL ASYMPTOTIC BASES

Let h be a positive integer. A subset B of the set $\mathbf{N}$ of nonnegative integers is said to be an ***asymptotic basis of order*** h if all but finitely many natural numbers are sums of h, not necessarily distinct, elements of B. An asymptotic basis B of order h is called ***minimal*** if no proper subset of B is an asymptotic basis of order h. Minimal additive bases have been studied by Erdös, Härtter, Nathanson, and Stöhr [3, 4, 6, 7, 9-12, 14, 15, 17, 20].

2. THE ORDER OF A MINIMAL BASIS

Erdös and Nathanson [12] raised the question whether it is possible for a set A to be simultaneously a minimal asymptotic basis of two different orders. If the set A is a minimal asymptotic basis of order h and an asymptotic basis of order $k < h$, then A is certainly a minimal asymptotic basis of order k. It is proved in [12] that if a set A is an asymptotic basis of order 2, then A cannot be a minimal asymptotic basis of order 4; in other words, there is no minimal asymptotic basis of orders 2, 3 and 4. We prove the following result.

THEOREM. ***There is no minimal asymptotic basis of orders*** 2 ***and*** 3.

This means that a minimal asymptotic basis of order 2 is not a minimal asymptotic basis of any order $k > 2$. To prove the theorem we use the following lemma.

LEMMA. ***Let*** A ***be a set of nonnegative integers such that***
(A.1) 0 ***belongs to*** A, ***and***
(A.2) ***for any two elements*** a, a' ***of*** A ***with*** $a > a' > 0$, ***we have that*** $a - a'$ ***does not belong to*** A.
If A ***is an asymptotic basis of order*** 2, ***then for every nonzero element*** a^* ***of*** A ***the set*** $A \setminus \{a^*\}$ ***is an asymptotic basis of order*** 3.

Notation. If A, B are nonempty sets of integers, let $A + B$

denote the set consisting of all sums of the form $a+b$ where $a \in A$ and $b \in B$. If B has a single element, say $B = \{b\}$, we write $A+b$ instead of $A+\{b\}$. We also define $2A = A+A$ and, for any integer $h > 2$, $hA = A+(h-1)A$.

Proof of the lemma. Let a^* be a positive element of A. We shall show that the set

$$B = A \setminus \{a^*\}$$

is an asymptotic basis of order 3. Since 0 belongs to B, we have that $2B$ is a subset of $3B$. Hence, it suffices to show that all but finitely many elements of the set

$$C = \mathbf{N} \setminus (2B)$$

belong to $3B$. If C is finite, there is nothing to prove. Let

$$C = \{c_1, c_2, \ldots, c_i, \ldots\}$$

where

$$c_1 < c_2 < \ldots < c_i < \ldots .$$

Since A is an asymptotic basis of order 2, every sufficiently large c_i belongs to $2A$. But c_i does not belong to $2B$. Hence, for all $i > i_1$, we have that c_i is of the form

$$c_i = a^{(i)} + a^*$$

where $a^{(i)} \in A$, $a^{(i)} > 0$.

Now, if c_i is large enough, we can choose an element b of B with $0 < b < c_i$ and such that $c_i - b > c_{i_1}$. If $c_i - b$ belongs to C, then $c_i - b = c_j$ for an index j, that is

$$a^{(i)} + a^* - b = a^{(j)} + a^* ,$$

or $a^{(i)} = b + a^{(j)}$, where $a^{(i)}$, b, $a^{(j)}$ are nonzero elements of A. But this is not true. So $c_i - b$ is not in C. The complement of C in $\mathbf{N}$ is $2B$; therefore

$$c_i - b = b_1 + b_2$$

where b_1, b_2 are elements of B. It follows that

$$c_i = b + b_1 + b_2 \in 3B$$

for all large i, and B is an asymptotic basis of order 3. Thus the lemma is proved.

Proof of the theorem. Let A be an asymptotic basis of order 2. We suppose that A is also a minimal asymptotic basis of order 3 and we shall arrive at a contradiction.

Let us first show that, without loss of generality, we can assume that the smallest element of A is zero. Consider a positve integer k, an integer n and a set X of integers. One can easily see that

$$k(X+n) = kX + kn .$$

Clearly X is an asymptotic basis of order k if and only if $X+n$ is an asymptotic basis of order k. Moreover, for any $x \in X$,

$$k((X+n) \setminus \{x+n\}) = k(X \setminus \{x\}) + kn .$$

We conclude that X is a minimal asymptotic basis of order k if and only if $X+n$ is a minimal asymptotic basis of order k. It becomes clear now that if 0 does not belong to A, one can consider equivalently the set $A' = A + (-\min A)$ which contains zero. So *we assume that zero belongs to* A.

The set A satisfies also property (A.2) of the lemma. To prove this, let a, a' be two elements of A such that $a > a' > 0$. Suppose that $a - a'$ belongs to A. The set A is a minimal asymptotic basis of order 3, so there are infinitely many natural numbers $n_1 < n_2 < \ldots$ such that

$$n_i \notin 3(A \setminus \{a\}) , \qquad (i = 1,2,\ldots) .$$

But A is an asymptotic basis of order 2; therefore for all $i > i_2$, $n_i \in 2A$. Consider an index $i_3 \geq i_2$ such that $n_{i_3} > 2a$. For all $i > i_3$ we have that

$$n_i = a + a^{(i)}$$

where $a^{(i)} \in A \setminus \{a\}$. It follows that

$$n_i = a' + (a-a') + a^{(i)} \in 3(A \setminus \{a\})$$

for all $i > i_3$. This contradiction proves that property (A.2) holds.

Thus, the set A satisfies all hypotheses of the lemma; hence, it is not a minimal asymptotic basis of order 3 . This contradiction proves the theorem.

3. MAXIMAL ASYMPTOTIC NONBASES

Nathanson [17] introduced the dual concept of minimal basis, that of maximal nonbasis. If the set A is not an asymptotic basis of order h, but, for every nonnegative integer $b \notin A$, the set $A \cup \{b\}$ is an asymptotic basis of order h, then we say that A is a ***maximal asymptotic nonbasis of order*** h. Maximal nonbases have been studied by Erdös, Hennefeld, Nathanson, and Turjányi [5-8, 10, 16-18, 21, 22].

Let A be a set of nonnegative integers and x a positive real number. We denote by $A(x)$ the number of positive elements of A not exceeding x. In a previous paper [1], written in common with J.-M. Deshouillers, we constructed, for every natural number $h \geq 2$, a class of maximal asymptotic nonbases of order h having the smallest possible density: each set A in this class satisfies

$$A(x) = \mathcal{O}(x^{1/h}) .$$

4. COMBINATORIAL ANALOGUES

Let $\mathcal{F}$ denote the collection of all finite subsets of $\mathbb{N}$. It has been proved (see, for instance, [2]) that many properties of the semigroup $(\mathbb{N}, +)$ hold also in $(\mathcal{F}, \cup)$ and in $(\mathcal{F}, \cap)$. We now state some results concerning union and intersection bases for $\mathcal{F}$.

4.1 Union bases

A subcollection $\mathcal{B}$ of $\mathcal{F}$ is called an ***asymptotic union basis of order*** h if all but finitely many sets in $\mathcal{F}$ are unions of h not necessarily distinct sets of $\mathcal{B}$; otherwise, $\mathcal{B}$ is an ***asymptotic union nonbasis of order*** h. An asymptotic union basis $\mathcal{B}$ of order h is called ***minimal*** if every proper subcollection of $\mathcal{B}$ is an asymptotic union nonbasis of order h. An asymptotic union nonbasis $\mathcal{B}$ of order h is called ***maximal*** if every subcollection of $\mathcal{F}$ that contains properly $\mathcal{B}$ is an asymptotic union basis of order h.

For each $h \geq 2$, Nathanson [19, p.223] constructed 'trivial' examples of minimal asymptotic union bases of order h: let

$T_1, T_2, \ldots, T_h$ be a partition of $\mathbb{N}$ into h nonempty sets at least two of which are infinite; then the collection of all finite subsets of the T_i's $(1 \le i \le h)$ is a minimal asymptotic union basis of order h. He also constructed 'nontrivial' minimal asymptotic union bases of order 2.

We proved [13] that, for any $h \ge 2$, there are no maximal asymptotic union nonbases of order h.

4.2 Intersection bases

A subcollection $\mathcal{B}$ of $\mathcal{F}$ is called an ***asymptotic intersection basis of order*** h if all but finitely many sets in $\mathcal{F}$ can be represented as the intersection of h not necessarily distinct sets in $\mathcal{B}$. Otherwise, $\mathcal{B}$ is an ***asymptotic intersection nonbasis of order*** h [19, p.229].

If $\mathcal{B}$ is an asymptotic intersection basis of order h then, for every B belonging to $\mathcal{B}$, $\mathcal{B} \setminus \{B\}$ is an asymptotic intersection basis of order h too. We conclude that ***there are neither minimal asymptotic intersection bases nor maximal asymptotic intersection nonbases of any order*** h.

5. OPEN PROBLEMS

5.1 The question whether there exist minimal asymptotic bases of two different orders h and k remains open, when $\min(h,k) \ge 3$.

5.2 Is there a minimal asymptotic basis A of order h satisfying the minimal density condition: $A(x) = O(x^{1/h})$?

5.3 Are there 'nontrivial' minimal asymptotic union bases of order $h \ge 3$? This is a question of Nathanson [19].

5.4 Can we find more general sufficient conditions on the structure of a commutative semigroup $(M, *)$ that would guarantee the existence or the nonexistence of minimal bases or maximal nonbases?

For other open problems we refer to the papers of Erdös and Nathanson, for instance [11] or [19].

REFERENCES

1. J.-M. Deshouillers et G. Grekos, Propriétés extrémales de bases additives, Bull.Soc.Math. France 107 (1979), 319-335.

2. M. Deza et P. Erdos, Extension de quelques théorèmes sur les densités de séries d'éléments de $\mathbb{N}$ à des séries de sous-ensembles finis de $\mathbb{N}$, Discrete Math. 12 (1975), 295-308.

3. P. Erdös, Einige Bemerkungen zur Arbeit von A. Stöhr 'Gelöste und ungelöste Fragen über Basen der natürlichen Zahlenreihe', J. Reine Angew.Math. 197 (1957), 216-219.

4. P. Erdös and E. Härtter, Konstruktion von nichtperiodischen Minimalbasen mit der Dichte 1/2 für die Menge der nichtnegativen

ganzen Zahlen, J. Reine Angew.Math. 221 (1966), 44-47.

5. P. Erdös and M.B. Nathanson, Maximal asymptotic nonbases, Proc. Amer.Math.Soc. 48 (1975), 57-60.

6. P. Erdös and M.B. Nathanson, Oscillations of bases for the natural numbers, Proc.Amer.Math.Soc. 53 (1975), 253-258.

7. P. Erdös and M.B. Nathanson, Partitions of the natural numbers into infinitely oscillating bases and nonbases, Comment.Math.Helvet. 51 (1976), 171-182.

8. P. Erdös and M.B. Nathanson, Nonbases of density zero not contained in maximal nonbases, J. London Math.Soc. (2), 15 (1977), 403-405.

9. P. Erdös and M.B. Nathanson, Sets of natural numbers with no minimal asymptotic bases, Proc.Amer.Math.Soc. 70 (1978), 100-102.

10. P. Erdös and M.B. Nathanson, Bases and nonbases of square-free integers, J. Number Theory 11 (1979), 197-208.

11. P. Erdös and M.B. Nathanson, Systems of distinct representatives and minimal bases in additive number theory, in: *Number Theory, Carbondale 1979*, Springer-Verlag Lecture Notes in Mathematics, Vol. 751, 1979, pp.89-107.

12. P. Erdös and M.B. Nathanson, Minimal asymptotic bases for the natural numbers, J. Number Theory, to appear.

13. G. Grekos, Nonexistence of maximal asymptotic union nonbases, Discrete Math., to appear.

14. E. Härtter, Ein Beitrag zur Theorie der Minimalbasen, J. Reine Angew.Math. 196 (1956), 170-204.

15. E. Härtter, Eine Bemerkung über periodische Minimalbasen für die Menge der nichtnegativen ganzen Zahlen, J. Reine Angew.Math. 214/215 (1964), 395-398.

16. J. Hennefeld, Asymptotic nonbases which are not subsets of maximal asymptotic nonbases, Proc.Amer.Math.Soc. 62 (1977), 23-24.

17. M.B. Nathanson, Minimal bases and maximal nonbases in additive number theory, J. Number Theory 6 (1974), 324-333.

18. M.B. Nathanson, s-maximal nonbases of density zero, J. London Math.Soc. (2), 15 (1977), 29-34.

19. M.B. Nathanson, Oscillations of bases in number theory and combinatorics, in: *Number Theory Day, New York 1976*, Springer-Verlag Lecture Notes in Mathematics, Vol.626, 1977, pp.217-231.

20. A. Stöhr, Gelöste und ungelöste Fragen über Basen der natürlichen Zahlenreihe, I, II, J. Reine Angew.Math. 194 (1955), 40-65, 111-140.

21. S. Turjányi, On maximal asymptotic nonbases of zero density, J. Number Theory 9 (1977), 271-275.

22. S. Turjányi, Note on maximal asymptotic nonbases of zero density, Publ.Math. Debrecen 26 (1979), 229-236.

MEAN VALUES FOR FOURIER COEFFICIENTS OF CUSP FORMS AND SUMS OF KLOOSTERMAN SUMS

Henryk Iwaniec
Mathematics Institute
Polish Academy of Sciences
ul. Śniadeckich 8, Warszawa, Poland

In this paper we prove an analogue of the large sieve inequality for Fourier coefficients both of holomorphic and of real-analytic cusp forms in respect of the modular group $SL(2,\mathbb{Z})$. The results will be applied for estimating trilinear forms of Kloosterman sums $S(n,m;c)$ over coefficients n,m and over modul c counted with a smooth weight function. For such sums it is shown that the Linnik-Selberg conjecture holds on average.

1. PRELIMINARIES

The group $\Gamma = PSL(2,\mathbb{Z})$ acts on the upper half-plane $H = \{z;\ z = x + iy,\ y > 0\}$ as linear fractional transformations

$$\gamma z = \frac{az + b}{cz + d}$$

for $\gamma = \begin{pmatrix} a & b \\ c & d \end{pmatrix} \in \Gamma$, $ad - bc = 1$, a,b,c,d-rational integers. Let M_k denote the space of holomorphic cusp forms for Γ of weight $k \geq 2$, $k \equiv 0 \pmod 2$. M_k is a Hilbert space with inner product given by

$$\langle f,g\rangle = \iint_{\mathcal{D}} f(z)\overline{g(z)}\, y^k dz$$

where $\mathcal{D} = \{z \in H;\ \frac{1}{2} \leq x \leq \frac{1}{2},\ |z| \geq 1\}$ is a fundamental domain and $dz = y^{-2}dxdy$ is Γ-invariant measure on H. It is known that M_k is finite dimensional generated by Poincaré series

$$P_m(z;k) = \frac{(4\pi m)^{k-1}}{\Gamma(k-1)} \sum_{\gamma \in \Gamma_\infty\backslash\Gamma} (cz + d)^{-k} e(m\gamma z), \quad m \geq 1$$

where Γ_∞ is the cyclic group generated by $\gamma z = z + 1$, the stabilizer of the cusp ∞. For $k = 4, 6, 8, 10$ and 14, $P_m(z;k)$ are identically equal to zero, thus M_k are empty. For $k = 12$ and $k \geq 16$ we have

$$\nu_k = \dim M_k = \begin{cases} [\frac{k}{12}] & \text{if } k \not\equiv 2 \pmod{12} \\ [\frac{k}{12}] - 1 & \text{if } k \equiv 2 \pmod{12} \end{cases}$$

and the first Poincaré series $P_m(z;k)$, $1 \le m \le \nu_k$ form a basis for M_k .

Every cusp form $f(z)$ has Fourier expansion around ∞

$$f(z) = \sum_{n=1}^{\infty} a_n e(nz) \quad .$$

The problem of estimating the coefficients a_n received great attention. In case of

$$P_m(z,k) = \sum_{n=1}^{\infty} p_n(m,k)e(nz)$$

it is known that (Petersson)

$$p_n(m,k) = \frac{(4\pi\sqrt{mn})^{k-1}}{\Gamma(k-1)} \{\delta_{nm} + 2\pi i^k \sum_{c=1}^{\infty} \frac{1}{c} S(n,m;c) J_{k-1}(\frac{4\pi\sqrt{mn}}{c})\} \tag{1}$$

where $J_{k-1}(x)$ is Bessel's function of order $k-1$, therefore the problem can be reduced to estimating sums of Kloosterman sums. By A. Weil's result

$$|S(n,m;c)| \le (n,m,c)^{\frac{1}{2}} \tau(c) c^{\frac{1}{2}} \tag{2}$$

it follows that $a_n \ll n^{\frac{k}{2} - \frac{1}{4} + \varepsilon}$ while the famous conjecture of Petersson proved recently by P. Deligne [2] says that

$$a_n \ll \tau(n) n^{\frac{k-1}{2}} \quad . \tag{3}$$

This shows that there is a great cancellation of terms in (1). At the Stockholm Congress Yu.V. Linnik [8] stated

<u>CONJECTURE</u>. For $T > (mn)^{\frac{1}{2}+\varepsilon}$ we have

$$\sum_{c \le T} \frac{1}{c} S(n,m;c) \ll T^{\varepsilon} \quad . \tag{4}$$

Shortly afterwards A. Selberg [13] extended the conjecture to all $T > (m,n)^{\frac{1}{2}+\varepsilon}$. In a first attack on (4) one is tempted to combine Deligne's (3) with (1) but one fails. The point is that the Bessel functions $J_{k-1}(x)$ of integral order are not sufficient to generate nice weights like the characteristic functions of intervals for example.

Very recently N.V. Kuznietsov [7] obtained the striking result

$$\sum_{c \le T} \frac{1}{c} S(n,m;c) \ll_{n,m} T^{\frac{1}{6}} (\log T)^{\frac{2}{3}} \quad . \tag{5}$$

This is one among many important consequences of his formula (see Lemma 2) which connects sums of Kloosterman sums weighted by Bessel's functions of imaginary order with Fourier coefficients of non-holomorphic cusp forms.

The theory of real-analytic cusp forms has been originated by A. Selberg [14] and H. Maass [9]. Let $L^2(\Gamma\backslash H)$ be the space of all Γ-invariant functions on H, square-integrable on $\mathcal{D}$ in $dz = y^2 dxdy$ measure. The Γ-invariant Laplacian

$$L = -y^2\left(\frac{\partial^2}{\partial x^2} + \frac{\partial^2}{\partial y^2}\right)$$

has a self-adjoint extension into $L^2(\Gamma\backslash H)$, it has a point spectrum $\lambda_0 = 0$, $1/4 < \lambda_1 \le \lambda_2 \le \dots$, $\lambda_j \sim j/12$ and it has a continuous spectrum. Constant functions have L-eigenvalue $\lambda_0 = 0$. The eigenfunctions $u_j(z)$ with eigenvalues $\lambda_j > 1/4$ are called cusp forms or Maass wave forms. They have Fourier-Bessel expansions

$$u_j(z) = \sqrt{y} \sum_{n\neq 0} \rho_j(n) K_{i\kappa_j}(2\pi|n|y)e(nx) \tag{6}$$

where $\lambda_j = 1/4 + \kappa_j^2$ and $\rho_j(n)$ depend on j and n only. The eigenfunctions of the continuous spectrum consist of Eisenstein series

$$E(z,s) = \sum_{\gamma\in\Gamma_\infty\backslash\Gamma} (\text{Im } \gamma z)^s \tag{7}$$

on the line $s = \frac{1}{2} + it$. The analytic continuation of $E(z,s)$ on $s = \frac{1}{2} + it$ is given by Fourier-Bessel expansion

$$\xi(s)E(z,s) = \xi(s)y^s + \xi(1-s)y^{1-s} + 2\sqrt{y} \sum_{n\neq 0} |n|^{\frac{1}{2}-s}\sigma_{2s-1}(n)$$

$$K_{s-\frac{1}{2}}(2\pi|n|y)e(nx) \tag{8}$$

where $\xi(s) = \pi^{-s}\Gamma(s)\zeta(2s)$, $\sigma_\nu(n) = \sum_{d|n} d^\nu$ and $K_\nu(x)$ is the modified Bessel function.

Usually the basis $\{u_j\}$ is chosen so that u_j are eigenfunctions of all the Hecke operators

$$T_n f(z) = \frac{1}{\sqrt{n}} \sum_{ad=n} \sum_{0<b\le d} f\left(\frac{az+b}{d}\right)$$

and of

$$T_{-1}f(z) = f(-\bar{z}) \quad .$$

This is possible because L, T_n, $n = \pm 1,2,3,\dots$ mutually commute and are hermetian. Letting $T_n u_j = \tau_j(n) u_j$ for $n = 1,2,\ldots$ and $T_{-1} u_j = \varepsilon_j u_j$ with $\varepsilon_j = \pm 1$ one gets $\rho_j(n) = \rho_j(1)\tau_j(n)$ and $\rho_j(-n) = \varepsilon_j \rho_j(n)$. The eigenvalues $\tau_j(n)$ are multiplicative. More precisely we have

$$Z_j(s) := \sum_{n=1} \tau_j(n) n^{-s} = \prod_p \left(1 - \frac{\tau_j(p)}{p^s} + \frac{1}{p^{2s}}\right)^{-1} .$$

It is conjectured that

$$|\tau_j(n)| \leq \tau(n) \quad . \tag{9}$$

The above statement is not yet proved. Several results are known for mean values of a different kind. In [7] Kuznietsov proved

$$\sum_{\kappa_j \leq X} \frac{|\rho_j(n)|^2}{\operatorname{ch} \pi \kappa_j} = \frac{X^2}{\pi^2} + O(X \log X + n^{\varepsilon} X + n^{\varepsilon + \frac{1}{2}})$$

which constitutes an improvement of some results of Bruggeman [1]. The sharpest estimate of $\tau_j(n)$ which occurred in print is due to N.V. Proskurin [12]

$$|\tau_j(n)| \leq \tau(n) n^{\frac{1}{4}} \quad .$$

2. STATEMENT OF RESULTS

Since T_n are symmetric $\tau_j(n)$ are real, positive or negative infinitely often. A number of questions arise here about variation of the sign of $\tau_j(n)$. In connection with the Riemann conjecture for $Z_j(s)$ (that complex zeros lie on $\operatorname{Re} s = \frac{1}{2}$) it is important to estimate sums over primes like

$$\sum_{p \leq X} \tau_j(p) p^{it} \quad . \tag{10}$$

We shall show that (10) is $\ll X^{\frac{1}{2}+\varepsilon}$ for almost all j. Precisely, we prove

THEOREM 1. *Let* $K \geq 1$, $N \geq 1$ *and let* a_n *be complex numbers. Then for any* $\varepsilon > 0$

$$\sum_{\kappa_j \leq K} \frac{1}{\operatorname{ch} \pi \kappa_j} \left| \sum_{N < n \leq 2N} a_n \rho_j(n) \right|^2 \ll (K^2 + N^{1+\varepsilon}) \sum_{N < n \leq 2N} |a_n|^2 \tag{11}$$

the constant implied in the symbol $\ll$ *depending on* ε *alone.*

A similar inequality will be proved for Fourier coefficients of holomorphic cusp forms. Let $f_{1,k}, \ldots, f_{\nu_k,k}$ be an orthonormal basis of M_k with Fourier expansion

$$f_{j,k}(z) = \sum_{n=1}^{\infty} \psi_{j,k}(n)e(nz) \quad .$$

THEOREM 2. *Under the assumptions of Theorem* 1 *we have*

$$\sum_{k\equiv 0 (\mathrm{mod}\ 2)} e^{-\frac{k-1}{K}} \frac{(k-1)!}{(4\pi)^{k-1}} \sum_{1\le j\le \nu_k} \Big| \sum_{N<n\le 2N} a_n n^{-\frac{k-1}{2}} \psi_{j,k}(n)\Big|^2$$
$$= (\tfrac{1}{2}K^2 + O(N^{1+\varepsilon})) \sum_{N<n\le 2N} |a_n|^2 \quad . \tag{13}$$

Both (11) and (13) remind one of the large sieve inequality for Dirichlet's characters. A corresponding result for the Fourier coefficients $n^{-ir}\sigma_{2ir}(n)$ of Eisenstein series is just a kind of mean-value theorem for Dirichlet polynomials (see [10]).

THEOREM 3. *Under the assumptions of Theorem* 1 *we have*

$$\int_0^K \Big| \sum_{N<n\le 2N} a_n n^{-ir}\sigma_{2ir}(n)\Big|^2 dr \ll (K^2 + N^{1+\varepsilon}) \sum_{N<n\le 2N} |a_n|^2 \tag{14}$$

Next we apply Theorems 1, 2 and 3 for estimating trilinear forms of Kloosterman sums

$$G^{\pm}(N,M,C) = \sum_{N<n\le 2N} \sum_{M<m\le 2M} \sum_{C<c\le 2C} a_n b_m g(n,m,c) S(n,\pm m;c)$$

with a weight function $g(n,m,c)$ having the properties

$$\mathrm{Supp}\ g(n,m,c) \subset [N,2N]\times[M,2M]\times[C,2C] \quad , \tag{15}$$

$g(n,m,c)$ is of C^2 class such that for $0 \le q_1,q_2,q_3 \le 2$

$$\Big|\frac{\partial^{q_1+q_2+q_3}}{\partial n^{q_1}\partial m^{q_2}\partial c^{q_3}} g(n,m,c)\Big| \le N^{-q_1} M^{-q_2} C^{-q_3} \quad . \tag{16}$$

THEOREM 4. *Let* $N,M,C \ge 1$ *and* $g(n,m,c)$ *be as above. Then for any* $\varepsilon > 0$ *we have*

$$G^{\pm}(N,M,C) \ll C^{1+\varepsilon}(MN)^{\frac{1}{2}}(\sum |a_n|^2)^{\frac{1}{2}}(\sum |b_m|^2)^{\frac{1}{2}}$$

the constant implied in $\ll$ *depending on* ε *alone.*

Remark $G^{\pm}(N,M,C)$ can be treated by the large sieve inequality

$$\sum_{c<C} \sum_{\substack{d(\text{mod } c)\\(d,c)=1}} \Big| \sum_{N<n\le 2N} a_n e(n\tfrac{d}{c})\Big|^2 \le (C^2+N) \sum_{N<n\le 2N} |a_n|^2 .$$

To make use of it one must separate the variables n, m and c in $g(n,m,c)$. For, we write

$$g(n,m,c) = \iiint \hat{g}(\lambda_1,\lambda_2,\lambda_3)e(\lambda_1 n+\lambda_2 m+\lambda_3 c)d\lambda_1 d\lambda_2 d\lambda_3$$

with

$$\hat{g}(\lambda_1,\lambda_2,\lambda_3) = \iiint g(n,m,c)e(-\lambda_1 n-\lambda_2 m-\lambda_3 c)dndmdc \ll$$

$$\ll (N^2\lambda_1^2+1)^{-1}(M^2\lambda_2^2+1)^{-1}(C^2\lambda_3^2+1)^{-1}NMC$$

by partial integration. Hence

$$G^{\pm}(N,M,C) \ll (C+\sqrt{M})(C+\sqrt{N})(\sum |a_n|^2)^{\frac{1}{2}}(\sum |b_m|^2)^{\frac{1}{2}} \tag{18}$$

which is sharper than (17) if $C \ll \sqrt{MN}$.

3. KUZNIETSOV'S FORMULAS

Proofs of Theorems 1, 2 and 4 depend on several formulas of Kuznietsov [6],[7] (see also Bruggeman [1] and Proskurin [11]).

LEMMA 1. *Let* $m,n \ge 1$. *Then for any* $t \in \mathbb{R}$ *we have*

$$\sum_{j=1}^{\infty} \frac{\rho_j(n)\overline{\rho_j(m)}}{\text{ch } \pi(\kappa_j+t)\text{ch } \pi(\kappa_j-t)} + \frac{1}{\pi}\int_{-\infty}^{\infty} (\frac{m}{n})^{it}\sigma_{2ir}(n)\sigma_{-2ir}(m) \frac{\text{ch } \pi r \ |\zeta(1+2ir)|^{-2}}{\text{ch } \pi(r+t)\text{ch } \pi(r-t)} dr \tag{19}$$

$$= \delta_{nm}\pi^{-2} \frac{t}{\text{sh } \pi t} - \frac{2it}{\pi \text{ sh } \pi t} \sum_{c=1}^{\infty} \frac{4\pi\sqrt{nm}}{c^2} S(n,m;c) \int_{-i}^{i} K_{2it}(\frac{4\pi\sqrt{nm}}{c} \upsilon) \frac{d\upsilon}{\upsilon}$$

where the path of the integration is a half unit circle $|\upsilon| = 1$, $\text{Re } \upsilon > 0$.

Kuznietsov's idea of the proof is based on computing scalar product $\langle U_n(\cdot,1+it),U_m(\cdot,1-it)\rangle$ of two Poincare series

$$U_n(z,s) = \sum_{\gamma\in\Gamma_\infty\backslash\Gamma} (\text{Im } \gamma z)^s e(n\gamma z)$$

in two ways. Apart from a simple factor the right-hand side of (19) results from computing this product by expanding $U_n(z,s)$ into Fourier series and applying Rankin's method while the left-hand side of (19) is a Bessel identity for $\langle U_n(\cdot,1+it),U_m(\cdot,1-it)\rangle$ in respect of the orthogonal basis of cusp forms $u_j(z)$ and the Eisenstein series $E(z,\frac{1}{2}+it)$. Another independent proof of a similar relation was given by Bruggeman [1].

Very clever manipulation with the continuous variable t led Kuznietsov [7] to

LEMMA 2. *Let* $\phi(x)$ *be of* C^3 *class on* $(0,\infty)$ *such that* $\phi(0) = \phi'(0) = 0$, $\phi^{(q)}(x) \ll x^{-2-\varepsilon}$ *as* $x \to \infty$ *for* $q = 0,1,2,3$. *Let* ϕ_H *be the component of* $\phi(x)$ *in* $L^1(\mathbb{R}^+,x^{-1}dx)$ *orthogonal to all Bessel functions of odd order, i.e.* $\phi = \phi_H + \phi_B$ *where*

$$\phi_B(x) = 2\sum_{\ell\equiv 1 (\mathrm{mod}\,2)} \ell\, J_\ell(x)\tilde{\phi}(\ell) \quad \textit{with} \quad \tilde{\phi}(\ell) = \int_0^\infty J_\ell(y)\phi(y)\,\frac{dy}{y}\,.$$

Define

$$\hat{\phi}(r) = \frac{\pi i}{2\,\mathrm{sh}\,\pi r}\int_0^\infty (J_{2ir}(x) - J_{-2ir}(x))\phi(x)\,\frac{dx}{x}\,.$$

For $n,m \geq 1$ *we have*

$$\sum_{c=1}^\infty \frac{1}{c} S(n,m;c)\phi_H\Big(\frac{4\pi\sqrt{nm}}{c}\Big) + \delta_{nm}\frac{1}{2\pi}\int_0^\infty J_0(u)\phi(u)du$$
$$= \sum_{\kappa_j} \rho_j(n)\overline{\rho_j(m)}\,\frac{\hat{\phi}(\kappa_j)}{\mathrm{ch}\,\pi\kappa_j} + \frac{1}{\pi}\int_{-\infty}^\infty \Big(\frac{m}{n}\Big)^{ir}\sigma_{2ir}(n)\sigma_{-2ir}(m)\,\frac{\hat{\phi}(r)dr}{|\zeta(1+2ir)|^2}. \qquad (20)$$

A sum with ϕ_B in place of ϕ_H can be expressed in terms of Fourier coefficients of holomorphic cusp forms in much similar form.

LEMMA 3. *For* $n,m \geq 1$ *we have*

$$\sum_{c=1}^\infty \frac{1}{c} S(n,m;c)\phi_B\Big(\frac{4\pi\sqrt{nm}}{c}\Big) = \delta_{nm}\frac{1}{2\pi}\int_0^\infty J_0(u)\phi(u)du$$
$$+ \frac{1}{\pi}\sum_{k\equiv 0(\mathrm{mod}\,2)}\ \sum_{1\leq j\leq \nu_k} \frac{(k-1)!}{(4\pi\sqrt{nm})^{k-1}}\,\psi_{j,k}(n)\overline{\psi_{j,k}(m)}\,\tilde{\phi}(k-1) \qquad (21)$$

Proof If $f(z)$ is a cusp form of weight k then its n-th Fourier coefficient is given by (Petersson)

$$a_n = \langle f,P_n(\cdot;k)\rangle\ .$$

Hence we first deduce that

$$P_n(z;k) = \sum_{1\le j\le \nu_k} \overline{\psi_{j,k}(n)}\, f_{j,k}(z)$$

and then by Parseval's identity and (1) that

$$\sum_{j=1}^{\nu_k} \psi_{j,k}(n)\overline{\psi_{j,k}(m)} = \langle P_m(\cdot,k),P_n(\cdot,k)\rangle = p_n(m,k)$$

$$= \frac{(4\pi\sqrt{nm})^{k-1}}{\Gamma(k-1)} \{\delta_{nm} + 2\pi i^k \sum_{c=1}^{\infty} \frac{1}{c} S(n,m;c) J_{k-1}(\frac{4\pi\sqrt{nm}}{c})\} \quad . \qquad (22)$$

Multiply both sides by $\frac{1}{\pi}(k-1)!\, i^k (4\pi\sqrt{nm})^{1-k} J_{k-1}(y)\phi(y)$, integrate over y from 0 to ∞ in respect of the logarithmic measure $y^{-1}dy$ and sum over $k \equiv 0 \pmod 2$, $k \ge 2$ getting (21). For δ_{nm}-terms we appealed to the identity

$$2\sum_{k=1}^{\infty} (2k-1)(-1)^k J_{2k-1}(y) = yJ_0(y)$$

which easily follows from the recurrence relations

$$J_{n-1}(z) + J_{n+1}(z) = \frac{2n}{z} J_n(z) \quad .$$

For $k = 2$ one must be careful because $P_n(z;k)$ are defined in a different manner (Hecke), however the relation (22) remains valid, both sides being equal to zero.

Lemmas 2 and 3 will be combined for estimating $G^+(N,M,C)$. In the case of $G^-(N,M,C)$ we need another formula of Kuznietsov [6].

LEMMA 4. *Let* $n,m \ge 1$ *and* $\phi(x)$ *be a function as in Lemma 2. We have*

$$\sum_{c=1}^{\infty} \frac{1}{c} S(n;-m;c)\phi(\frac{4\pi\sqrt{nm}}{c}) = \sum_{\kappa_j} \rho_j(n)\rho_j(m)\check{\phi}(\kappa_j)$$

$$+ \frac{1}{\pi}\int_{-\infty}^{\infty} (nm)^{-ir}\sigma_{2ir}(n)\sigma_{2ir}(m)\,\mathrm{ch}\,\pi r\,\check{\phi}(r)dr \qquad (23)$$

where

$$\check{\phi}(r) = \frac{4}{\pi}\int_0^{\infty} K_{2ir}(y)\phi(y)\frac{dy}{y} \quad .$$

4. BILINEAR FORMS OF KLOOSTERMAN SUMS

In this section we investigate sums of $S(n,m;c)$ over the coefficients n and m . Our arguments do not depend on the theory of automorphic functions and they are mostly elementary. Define

$$B(c,N) = \sum_{N<n,m\le 2N} b_n\overline{b_m}S(n,m;c)e(\frac{2\sqrt{nm}}{c}\,\Theta) .$$

LEMMA 5. *Let* $\Theta > 0$, $\varepsilon > 0$ *and* $N \ge 1$. *Then we have*

$$B(c,N) \ll c^{\frac{1}{2}+\varepsilon}N \sum|b_n|^2 \qquad \textit{for all} \quad c \ge 1 , \tag{24}$$

$$B(c,N) \ll (c + N + \sqrt{\Theta cN}) \sum|b_n|^2 \quad \textit{for all} \quad c \ge 1 , \tag{25}$$

$$B(c,N) \ll \Theta^{-\frac{1}{2}}c^{\frac{1}{2}}N^{\frac{1}{2}+\varepsilon} \sum|b_n|^2 \qquad \textit{for} \quad c \le N \quad \textit{and} \quad \Theta \le 2 , \tag{26}$$

the constant implied in $\ll$ *depending on* ε *alone.*

Of the three results above the last one is most crucial. The first follows trivially from A. Weil (2). The second is an easy consequence of the hybrid large sieve inequality:

$$\int_{-T}^{T} \sum_{d(\mathrm{mod}\ c)} \Big|\sum_{N<n\le 2N} b_n n^{it} e(n\,\frac{d}{c})\Big|^2 \ll (cT + N)\sum|b_n|^2 \quad . \tag{27}$$

A minor difficulty arises when separating the variables n and m in $e(2\Theta\sqrt{nm}/c)$. To this end we appeal to Mellin's transform. Notice that in the range of the summation we have $N < \sqrt{nm} \le 2N$, therefore $B(c,N)$ will not be affected if weights $\eta(\sqrt{nm}/N)$ with $\eta(x) = 1$ for $x \in (1,2]$ are attached at each term. In what follows we demand $\eta(x)$ to be of C^∞ class with $\mathrm{Supp}\ \eta(x) = [\frac{1}{2},3]$. Then

$$\eta(x)e(2\Theta c^{-1}xN) = \frac{1}{2\pi i}\int_{1-i\infty}^{1+i\infty} R(s)x^{-s}ds$$

where by Mellin's inversion formula

$$R(s) = \int_0^\infty \eta(x)e(2\Theta c^{-1}xN)x^{it}dx \ll \begin{cases} (|t|+1)^{-\frac{1}{2}} & \text{for all} \quad t \\ t^{-2} & \text{for } |t|>16\pi\Theta c^{-1}N \end{cases} \tag{28}$$

Thus

$$B(c,N) = \frac{1}{2\pi i}\int R(s) \sum_{m,n} b_n\overline{b_m}(\frac{\sqrt{nm}}{N})^{-s} S(n,m;c)ds$$

whence (25) by (27) and (28).

Now we proceed to prove (26) for $c \le N^{1-\varepsilon}$, the remaining case $N^{1-\varepsilon} < c \le N$ following from (25). By the Cauchy-Schwarz inequality

$$|B(c,N)|^2 \le \left(\sum_n |b_n|^2\right) \sum_n \eta\left(\frac{n}{N}\right) \left|\sum_m b_m S(n,m;c)e(2\Theta\sqrt{nm}/c)\right|^2$$

$$= \left(\sum_n |b_n|^2\right) \sum_{\substack{m_1,m_2\\ d_1,d_2}} b_{m_1}\overline{b_{m_2}}\, e\left(\frac{m_1\overline{d}_1 - m_2\overline{d}_2}{c}\right) \sum_n f(n)$$

where $f(n) = \eta\left(\frac{n}{N}\right)e\left(\frac{d_1-d_2}{c}n + \frac{2\Theta(\sqrt{m_1}-\sqrt{m_2})}{c}\sqrt{n}\right) = \eta\left(\frac{n}{N}\right)e(An + B\sqrt{n})$, say.
For the inner sum we apply the Poisson summation formula

$$\sum_n f(n) = \sum_h \hat{f}(h) , \quad h \in \mathbb{Z}$$

where

$$\hat{f}(h) = \int \eta\left(\frac{t}{N}\right)e((A-h)t + B\sqrt{t})dt \quad .$$

If $h \ne A$ then $|A-h| \ge \frac{1}{c}$ while $B/2\sqrt{t} \le (\sqrt{2}-1)\Theta/c \le 2(\sqrt{2}-1)/c$. Therefore by partial integration $q = [2/\varepsilon]$ times $\hat{f}(h) \ll N((A-h)N)^{-q}$ giving

$$\sum_{h \ne A} \hat{f}(h) \ll N\left(\frac{c}{N}\right)^q \ll c^{-1} \quad .$$

Now consider the case $h = A$. Since h is an integer this implies $d_1 = d_2$. If $m_1 \ne m_2$ we integrate by parts getting

$$\hat{f}(A) \ll |B|^{-1}N^{\frac{1}{2}} \ll \frac{cN}{\Theta|m_1-m_2|}$$

and if $m_1 = m_2$ we take the trivial estimate $\hat{f}(A) \ll N$. The summation over $d = d_1 = d_2$, $(d,c) = 1$ yields Ramanujan's sum for which we have

$$\left|\sum_{\substack{d(\text{mod } c)\\ (d,c)=1}} e\left(\frac{m_1-m_2}{c}\overline{d}\right)\right| \le (m_1-m_2,c) \quad .$$

Gathering all the above estimates together we complete the proof of (26).

5. PROOF OF THEOREM 2

By (22) the left-hand side of (13) is equal to

$$D_K \sum |a_n|^2 + \pi \sum_{c=1}^{\infty} \frac{1}{c} \sum_{n,m} a_n\overline{a_m}S(n,m;c)E_K\left(\frac{4\pi\sqrt{nm}}{c}\right) \qquad (29)$$

where

$$D_K = \sum_{\ell=1}^{\infty} (2\ell-1)e^{-\frac{2\ell-1}{K}} = \frac{1}{2}\,\frac{\operatorname{ch}\frac{1}{K}}{(\operatorname{sh}\frac{1}{K})^2} = \frac{K^2}{2} + O(1) \tag{30}$$

and

$$E_K(x) = 2\sum_{\ell=1}^{\infty} (2\ell-1)(-1)^{\ell} e^{-\frac{2\ell-1}{K}} J_{2\ell-1}(x) \quad .$$

It remains to estimate bilinear forms

$$F_K(c,N) = \sum_{N<n,m\le 2N} a_n \overline{a_m} S(n,m;c) E_K\left(\frac{4\pi\sqrt{nm}}{c}\right) \quad .$$

To this end we first prove

LEMMA 6. *For* $x > 0$ *and* $K > 0$ *we have*

$$E_K(x) = \tfrac{1}{2}(\operatorname{sh}\tfrac{2}{K}) \int_0^1 \frac{\xi x J_0(\xi x)\,d\xi}{[(\operatorname{ch}\frac{1}{K})^2 - \xi^2]^{3/2}} \quad . \tag{31}$$

Proof For $b > 1$ it holds (see [3], p.106) that

$$\int_0^{\infty} e^{iby} J_n(y)\,dy = i^n (b^2-1)^{-\frac{1}{2}} (b + \sqrt{b^2-1})^{-n} \quad .$$

In particular for $n = 0$ differentiating over b gives

$$\int_0^{\infty} e^{iby} J_0(y)\,y\,dy = ib(b^2-1)^{-3/2} \quad .$$

Moreover for $x,y > 0$ we have (see [7])

$$2\sum_{\ell=1}^{\infty} (2\ell-1) J_{2\ell-1}(x) J_{2\ell-1}(y) = y\int_0^1 \xi x J_0(\xi x) J_0(\xi y)\,d\xi \quad .$$

Hence

$$2\sum_{\ell=1}^{\infty} (2\ell-1)(-1)^{\ell} (b+\sqrt{b^2-1})^{-(2\ell-1)} J_{2\ell-1}(x)$$
$$= -b\sqrt{b^2-1} \int_0^1 \frac{\xi x J_0(\xi x)\,d\xi}{[b^2-\xi^2]^{3/2}}$$

which for $b = \operatorname{ch}\frac{1}{K}$ becomes (31).

Combining Lemmas 5 and 6 we shall prove that

$$F_K(c,N) \ll c^{-\varepsilon} N^{1+2\varepsilon} \sum_{N<n\le 2N} |a_n|^2 \tag{32}$$

If $c > N^2$ or $N < c \le N^2$ the result easily follows from the integral representation

$$J_0(\xi x) = \frac{1}{\pi}\int_0^{\pi} \cos(\xi x \sin\Theta)d\Theta \tag{33}$$

from the crude estimate

$$(\mathrm{sh}\,\frac{2}{K})\int_0^1 \frac{\xi d\xi}{[(\mathrm{ch}\,\frac{1}{K})^2 - \xi^2]^{3/2}} \ll 1$$

and from (24) and (25) respectively. For $c \le N$ we take $\varepsilon = \sqrt{c/N}$ and split up the integral (33) as follows

$$J_0(\xi x) = \frac{1}{\pi}\int_{-\varepsilon}^{\varepsilon} \cos(\xi x \cos\Theta)d\Theta + \frac{2}{\pi}\,\frac{\sin(\xi x \cos\varepsilon)}{\xi x \sin\varepsilon}$$
$$+ \frac{2}{\pi\xi x}\int_0^{\frac{\pi}{2}-\varepsilon} \frac{\mathrm{tg}\Theta}{\cos\Theta}\sin(\xi x \sin\Theta)d\Theta \tag{34}$$

the last two terms arising by partial integration. If $\xi > \varepsilon^2$ we apply (26) and (34) and if $\xi \le \varepsilon^2$ we apply (25) and (33) in either case getting a contribution to $F_K(c,N)$ less than $O(N^{1+\varepsilon}\sum |a_n|^2)$ as claimed.

Finally Theorem 2 follows from (29), (30) and (32).

6. PROOF OF THEOREM 1

Multiply both sides of (19) by $t(\mathrm{sh}\,\pi t)e^{-(t/K)^2}a_n\overline{a_m}$, sum over $n,m \in (N,2N]$ and integrate from $t = 0$ to $t = \infty$. Since

$$\int_0^{\infty} \frac{t\ \mathrm{sh}\ \pi t}{\mathrm{ch}\ \pi(t+r)\mathrm{ch}\ \pi(t-r)}\, e^{-(t/K)^2}\,dt \gg \frac{|r|}{\mathrm{ch}\ \pi r}\, e^{-(r/K)^2}$$

and the integral in (19) contributing nonnegative amount can be discarded it yields

$$\sum_{\kappa_j} \frac{\kappa_j}{\mathrm{ch}\ \pi\kappa_j}\, e^{-(\kappa_j/K)^2} \left|\sum_n a_n\rho_j(n)\right|^2 \ll K^3\sum_n |a_n|^2 + \Delta(K,N) \tag{35}$$

where we put

$$\Delta(K,N) = \sum_{c=1}^{\infty}\sum_{n,m} a_n\overline{a_m}\,\frac{\sqrt{nm}}{c^2}\,S(n,m;c)\Phi\left(\frac{4\pi\sqrt{nm}}{c}\right)$$

and for $x = 4\pi\sqrt{nm}/c$ we put

$$\Phi(x) = \int_0^{\infty} t^2 e^{-(t/K)^2}\int_{-i}^{i} K_{2it}(xu)\,\frac{du}{u}\,dt$$

By (7.12.21) of [3] we have for $\mathrm{Re}\ u > 0$

$$K_{2it}(xu) = \int_0^{\infty} e^{-xu\,\mathrm{ch}\,\xi}\cos(2t\xi)d\xi$$
$$= \frac{xu}{2t}\int_0^{\infty} e^{-xu\,\mathrm{ch}\,\xi}\,\mathrm{sh}\,\xi\,\sin(2t\xi)d\xi$$

Also we have (see [4], p.214)

$$\int_0^\infty te^{-(t/K)^2}\sin(2t\xi)dt = \frac{\sqrt{\pi}}{2}\ K^3e^{-(\xi K)^2}$$

Hence

$$\Phi(x) = i\ \frac{\sqrt{\pi}}{2}K^3\int_0^\infty e^{-(\xi K)^2}\xi\ \mathrm{th}\ \xi\ \sin(x\ \mathrm{ch}\ \xi)d\xi$$

If $c < NK^{-2}$ (i.e. $x >> K^2$) we may do better when integrating by parts

$$\Phi(x) = i\ \frac{\sqrt{\pi}}{2}\ x^{-1}K^3\int_0^\infty e^{-(\xi K)^2}(1 - \xi\ \mathrm{th}\ \xi - 2\xi^2K^2)\cos(x\ \mathrm{ch}\ \xi)\frac{d\xi}{\mathrm{ch}\ \xi} \qquad (37)$$

Let $\Delta_\nu(K,N)$, $\nu = 1,2,3,4$ denote partial sums of $\Delta(K,N)$ in respect of the variable c from the intervals $c_1 \le NK^{-2} < c_2 \le N < c_3 \le N^2 < c_4$ respectively. For $\Delta_4(K,N)$ apply (36) and (24) giving

$$\Delta_4(K,N) << \sum_{c>N^2} c^{-\frac{3}{2}+\varepsilon}N^2\sum|a_n|^2 << N^{1+\varepsilon}\sum|a_n|^2\ . \qquad (38)$$

For $\Delta_3(K,N)$ apply (36) and (25) giving

$$\Delta_3(K,N) << \sum_{N<c\le N^2} c^{-1}N\sum|a_n|^2 << N^{1+\varepsilon}\sum|a_n|^2\ . \qquad (39)$$

For $\Delta_2(K,N)$ apply (36), (26) if $\xi \le 1$ and (25) if $\xi > 1$ giving

$$\Delta_2(K,N) << \sum_{NK^{-2}<c\le N} c^{-\frac{3}{2}}N^{\frac{3}{2}+\varepsilon}\sum|a_n|^2 << KN^{1+\varepsilon}\sum|a_n|^2\ . \qquad (40)$$

For $\Delta_1(K,N)$ apply (37), (26) if $\xi \le 1$ and (25) if $\xi > 1$ giving

$$\Delta_1(K,N) << \sum_{c\le NK^{-2}}(c^{-\frac{1}{2}}N^{\frac{1}{2}+\varepsilon}K^2 + c^{-2}N^2K^3e^{-K^2})\sum|a_n|^2$$

$$<< (KN^{1+\varepsilon} + K^3N^2e^{-K^2})\sum|a_n|^2\ . \qquad (41)$$

Gathering (35), (38)-(41) together we get

$$\sum_{\kappa_j\le K}\frac{\kappa_j}{\mathrm{ch}\ \pi\kappa_j}\Big|\sum_{N<n\le 2N}a_n\rho_j(n)\Big|^2 << K(K^2+N^{1+\varepsilon}+K^2N^2e^{-K^2})\sum_{N<n\le 2N}|a_n|^2\ .$$

Since the left-hand side is nondecreasing in K we may replace K in the right-hand side by any number $K_1 \ge K$. On taking $K_1 = K + \log N$ we first ignore the term $K^2N^2e^{-K^2}$ and then we derive (11) by partial summation.

7. PROOF OF THEOREM 4

As we pointed out in Section 3 it is sufficient to prove (17) for $C > 8\pi\sqrt{MN}$. Let $\phi(x)$ be a function of C^2 class with compact support in $(0,1)$ such that

$$\int |\phi''(x)| x \, dx \le 1 .$$

We first show that

$$H^{\pm}(N,M,\phi) = \sum_{N<n\le 2N} \sum_{M<m\le 2M} \sum_{c} a_n b_m \frac{1}{c} S(n,\pm m;c)\phi\left(\frac{4\pi\sqrt{nm}}{c}\right)$$
$$\ll (MN)^{\frac{1}{2}+\varepsilon}\left(\sum |a_n|^2\right)^{\frac{1}{2}}\left(\sum |b_m|^2\right)^{\frac{1}{2}} . \quad (42)$$

To this end we apply (20) and (21) in case of $H^{+}(N,M,\phi)$ and (23) in case of $H^{-}(N,M,\phi)$. We have to estimate the transforms $\hat{\phi}$, $\tilde{\phi}$ and $\check{\phi}$.

LEMMA 7. *For all* $r > 0$ *we have*

$$\hat{\phi}(r), \tilde{\phi}(r), (\operatorname{ch} \pi r)\check{\phi}(r) \ll (r^2+1)^{-1} \quad (43)$$

<u>Proof</u> All results are trivial for $r \le 1$. If $r \ge 1$ we utilize power series expansions (note that x we deal with is in $(0,1)$)

$$J_\nu(x) = \sum_{\ell=0}^{\infty} \frac{(-1)^\ell}{\ell!\Gamma(\ell+1+\nu)} \left(\frac{x}{2}\right)^{2\ell+\nu} , \quad \nu = \pm 2ir, \ \nu = r-1$$

$$K_{2ir}(x) = \frac{\pi}{2 \operatorname{sh} 2\pi r} \operatorname{Im} \sum_{\ell=0}^{\infty} \frac{1}{\ell!\Gamma(\ell+1+2ir)} \left(\frac{x}{2}\right)^{2\ell+2ir}$$

and integrate termwise. A typical term to be estimated is

$$\int \phi(y) y^{2\ell+\nu-1} dy = \frac{1}{(2\ell+1+\nu)(2\ell+\nu)} \int \phi''(y) y^{2\ell+\nu+1} dy \ll (\ell + |\nu|)^{-2}$$

An application of Stirling's formula for $\Gamma(\ell+1+\nu)$ and $\Gamma(\ell+1+2ir)$ completes the proof of Lemma 7.

By (20), (21), (43), (11), (13), (14) and the Cauchy-Schwarz inequality we obtain (42) for $H^{+}(N,M,\phi)$. Analogously, by (23), (43), (11), (14) and the Cauchy-Schwarz inequality we obtain (42) for $H^{-}(N,M,\phi)$. It remains to derive (17) from (42). Writing $h(x_1,x_2;x) = g(x_1,x_2,4\pi\sqrt{x_1x_2}/x)$ we get $g(n,m,c) = h(n,m;4\pi\sqrt{nm}/c)$. Just for separating the variables x_1,x_2,x we write

$$h(x_1,x_2;x) = \iint \hat{h}(\lambda_1,\lambda_2;x)e(-\lambda_1x_1-\lambda_2x_2)d\lambda_1d\lambda_2$$

with

$$\hat{h}(\lambda_1,\lambda_2;x) = \iint h(x_1,x_2;x)e(\lambda_1x_1+\lambda_2x_2)dx_1dx_2 \ .$$

By partial integration twice in each variable λ_1,λ_2 we deduce that

$$x^2 \frac{\partial^2}{\partial x^2} \hat{h}(\lambda_1,\lambda_2;x) \ll (N^2\lambda_1^2+1)^{-1}(M^2\lambda_2^2+1)^{-1}NM \ .$$

Therefore (17) follows from (42) for $\phi(x) = \hat{h}(\lambda_1,\lambda_2;x)$.

We note that Theorem 1 is applied in [5] to prove that

$$\int_T^{T+T^{2/3}} |\zeta(\tfrac{1}{2}+it)|^4dt \ll T^{2/3+\varepsilon} \ .$$

Other applications and generalizations of the above methods for congruence subgroups will be discussed in a forthcoming paper by J.-M. Deshouillers and the author.

REFERENCES

1. R.W. Bruggeman, Fourier coefficients of cusp forms, Inventiones math. 45 (1978), 1-18.

2. P. Deligne, La conjecture de Weil I, Publ.Math. I.H.E.S., 43 (1974), 273-307.

3. A. Erdélyi, W. Magnus and F. Oberhettinger and F.G. Tricomi, Higher Transcendental Functions II, McGraw-Hill, New York, 1953.

4. I.S. Gradsztejn and I.M. Ryzyk, Tables of integrals, sums, series and products (in Polish), PWN Warszawa, 1964.

5. H. Iwaniec, Fourier coefficients of cusp forms and the Riemann zeta-function, Séminaire de Théorie des Nombres, Bordeaux 1979/80.

6. N.V. Kuznietsov, Petersson hypothesis for forms of weight zero and Linnik hypothesis (in Russian), Preprint No.02, Khab. K.H.I.I., Khabarovsk (1977).

7. N.V. Kuznietsov, Petersson hypothesis for parabolic forms of weight zero and Linnik hypothesis. Sums of Kloosterman sums, Math. Sbornik 111 (153), No.3 (1980), 334-383.

8. Y.V. Linnik, Additive problems and eigenvalues of the modular operators, Proc.Internat.Congr.Math. (Stockholm 1962), 270-284.

9. H. Maass, Über eine neue Art von nichtanalitischen automorphen Funktionen, Math.Ann. 121, No.2 (1949), 141-183.

10. H.L. Montgomery, Topics in Multiplicative Number Theory, Lect. Notes in Math. 227 (1971), Springer-Verlag, Berlin-New York.

11. N.V. Proskurin, Summation formulas for generalised Kloosterman sums (in Russian), Zap.Naučn.Sem. Leningrad. Otdel.Mat.Inst.Steklov. (LOMI) (1979) Vol.82, 103-135.

12. N.V. Proskurin, Estimates for eigenvalues of Hecke operators in the space of parabolic forms of weight zero (in Russian), ibid. 136-143.

13. A. Selberg, On the estimation of Fourier coefficients of modular forms, Proc. of Symposia in Pure Math. VIII, AMS, Providence (1965), 1-15.

14. A. Selberg, Harmonic analysis and discontinuous groups in weakly symmetric Riemannian spaces with applications to Dirichlet series, J. Indian Math.Soc. 20 (1956), 47-87.

NONSTANDARD METHODS IN DIOPHANTINE GEOMETRY

by
Ernst Kani

The basic methods and results in the subject of Diophantine Geometry (as set forth in the book of Lang[11]) were developed by Mordell, Weil, Siegel and others in the 1920's and '30's. The first such result is the celebrated Theorem of Mordell-Weil which may be stated as follows.

<u>Theorem</u> 1 (Mordell[17],Weil[30]): Let C be a curve defined over a number field K. Then the group $J_C(K)$ of K-rational points of the Jacobian variety J_C of C is a finitely generated group.

Siegel subsequently (in 1929) showed how one can combine this theorem with a theorem on diophantine approximations (the Thue-Siegel Theorem) to obtain:

<u>Theorem</u> 2 (Siegel[28],Mahler[14]): Let $C \subset \mathbb{P}^n$ be a projective curve of genus $g \geq 1$ which is defined over a number field K, and let S be a finite set of places of K. Then there exist only finitely many S-integral points $P \in C(K)$.

Recall that a point $P \in \mathbb{P}^n$ is called S-<u>integral</u> if all its coordinates (for some choice of homogeneous coordinates) are $\mathfrak{p}$-integral for all finite $\mathfrak{p} \notin S$.

Actually, Siegel proved a bit more: he also characterized those curves of genus 0 for which the assertion fails. By using this characterization, he was able to strengthen the classical Irreducibility Theorem of Hilbert considerably. Recall that Hilbert's Theorem is the following.

Theorem 3 (Hilbert[7]): Every number field K is hilbertian in the sense that the following property holds:

(H) For each absolutely irreducible polynomial $f(T,X)$ with coefficients in K there exist infinitely many $t \in K$ such that $f(t,X) \in K[X]$ is irreducible.

Siegel's sharpening of this theorem is as follows.

Corollary: Let $f(T,X) \in K[T,X]$ be a polynomial with S - integral coefficients. Then there exists a number field K' in which $f(t,X)$ becomes reducible for infinitely many S-integral $t \in K'$ if and only if, after a suitable substitution of the form

$$T = a_n Y^n + a_{n-1} Y^{n-1} + \dots + a_{-n} Y^{-n},$$

f becomes reducible as a polynomial in Y.

Coming back to Siegel's theorem, for curves of genus 1 it is, in a sense, best possible, since there exist such curves with infinitely many K-rational points. For curves of genus $g \geq 2$, however, it falls somewhat short of what one conjectures to be true, namely:

Mordell's Conjecture (Mordell[17]): Every curve of genus $g \geq 2$ defined over a number field K has only finitely many K-rational points.

While this conjecture is at present still far from being settled, there exist several partial results in this direction.

Theorem 4 (Mumford[18]): Let $C \subset \mathbb{P}^n$ be a curve of genus $g \geq 2$ defined over a number field K, and suppose that the K-rational points $P_i \in C(K)$ are ordered according to increasing (logarithmic) height: $h(P_1) \leq h(P_2) \leq \dots$ Then there exist constants a and b with $a > 0$ such that

$$h(P_n) \geq e^{an+b}, \qquad \forall\, n \geq 1.$$

Here, the logarithmic height of a point $P = (a_0,\ldots,a_n) \in \mathbb{P}^n_K$ is defined by

$$h(P) = -\sum_{\mathfrak{p}\in M_K} \operatorname{Min}_i v_{\mathfrak{p}}(a_i),$$

where the sum runs over all archimedean and non-archimedean primes of K, and $v_{\mathfrak{p}}$ denotes the normalized valuation associated to $\mathfrak{p}$ (written additively):

$$v_{\mathfrak{p}}(a) = -\log |a|_{\mathfrak{p}}^{N_{\mathfrak{p}}}.$$

Theorem 5 (Dem'janenko[1],Manin[16]): Suppose that C and C' are curves defined over K such that

$$\operatorname{rank} \operatorname{Hom}_K(J_C,J_{C'}) > \operatorname{rank} J_C(K).$$

Then C' has only finitely many K-rational points.

Corollary (Manin[16]): Let $N > 1$ be an integer and suppose that $k \gg 0$. Then the modular curve $X_0(N^k)$ has only finitely many K-rational points.

In this lecture I would like to outline how one can use the nonstandard methods of A. Robinson (cf. Robinson[22], Stroyan-Luxemburg[29]) to prove the above results. For lack of time, however, I shall mainly concentrate on the proof of the Mordell-Weil Theorem, or, more precisely, on the proofs of the following two theorems from which (by using an "infinite descent argument") the Mordell-Weil Theorem may easily be deduced (cf. Lang[13], Mumford[19]).

Theorem 6 (weak Mordell-Weil): There exists an integer $m > 1$ and a finite extension K' of K such that $J_C(K')/mJ_C(K')$ is finite.

Theorem 7 (Néron[21],Tate[12],[15]): There exists a real-valued, positive definite quadratic form h_J on $J_C(K)$ with the following property:

(*) Every h_J-bounded subset of $J_C(K)$ is finite.

In particular, $h_J(P) = 0$ if and only if $P \in J_C(K)$ is a point of finite order.

The quadratic form constructed in Theorem 7 is called the <u>Néron-Tate height</u> on J_C. As a consequence of the explicit construction of h_J by means of "intersection products", the theorems of Mumford and Dem'janenko-Manin become easy corollaries, as I will briefly indicate.

§1. <u>Nonstandard Methods</u>

a) Enlargements

Roughly speaking, the nonstandard methods consist of embedding each object X which we are considering in a (much) larger object *X, called an <u>enlargement</u> of X, such that

1) the embedding is compatible with the basic operations of set theory
2) the embedding is "functorial": each map $f:X \to Y$ extends uniquely to a map $*f:*X \to *Y$
3) the objects *X (resp. maps *f) have similar properties as X (resp. f).

Let me explain this in some more detail in the case that X = K is a number field, which I regard as coming equipped with its canonical set M_K of places. As Roquette has suggested, one might view the enlargement *K, together with its canonical set $*M_K$ of places, as a <u>global completion</u> of K since it possesses the following properties:

(A) *K is a valued field with respect to the set $*M_K$ of places, and there exists an embedding $K \to *K$ such that every $v \in M_K$ has a unique extension $*v \in *M_K$ to *K.

(B) *K is an elementary extension of K (i.e. the elementary properties of K are inherited by *K).

(C) *K is <u>saturated</u>: every binary relation on K which satisfies a "Cauchy condition" (more precisely: is concurrent) has a "limit" in *K.

(D) A statement about K is true if and only if the "corresponding" statement about *K is true.

Note that the properties (A) - (C) are analogous to the following well-known properties of a local field $\hat{K}_v$:

(a) $\hat{K}_v$ is a valued field with respect to a canonical valuation $\hat{v}$, and there is an embedding $K \to \hat{K}_v$ such that $\hat{v}$ extends v.

(b) K is dense in $\hat{K}_v$ (hence, many properties of K are inherited by $\hat{K}_v$ by continuity).

(c) $\hat{K}_v$ is complete with respect to $\hat{v}$.

On the other hand, property (D) has no local analogue, since it reflects the global nature of *K.

By property (D), in order to prove a statement about K, it is enough to prove the corresponding statement about *K. But this seems to have made the problem harder rather than easier. Nevertheless, by using the following principle of relativization, we have added a new dimension to the problem which allows us (in certain cases) to get a better handle on it.

b) Principle of Relativization

Vaguely speaking, this principle consists of considering the object *X <u>relative</u> <u>to</u> X. Depending upon the nature of the object X, there are several variants to this principle:

I) X = A is an abelian group: consider the factor group *A/A.

II) $X = \Gamma$ is an ordered abelian group: consider the factor group $\dot{\Gamma} = {}^*\Gamma/\langle\Gamma\rangle$, where $\langle\Gamma\rangle$ denotes the convex hull of Γ in ${}^*\Gamma$. Note that the ordering on ${}^*\Gamma$ induces an ordering on $\dot{\Gamma}$.

III) $X = f$ is a real-valued function on a set S: consider the function $\dot{f}:{}^*S \to \dot{\mathbb{R}}$, defined by the composition $\dot{f} = pr \circ {}^*f$.

By using these "relative objects", many statements about X can be rather elegantly reformulated. For example, we have the following <u>Nonstandard Dictionary</u>:

standard	nonstandard
1) A is finite	${}^*A/A = 0$
2) A/mA is finite	${}^*A/A$ is m-divisible
3) f is bounded on S	$\dot{f} = 0$ on ${}^*S$
4) f-bounded subsets of S are finite	$\dot{f}(s) = 0 \Leftrightarrow s \in S$
5) f is quasi-quadratic (cf. Lang[13], p.86) on A	$\dot{f}$ is quadratic on ${}^*A$

Let us apply this principle to the number field K: in this case, we are interested in an arithmetic of ${}^*K$ relative to K. For this we shall use the following construction (basically due to Robinson[23],[24]).

Let $v \in {}^*M_K$; by definition, v is a map from ${}^*K^\times$ onto some subgroup $\Gamma_v \subset {}^*\mathbb{R}$. Set $\dot{\Gamma}_v = \Gamma_v/\langle v(K^\times)\rangle$, and define

$$\dot{v} : {}^*K^\times \to \dot{\Gamma}_v$$

by $\dot{v} = pr \circ v$, where $pr:\Gamma_v \to \dot{\Gamma}_v$ is the projection. One easily checks that $\dot{v}$ is a non-trivial valuation in the sense of Krull (i.e. $\dot{v}$ satisfies the strong triangle inequality) - even if v is archimedean!

<u>Remark</u>: If v is <u>standard</u>, i.e. $v = {}^*w$ is the canonical

extension of a valuation w of K, then $\dot{v}$ coincides with $\dot{w}$ as defined in III) above. On the other hand, if v is nonstandard, then v is already trivial on K and hence $\dot{v} = v$.

Since, by construction, $\dot{v}$ is trivial on K, the set $\dot{M}_K = \{\dot{v} : v \in {}^*M_K\}$ defines an arithmetic of *K relative to K. We now come to

Basic Philosophy: $\dot{M}_K$ gives *K the structure of a "function field" (over K).

§2. The Geometry of *K/K

To explain the "geometry" of *K/K, let me list some properties of the arithmetic object $({}^*K, \dot{M}_K)$ which are analogous to geometric properties shared by (certain) varieties.

The first of these may be interpreted as giving *K a "projective structure". To explain this, let me recall (cf. Lang[11]) that if V is a normal projective variety over K, and $\varphi : V \to \mathbb{P}^n$ an embedding in projective space, then the degree $\deg_\varphi(D)$ of a divisor D on V is defined, and hence we have a function $\deg_\varphi : \mathrm{Div}(V) \to \mathbb{Z}$ on the divisor group Div(V) which is additive, monotone and vanishes on principal divisors. Furthermore, it is known that for $f \in K(V)$, $\deg_\varphi((f)_\infty) = 0 \Leftrightarrow f \in K$. (Here, $(f)_\infty$ denotes the pole divisor of (f).) Let me sum these properties up by saying: $\deg_\varphi$ is a non-degenerate degree function on V.

Now a similar property is shared by *K: if $\mathfrak{D}({}^*K/K)$ denotes the divisor group of *K with respect to $\dot{M}_K$ in the sense of Weil[32] (i.e. $\mathfrak{D}({}^*K/K)$ is the smallest subgroup of $\prod \dot{\Gamma}_v$ (the product extending over all $v \in {}^*M_K$) which contains all principal divisors and is closed under the formation of minima and maxima) then we have:

Property 1: $\mathfrak{D}({}^*K/K)$ has a non-degenerate, $\dot{\mathbb{R}}$-valued degree function $\dot{\delta}$.

Note that this immediately implies

<u>Property</u> 1': The only $\dot{M}_K$-holomorphic functions of *K are the constants; i.e. $\bigcap_{\vartheta} \mathfrak{O}_{\vartheta} = K$.

To construct $\dot{s}$, use the usual (logarithmic) size function on the divisor group $\mathfrak{D}(K)$ of K which is defined by

$$s(D) = \sum_{v \in M_K} v(D).$$

(Note that v is normalized!) Relativizing s yields an $\dot{\mathbb{R}}$-valued homomorphism $\dot{s}$ on $\dot{\mathfrak{D}}(K) = {}^*\mathfrak{D}(K)/\langle\mathfrak{D}(K)\rangle$. By making use of the canonical identification $\dot{\mathfrak{D}}(K) \cong \mathfrak{D}({}^*K/K)$, we can transport $\dot{s}$ to $\mathfrak{D}({}^*K/K)$. It is then clear that $\dot{s}$ is a degree function. The fact that $\dot{s}$ is non-degenerate, on the other hand, is equivalent to the assertion that a number field K (or, more precisely, the projective line $\mathbb{P}^1_K$) has only finitely many elements of bounded height.

Now *K/K is a rather special function field, as the next property shows.

<u>Property</u> 2: *K/K is <u>simply connected</u> in the sense that there is no proper finite extension L of $^*K\cdot\overline{K}$ ($\overline{K}$ = the algebraic closure of K) which is unramified at all valuations of $^*K\cdot\overline{K}/\overline{K}$.

This property is (basically) the nonstandard equivalent of the <u>Theorem of Hermite</u>[6] which states that there exist only finitely many extension fields of a number field K with bounded discriminant and degree.

Property 2 seems to suggest that *K/K is "rational", but this contention has serious drawbacks. For then, by an extended version of Lüroth's theorem, one would expect that every subfield $F \subset {}^*K$ which is a function field of <u>one</u> variable over K is also rational, but this is not so. To see this, we will use the next property of *K/K.

<u>Property</u> 3: A function field F/K of one variable may be embedded in *K ⇔ F has infinitely many K-rational places.

Since it is easy to construct elliptic curves with infinitely many K-rational points, *K has many elliptic subfields and hence cannot be considered to be "rational". On the other hand, Mordell's Conjecture suggests that *K/K is "almost rational" since we can reformulate the conjecture as:

(M) Every subfield $F \subset$ *K which is a function field of one variable over K has genus $g \leq 1$.

Property 3 may be generalized to function fields of higher dimension as follows.

<u>Property</u> 3': A function field F/K may be K-embedded in *K if and only if the set of K-rational places of F/K is (Zariski-) dense in the set of all places of F/K.

In particular, we have:

<u>Property</u> 3": Every finitely generated subfield of *K/K has a K-rational place.

The last "geometric" property of *K/K which we will consider is analogous to the <u>Lemma of Matsusaka</u> which is the key point in proving the <u>Theorem of Bertini</u>.

<u>Property</u> 4: There exists $t \in$ *K $\setminus$ K such that K(t) is algebraically closed in *K.

Again, this property is a reformulation of a classical theorem about number fields, namely the aforementioned Irreducibility Theorem of Hilbert (Theorem 3). This equivalence was established by Gilmore and Robinson[4] in 1955, and used by Roquette[27] to give a nonstandard proof of Hilbert's Theorem. More recently, Weissauer[33] noticed that it is possible to strengthen the Gilmore-Robinson Criterion as follows.

<u>Proposition</u> 1 (Weissauer[33]): A field K is hilbertian if and only if there exists $t \in {}^*K \setminus K$ such that the algebraic closure Ω_t of K(t) in *K possesses the following property:

(+) There exist only finitely many valuations v_P of Ω_t/K such that $v_P(t) < 0$.

Weissauer used this criterion to show that the class of hilbertian fields is very large. For example:

<u>Theorem</u> 8 (Weissauer[33]): Let R be a Krull ring of dimension ≥ 2. Then the quotient field of R is hilbertian.

Since, by the Theorem of Mori-Nagata (cf. Nagata[20], p. 118), the integral closure of a noetherian integral domain is a Krull ring (of the same dimension), we have:

<u>Corollary</u>: The quotient field of a noetherian integral domain of dimension ≥ 2 is hilbertian.

In particular, every power series field $k((t_1,\ldots,t_n))$ in $n \geq 2$ variables is hilbertian. (This is false for $n = 1$!)

Just to show the power of Weissauer's Criterion, let me give a proof of Hilbert's Theorem based on it.

<u>Proof of Theorem</u> 3: Let $t_o \in K$ (t_o not a root of unity), and let $t = t_o^{\omega}$, where $\omega \in {}^*\mathbb{N} \setminus \mathbb{N}$. It is claimed that t satisfies (+) above. To see this, consider the set $P_t = \{v \in {}^*M_K: v(t) < 0\}$. Since t is a power of t_o, we have $P_t = P_{t_o}$. But P_{t_o} is finite (since $t_o \in K$), and hence so is $\dot{P}_t = \{\vartheta \in \dot{M}_K: \vartheta(t) < 0\}$. To conclude the proof of the claim, we now use

<u>Lemma</u> (Robinson[24]): Let $F \supset K(t)$ be a subfield of Ω_t which is finite over K(t). Then every valuation v_P of F/K is induced by a valuation ϑ of *K/K.

Closely connected with this lemma (which can be deduced from Property 1' by using the Riemann-Roch Theorem) is the nonstandard formulation of the Siegel-Mahler Theorem (cf. Robinson-Roquette[25]):

<u>Theorem</u> *2: Let $F \subset {}^*K$ be a function field over K of genus $g \geq 1$. Then every valuation v_P of F/K is induced by a nonstandard valuation $v = \overset{*}{v}$ of ${}^*M_K$.

§3. <u>The Mordell-Weil Theorem</u>

Using the nonstandard dictionary of §1 and the fact that ${}^*J_C(K) = J_C({}^*K)$, we can reformulate Theorems 6 and 7 as follows.

<u>Theorem</u> *6: There exists an integer $m > 1$ and a finite extension K' of K such that $J_C({}^*K')/J_C(K')$ is m-divisible.

<u>Theorem</u> *7: There exists a real-valued function q on $J_C(K)$ such that the relativized function $\overset{*}{q}: J_C({}^*K) \to \overset{*}{\mathbb{R}}$ is a positive definite quadratic form which vanishes precisely on $J_C(K)$.

<u>Remark</u>: To establish the equivalence between Theorems 7 and *7, one also needs "Tate's trick" which enables one to replace a quasi-quadratic form by a quadratic form (cf. Lang[13], p. 87).

In keeping with the philosophy explained above, let us look at the function field analogue of these theorems; i.e. let us replace the field *K by a function field L/K, keeping in mind the "geometric" properties which we have established for *K/K.

Beginning with Theorem *6, note that the following is true: <u>if</u> L/K <u>is a simply connected function field, then</u> $J_C(L) = J_C(K)$. While this fact is no longer true if L/K is not finitely generated, one can prove the following (cf. [10]).

Proposition 2: Suppose that L/K is simply connected and that every finitely generated subfield of L/K has a K-rational place. Suppose further that C is a curve of genus g defined over K which has a K-rational point and a non-special divisor of degree g defined over K. Then the group $J_C(L)/J_C(K)$ is divisible.

Since L = *K satisfies the hypotheses of the proposition by Properties 2 and 3", Theorem *6 is indeed a consequence of Proposition 2. (Note that the additional hypotheses on C can always be realized after a finite extension K' of K.)

For the proof of Theorem *7, we are faced with the problem of constructing a quadratic form on $J_C(*K)$. Now in the geometric analogue, the existence of such a quadratic form is well-known, at least in the case when L/K is a function field of one variable. For then we are concerned with a surface S = C × C' which is a product of two curves (C' being a model of the function field L/K), and the divisor group of a surface carries a canonical symmetric bilinear form $(D.E) \in \mathbb{Z}$, given by counting the intersection indices of the two divisors D and E on S. It is easy to see that this pairing induces a pairing on $J_C(L)$, which is a quotient of a subgroup of the divisor class group Pic(S) of S. For historic reasons, let us write

$$\sigma(D) = -(D.D), \qquad \text{for } D \in J_C(L);$$

σ is called the Weil metric (or Weil trace) on $J_C(L)$.

The fundamental theorem concerning this metric is the celebrated Theorem of Castelnuovo-Severi (cf. Weil[31], Roquette[26]) which may be stated as follows.

Theorem 9 (Castelnuovo-Severi): The Weil metric is positive definite on $J_C(L)$ and vanishes only on $J_C(K)$.

There is a striking resemblance between Theorem *7

and the Theorem of Castelnuovo-Severi; what remains to be done is to make the analogy precise.

To begin with, we have to generalize the notion of an intersection product in such a way that it is applicable to the general situation which we have in mind. For this, we will use Roquette's interpretation (Roquette[26]) of the intersection product. That is, we view the surface as being fibered over the curve C', and define an <u>intersection divisor</u> D.E on C' (called <u>divisor residue</u> in Roquette[26]). By taking the degree of this divisor, one arrives at the previous concept of the intersection product.

This method can easily be generalized to an arbitrary base variety C' (not just a curve). Somewhat more difficult is the generalization to a "non-geometric base" since the local rings involved need no longer be noetherian. Here, the term "non-geometric base" refers to the following situation:

L/K a field extension (of transcendence degree ≥ 1)

M a set of valuations of L/K

C a curve defined over K

Then one has (cf. [9]):

<u>Theorem</u> 10: To each pair D, E of disjoint (= without common components) divisors of C which are defined over L, there exists an intersection divisor $D.E \in \mathfrak{D}(M,L)$ with the following properties:

(1) $(D_1 + D_2).E = D_1.E + D_2.E$

(2) $D \geq 0,\ E \geq 0 \Rightarrow D.E \geq 0$

(3) $D.E = E.D$

(4) D, E rational over K $\Rightarrow D.E = 0$

(5) $D_1 \sim D_2,\ \deg(E) = 0 \Rightarrow D_1.E \sim D_2.E$

(6) If $D = \min\max(f_{ij})$, with $f_{ij} \in F^{\times}$ (where $F = K(C)$), and $E = P$ is a prime divisor rational over L, then

$$D.P = \min_i \max_j (f_{ij}(P)).$$

Since for each pair D, E of divisors of C there exists a divisor $D' \sim D$ which is disjoint from E, and since, by (5) above, the divisor class of $D'.E$ does not depend on the choice of D' when $\deg(E) = 0$, the intersection product defines a symmetric bilinear map on the group of divisor classes of degree 0 on C which are rational over L. As this group may be identified with $J_C(L)$, we obtain:

<u>Corollary</u>: The intersection product induces a symmetric bilinear map

$$(\cdot.\cdot) : J_C(L) \times J_C(L) \to \mathfrak{C}(M,L),$$

where $\mathfrak{C}(M,L)$ denotes the divisor class group of L/K.

If, as in the case that $L = {}^*K$ and $M = \dot{M}_K$, the divisor (class) group possesses a degree function δ with values in some totally ordered group Γ, then we can define a <u>Weil metric</u> with respect to δ by setting, for $D, E \in J_C(L)$,

$$\langle D,E\rangle = \langle D,E\rangle_\delta = -\delta(D.E).$$

One can then generalize the Theorem of Castelnuovo-Severi as follows (cf. [9]).

<u>Theorem</u> 11: The Weil metric $\langle\ ,\ \rangle_\delta$ is positive definite on $J_C(L)$. If, in addition, the degree function δ is non-degenerate, then $\langle D,D\rangle_\delta = 0 \Leftrightarrow D \in J_C(K)$.

While in general the proof is somewhat involved, for divisor classes which are of the form $D \sim P - P_0$, the proof is considerably easier and follows directly from the following <u>Adjunction Formula</u>.

<u>Proposition</u> 3: Let P, P_0 be points on C which are rational over L and K, respectively. Then

$$\langle P-P_0,P-P_0\rangle_\delta = \delta(P.(W+2P_0)),$$

where W is a canonical divisor on C which is rational over K.

Let us now apply the above to the case that $L = {}^*K$ and $\delta = \mathring{\delta}$ is the degree function on $^*K/K$ as constructed in Property 1. Then, by Theorems 10 and 11, we have a positive definite quadratic form on $^*J_C(K)$ which vanishes precisely on $J_C(K)$. Thus we are through with the proof of Theorem *7 once we have verified that this quadratic form is of the form $\mathring{q}$ for some function $q: J_C(K) \to \mathbb{R}$. But this is an easy exercise, since the above quadratic form is defined "geometrically". Summing up, we therefore have a canonical quadratic form h_J on $J_C(K)$ such that

$$\mathring{h}_J(D) \quad = \quad \langle D, D\rangle_{\mathring{\delta}} \; ,$$

for all $D \in {}^*J_C(K) = J_C({}^*K)$. The form h_J is called the <u>Néron-Tate height</u> on $J_C(K)$.

We can also use intersection products to give a non-standard interpretation of the height functions h_A on the curve. Here, $h_A = h \circ \varphi_A$, where φ_A is a rational map associated to the divisor A (cf. Lang[11]).(Note that φ_A and h_A are not uniquely defined.) Using property (6) of Theorem 10, we have:

<u>Proposition</u> 4: If A is a K-rational divisor on C of the form $A = -\min\,(f_i)$, then

$$\mathring{h}_A(P) \quad = \quad \mathring{\delta}(P.A), \qquad \forall\; P \in {}^*C(K) = C({}^*K)$$

Since the above equation is valid for any height function associated to A, and the right hand side depends only on A (and P), we see that for any two choices h_A and h_A' of height functions the difference $h_A - h_A'$ is bounded on $C(K)$. In other words, h_A is <u>equivalent</u> to h_A': $h_A \sim h'_A$.

Using the Adjunction Formula, we can apply this to compute the restriction of h_J to the curve C which we view as embedded in J_C via the map $\Phi_{P_0} : P \to P - P_0$.

<u>Proposition</u> 5: $$h_J \circ \Phi_{P_0} \;\sim\; h_{W+2P_0} \; .$$

§4. The Theorems of Mumford and Dem'janenko-Manin

We shall prove the following equivalent form of Mumford's theorem.

<u>Theorem</u> 4': Let C be a curve of genus $g \geq 2$ defined over a number field K, and assume that the points $P_i \in C(K)$ are ordered according to increasing height $h_W(P_i)$, where W is a canonical divisor on C. Then there exist integers N and N_0 such that for all $n \geq N_0$

$$h_W(P_{N+n}) \geq \frac{16}{9}(g-1)^2 h_W(P_n).$$

The proof of this is based on the following elementary lemma about euclidean spaces (cf. Mumford[18]).

<u>Lemma</u> 1: Let V be a euclidean space with norm $\| \; \|$ and inner product $\langle \; , \; \rangle$, and let $v_1, v_2, \ldots$ be a sequence of vectors such that $\|v_1\| \leq \|v_2\| \leq \ldots$. Suppose that there exist constants $c > 0$ and $\kappa \geq 0$ such that

$$(1) \qquad \langle v_i, v_j \rangle \leq \frac{1}{c}(\|v_i\|^2 + \|v_j\|^2) + \kappa$$

holds for every pair $i \neq j$. Then there exists an integer N such that for all n with $\|v_n\|^2 \geq 3c\cdot\kappa$, we have

$$(2) \qquad \|v_{N+n}\| \geq \frac{2c-4}{3}\|v_n\|.$$

We shall apply this lemma to $V = J_C(K) \otimes \mathbb{R}$, which is finite dimensional by the Mordell-Weil Theorem and becomes a euclidean space under the Néron-Tate height h_J (cf. Lang[13] for a discussion of this). For v_n we shall take

$$v_n = \psi(P_n) = (2g-2)P_n - W.$$

Thus, the proof of Theorem 4' will be complete once we have:

<u>Lemma</u> 2: a) $h_J \circ \psi \sim 2g(2g-2)h_W$.

b) There is a constant $\varkappa \geq 0$ such that (1) holds with $c = 2g$.

The nonstandard version of this lemma is as follows.

Lemma *2: a) $\langle \Psi(P), \Psi(P) \rangle_{\mathfrak{s}} = 2g(2g-2)\mathfrak{s}(P.W)$, $\forall\, P \in C({}^*K)$.

b) $\langle \Psi(P_i), \Psi(P_j) \rangle_{\mathfrak{s}} \leq \frac{1}{2g}(\langle \Psi(P_i), \Psi(P_i) \rangle_{\mathfrak{s}} + \langle \Psi(P_j), \Psi(P_j) \rangle_{\mathfrak{s}})$ for $i \neq j$.

Now a) follows directly from the Adjunction Formula, and b) from the fact that $\mathfrak{s}(P_i.P_j) \geq 0$ by property (2) of the intersection product (cf. [8] or [10] for details).

We now turn to the theorem of Dem'janenko-Manin. To begin with, let me restate the theorem as follows.

Theorem 5': Suppose C' is a curve defined over a number field K which has infinitely many K-rational points. Then for every curve C defined over K we have

$$\text{rank } J_C(E)/J_C(K) \leq \text{rank } J_C(K),$$

where $E = K(C')$ is the function field of C' over K.

To see that Theorem 5' is equivalent to Theorem 5, note that by Deuring[2] or Weil[31]

$$\text{Hom}_K(J_C, J_{C'}) \cong J_C(E)/J_C(K),$$

both groups being isomorphic to the group of K-rational correspondences between C and C'.

By Property 3 of *K/K, the nonstandard reformulation of Theorem 5' is:

Theorem *5: Suppose that $E \subset {}^*K$ is a function field of one variable over K. Then for every curve C defined over K we have

$$\text{rank } J_C(E)/J_C(K) \leq {}^*\text{rank } {}^*J_C(K).$$

Next, we need a criterion for bounding the ranks of $J_C(E)/J_C(K)$ and $J_C(K)$. To do this, let

$\sigma(\ ,\ )$ be the bilinear form associated to the Weil metric on $J_C(E)$

$h_J(\ ,\ )$ be the bilinear form associated to the Néron-Tate height on $J_C(K)$.

<u>Lemma</u> 3: a) rank $J_C(E)/J_C(K) \geq n$ if and only if there exist $D_1, \ldots, D_n \in J_C(E)$ such that $\det(\sigma(D_i,D_j)) \neq 0$.

b) rank $J_C(K) \geq n$ if and only if there exist $D_1, \ldots, D_n \in J_C(K)$ such that $\det(h_J(D_i,D_j)) \neq 0$.

Statement a) resp. b) of the lemma follows from the Theorem of Castelnuovo-Severi resp. the Theorem of Néron-Tate which asserts that σ resp. h_J is a non-degenerate, positive definite quadratic form on $J_C(E)/J_C(K)$ resp. $J_C(K)/J_C(K)_{Tor}$.

<u>Corollary</u>: If $E \subset {}^*K$, then rank $J_C(E)/J_C(K) \leq {}^*\text{rank}\ {}^*J_C(K)$ if and only if $\det({}^*h_J(D_i,D_j)) \neq 0$ for all $D_1, \ldots, D_n \in J_C(E)$ with $\det(\sigma(D_i,D_j)) \neq 0$.

Thus, to conclude the proof of Theorem *5, we need to verify the second condition of the corollary. However, this follows from:

<u>Lemma</u> 4: There exists $\rho \in {}^*\mathbb{R}$ such that for all $D_1, \ldots, D_n \in J_C(E) \subset J_C({}^*K)$ we have

$$\frac{\det({}^*h_J(D_i,D_j))}{\rho^n} \simeq \det(\sigma(D_i,D_j)).$$

Here, as usual, the symbol $a \simeq b$ (for $a, b \in {}^*\mathbb{R}$) means that $a - b$ is <u>infinitesimal</u>.

<u>Proof</u>: Since det is homogeneous of degree n (and since $\sigma(D_i,D_j) \in \mathbb{Z}$), it is enough to show that $\sigma(D_i,D_j)$

$\sim {}^*h_J(D_i,D_j)/\rho$, for some ρ (independent of D_i and D_j).

By definition of the Weil metric, we have

(1) $$\sigma(D_i,D_j) = -\deg_E(D_i.D_j),$$

where $D_i.D_j$ denotes the intersection divisor of D_i and D_j in E. On the other hand, by construction of the Néron-Tate height,

(2) $$\dot{h}_J(D_i,D_j) = -\hat{s}(D_i.D_j),$$

where now the intersection divisor $D_i.D_j$ is viewed as a divisor of *K/K. Finally, by the "generalized lemma of Artin-Whaples" (Robinson-Roquette[25], p. 152), there exists an infinitely large $\rho \in {}^*\mathbb{R}$ such that for any divisor A of E/K

(3) $$\deg_E(A) \sim \hat{s}(A)/\rho.$$

(Note that since ρ is infinitely large, $\hat{s}(A)/\rho$ may be viewed as an element of $^*\mathbb{R}$ modulo infinitesimals.) Combining (1), (2) and (3) then yields the result.

* * *

References

[1] V.A. DEM'JANENKO, Rational points of a class of algebraic curves. Izv. Akad. Nauk SSSR Ser. Mat. 30 (1966), 1373-1396 = Am. Math. Soc. Transl. (2) 66 (1968), 246-272.

[2] M. DEURING, Arithmetische Theorie der Korrespondenzen algebraischer Funktionenkörper I. J. reine angew. Math. 177 (1937), 161-191.

[3] M. DEURING, Arithmetische Theorie der Korrespondenzen algebraischer Funktionenkörper II. J. reine angew. Math. 183 (1940), 25-36.

[4] P.C. GILMORE, A. ROBINSON, Metamathematical Considerations on the relative irreducibility of polynomials. Can. J. Math. 7 (1955), 483-489.

[5] R. HARTSHORNE, Algebraic Geometry. Springer Verlag, New York, 1977.

[6] C. HERMITE, Extrait d'une lettre de M. C. Hermite à M. Bourchardt sur le nombre limité d'irrationalités

auxquelles se réduisent les racines des équations à coefficients entiers complexes d'un degré et d'un discriminant donnés. J. reine angew. Math. 53 (1857), 182-192 = Oeuvres I (Gauthiers-Villars,Paris,1905),415-428.

[7] D. HILBERT, Über die Irreduzibilität ganzer rationaler Funktionen mit ganzzahligen Koeffizienten. J. reine angew. Math. 110 (1892), 104-129 = Gesammelte Abhandlungen II (Springer, Berlin, 1933), 264-286.

[8] E. KANI, Nonstandard diophantine geometry. In: Proc. Queen's Number Theory Conference, 1979 (P. Ribenboim, ed.), Queen's Papers in Pure and Applied Math. No. 54 (Queen's Univ. Press, Kingston, 1980), 129-172.

[9] E. KANI, Eine Verallgemeinerung des Satzes von Castelnuovo-Severi. J. reine angew. Math. 318 (1980),178-220.

[10] E. KANI, On the Néron-Tate height on the Jacobian of a curve. To appear.

[11] S. LANG, Diophantine Geometry. Interscience, New York, 1962.

[12] S. LANG, Les formes bilinéaires de Néron et Tate. Sem. Bourbaki 1963/1964, no. 274.

[13] S. LANG, Elliptic Curves: Diophantine Analysis. Springer Verlag, Berlin, 1978.

[14] K. MAHLER, Über die rationalen Punkte auf Kurven vom Geschlecht 1. J. reine angew. Math. 170 (1934),168-178.

[15] YU. I. MANIN, The Tate height of points on an abelian variety. Its variants and applications. Izv. Akad. Nauk SSSR Ser. Mat. 28 (1964), 1363-1390 = Am. Math. Soc. Transl. (2) 59 (1966), 82-110.

[16] YU. I. MANIN, The p-torsion of elliptic curves is uniformly bounded. Izv. Akad. Nauk SSSR 33 (1969), 459-465 = MAth. USSR Izv. 3 (1969), 433-438.

[17] L.J. MORDELL, On the rational solutions of the indeterminate equation of the third and fourth degrees. Proc. Cambridge Phil. Soc. 21 (1922), 179-192.

[18] D. MUMFORD, A remark on Mordell's Conjecture. Am. J. Math. 87 (1965), 1007-1016.

[19] D. MUMFORD, Abelian Varieties. (App. by C.P. Ramanujam and Yu.I. Manin). Oxford Univ. Press, Oxford, 1974.

[20] M. NAGATA, Local Rings. Krieger Publ. Co., Huntington N.Y., 1975.

[21] A. NERON, Quasi-fonctions et hauteurs sur les variétés abéliennes. Ann. Math. (2) 82 (1965), 249-331.

[22] A. ROBINSON, Non-standard Analysis. North-Holland Publ. Co., Amsterdam, 1966.

[23] A. ROBINSON, Algebraic function fields and non-standard arithmetic. In: Contributions to Non-standard Analysis (W.A.J. Luxemburg, A. Robinson, eds.) North-Holland Publ. Co., Amsterdam, 1972 = Selected Papers II (Yale Univ. Press, New Haven, 1979), 256-269.

[24] A. ROBINSON, Nonstandard points on curves. J. Number Th. 5 (1973), 301-327 = Selected Papers II, 306-332.

[25] A. ROBINSON, P. ROQUETTE, On the finiteness theorem of Siegel and Mahler concerning diophantine equations. J. Number Th. 7 (1975), 121-176 = Selected Papers II, 370-425.

[26] P. ROQUETTE, Arithmetischer Beweis der Riemannschen Vermutung in Kongruenzfunktionenkörper beliebigen Geschlechts. J. reine angew. Math. 191 (1953), 195-252.

[27] P. ROQUETTE, Nonstandard aspects of Hilbert's irreducibility theorem. In: Model Theory and Algebra - a Memorial Tribute to Abraham Robinson (D.H. Saracino, V.B. Weispfenning, eds.) Lecture Notes in Math. 498, (Springer Verlag, Berlin, 1975), 231-275.

[28] C.L. SIEGEL, Über einige Anwendungen diophantischer Approximationen. Abh. Preuss. Akad. Wiss. Phy.-Math.. Kl. 1929, Nr. 1 = Gesammelte Abhandlungen I (Springer Verlag, Berlin, 1966), 206-226.

[29] K.D. STROYAN, W.A.J. LUXEMBURG, Introduction to the Theory of Infinitesimals. Academic Press, New York, 1976.

[30] A. WEIL, L'arithmétique sur les courbes algébriques. Acta Math. 52 (1928), 281-315 = Oeuvres Scien. I (Springer Verlag, New York, 1979), 11-45.

[31] A. WEIL, Sur les Courbes Algébriques et les Variétés qui s'en deduisents. Hermann & Cie., Paris, 1948.

[32] A. WEIL, Arithmetic on algebraic varieties. Ann. Math. (2) 53 (1951), 412-444 = Oeuvres Sc. I,450-482.

[33] R. WEISSAUER, Hilbertsche Körper. Dissertation, Heidelberg, 1980.

AN ADELIC PROOF OF THE HARDY-LITTLEWOOD THEOREM ON WARING's PROBLEM

Gilles Lachaud
Université de Nice
7, Avenue Bieckert
06000 - Nice, France

1. THE GAUSS TRANSFORM

Let F be an homogeneous polynomial with n variables, of degree d, and with coefficients in the ring $\underline{Z}$ of rational integers.

Let ϕ be a standard function on the adelic vector-space $\underline{A}^n$ of dimension n, where $\underline{A}$ is the ring of adeles of the field $\underline{Q}$ of rational numbers, and let χ be the Tate character of $\underline{A}$ (cf. [12]). We define the *Gauss transform of* ϕ *(relative to* F *)* as

$$G_{\underline{A}}(\phi,\xi) = \int_{\underline{A}^n} \phi(x)\chi(\xi F(x))dx$$

where $\xi \in \underline{A}$ and where dx is the Haar measure on $\underline{A}^n$ such that the induced quotient measure satisfies

$$\int_{\underline{A}^n/\underline{Q}^n} dx = 1 \quad .$$

The set P is the set of prime numbers, and the set $\overline{P} = P \cup \{0\}$ is the set of places of the field $\underline{Q}$ (we denote by $|x|_0$ the usual archimedean absolute value of the number $x \in \underline{Q}$).

If K is a field and if F is an homogeneous polynomial with coefficients in K, we set

$$\Delta_K(F) = \{x \in K^n \mid dF(x) = 0\}$$

and we say that F is *strongly non-degenerate on* K if $\Delta_K(F) = \{0\}$.

This definition allows us to introduce the following assumptions:

(SS 1) One has $n > 2d$ and $d \geq 3$;

(SS 2) The homogeneous polynomial F is non-degenerate over $\underline{Q}_p$ for every $p \in \overline{P}$.

We note that the *Fermat form*

$$F(x_1,\dots,x_n) = x_1^d + \dots + x_n^d$$

satisfies the condition (SS 2).

We then have the following result of Igusa [5] depending on Deligne's theorem on trigonometrical sums [2,9].

THEOREM 1. *Under the assumptions* (SS 1) *and* (SS 2) *the integral* $G_{\underline{A}}(\phi,\xi)$ *converges and defines an integrable function on* $\underline{A}$.

2. THE SINGULAR SERIES

If $p \in \overline{P}$ and if $t \in \underline{Q}_p$ we set

$$U_p(t) = \{x \in \underline{Q}_p^n \mid F(x) = t\} \quad .$$

Since F is homogeneous, the hypersurface $U_p(t)$ is non-singular if $t \neq 0$. We denote by $(\omega_t)_p$ the *Leray form* on $U_p(t)$; we have

$$(\omega_t(x))_p \wedge dF(x) = dx$$

if $x \in U_p(t)$.

We denote now by $U_{\underline{A}}(t)$ (cf. [10]) the set of adelic points of the hypersurface $F(x) = t$, and by $\underline{\omega}_t$ the restricted product of the measures $(\omega_t)_p$. If ϕ belongs to the Schwartz-Bruhat space $S(\underline{A}^n)$, we set

$$S_{\underline{A}}(\phi,t) = \int_{U_{\underline{A}}(t)} \phi(x)\ \underline{\omega}_t(x) \quad .$$

We then have the

THEOREM 2 (Weil-Igusa). *Under the assumptions* (SS 1) *and* (SS 2), *one has*

$$S_{\underline{A}}(\phi,t) = \hat{G}_A(\phi,-t)$$

for every $t \neq 0$.

The formula of Theorem 2 implies that if we take for ϕ a decomposable function:

$$\phi(x) = \prod_{p \in \overline{P}} \phi_p(x_p) \quad ,$$

then for $x = (x_p)_{p \in \overline{P}} \in \underline{A}^n$, we have

$$S_{\underline{A}}(\phi,t) = \prod_{p \in \overline{P}} S_p(\phi_p,t)$$

with

$$S_p(\phi_p,t) = \int_{U_p(t)} \phi_p(x)\ (\omega_t(x))_p \quad .$$

We now make the assumption that ϕ_p is the characteristic function of $\underline{Z}_p^n$, and we just set, once this assumption is made,

$$S_p(\phi_p,t) = S_p(t)$$

and

$$S_f(t) = \prod_{p \in P} S_p(t) \quad .$$

The Theorem 2 then also implies that we then have

$$S_p(t) = 1 + \sum_{e \geq 1} A(p^e) \quad ,$$

with

$$A(q) = \sum_{(a,q)=1} \gamma_q(a)\exp(-2i\pi ta/q) \quad ,$$

where

$$\gamma_q(a) = q^{-n} \sum_{x \bmod q} \exp(2i\pi F(x)/q) \quad ;$$

we thus recover the usual expression of the Singular Series for $S_f(t)$ as in [1] and [4].

We have the following result:

PROPOSITION 2. *If* $U_{\underline{A}}(t)$ *is non-empty when* t *is sufficiently large, then*

$$1 \ll S_f(t) \ll 1 \quad .$$

We now assume that the homogeneous polynomial F is the Euler form and that d is even; we fix $t \in \underline{N}$ and set $R = [t^{1/d}] + 1$, in such a way that if

$$||x||_0 = \mathrm{Max}(|x_1|_0, \ldots, |x_n|_0) > R - 1$$

when

$x = (x_1 , \ldots , x_n) \in \underline{R}^n$, then $F(x) > t$.

Let ψ_o be a C^∞ function on $\underline{R}$ which is equal to 1 if $|x|_o \le R - (1/3)$, and equal to 0 if $|x|_o \ge R + (1/3)$. We set

$$\phi_o(x) = \psi_o(x_1) \ldots \psi_o(x_n)$$

for $x = (x_1 , \ldots , x_n) \in \underline{R}^n$. With this choice of ϕ_o , we have

$$S_o(\phi_o,\tau) = S_o(\tau)$$

for any $\tau \le t$, if we define the latter as

$$S_o(t) = t^{(n/d)-1} \int_{U_o(1)} (\omega_1)_o$$

(recall that $U_o(1)$ is compact). If we put

$$S_{\underline{A}}(t) = S_o(t)S_f(t) \quad ,$$

we then have

$$S_{\underline{A}}(\phi,\tau) = S_A(\tau)$$

for $\tau \le t$, when ϕ is the decomposable function on $\underline{A}^n$ constructed from the functions ϕ_o and ϕ_p defined above.

3. THE HARDY-LITTLEWOOD THEOREM

Here, we take for F the Fermat form and we assume that d is even; we set $D = 2^{d-1}$ and

$$N(t) = \# \{x \in \underline{Z}^n \mid F(x) = t\}$$

for $t \in \underline{N}$. We are now going to give a reading, following the adelic language, of the proof given in [1] of the

HARDY-LITTLEWOOD THEOREM. *If* $d \ge 3$ *and* $n > dD$ *, then*

$$N(t) = S_{\underline{A}}(t) \ (1 + O(t^{-\theta}))$$

with $\theta > 0$.

We have to specify that this assertion is equivalent to the following:

$$N(t) = V_o S_f(t) t^{(n/d)-1} + O(t^{(n/d)-1-\theta})$$

once the global Singular Series has been explicited. If t is sufficiently large, we know that $U_{\underline{A}}(t) \neq \phi$ (cf. [1], lemma 11); thus the Proposition 2, joined to the above assertion, implies that $N(t) > 0$ and consequently $U_{\underline{Q}}(t) \neq \phi$.

With the function ϕ chosen in n° 2, we have

$$N(t) = \# \{x \in \underline{Q}^n \mid \phi(x) = 1 \text{ and } F(x) = t\} ,$$

so that in fact the Hardy-Littlewood theorem establishes a relation between

$$N(t) = \sum_{U_{\underline{Q}}(t)} \phi(x)$$

and

$$S_{\underline{A}}(t) = \int_{U_{\underline{A}}(t)} \phi(x) \, \underline{\omega}_t(x) .$$

If $\xi \in \underline{A}$, we introduce the (finite) trigonometrical sum

$$f(\xi) = \sum_{x \in \underline{Q}} \psi(x) \chi(x^d \xi) ,$$

where $\psi = \Pi \psi_p$, denoting by ψ_p the characteristic function of $\underline{Z}_p$ if $p \in P$ and by ψ_o the C^∞ function with compact support on $\underline{R}$ introduced in n° 2. We thus have

$$f(\xi)^n = \sum_{x \in \underline{Q}^n} \phi(x) \chi(\xi F(x))$$

$$= \sum_t \chi(\xi t) N(t)$$

and consequently,

$$N(t) = \int_{\underline{A}/\underline{Q}} f(\xi)^n \chi(-t\xi) \, d\xi . \qquad (*)$$

We set also

$$g(\xi) = \int_{\underline{A}} \psi(x) \chi(x^d \xi) \, dx ,$$

so that

$$S_{\underline{A}}(t) = \int_{\underline{A}} g(\xi)^n \chi(-t\xi) \, d\xi . \qquad (**)$$

In order to establish a comparison between $N(t)$ and $S_{\underline{A}}(t)$, we are going to introduce a subset $M \subset \underline{A}$ such that f^n and g^n are negligibles out of M and such that the difference $f^n - g^n$ is negligible in M. For $\xi \in \underline{A}$, we put

$$\underline{Q}(\xi) = \prod_{p \in P} \mathrm{Max}(1, |\xi_p|_p) ,$$

and we define the ***major set*** in $\underline{A}$ (depending on $\delta > 0$) as

$$M = \{\xi \in \underline{A} \mid |\xi|_0 \le R^{-d+\delta} \text{ and } Q(\xi) \le R^{\delta}\} .$$

The restriction of the projection from $\underline{A}$ to $\underline{A}/\underline{Q}$ is injective on M; we denote its image by $\overline{M}$, and we define the ***minor set*** $\overline{m}$ as the complementary of $\overline{M}$ in $\underline{A}/\underline{Q}$.

Using Weyl's inequality, we have:

PROPOSITION 3. *If* $n > dD$ *and* $dD\, n < \delta < 1$, *then*

$$\int_{\overline{m}} |f(\xi)|^n \, d\xi \ll R^{n-d-\theta_1}$$

with $\theta_1 > 0$.

A careful estimation of the behaviour of the function g shows:

PROPOSITION 4. *If* $n > 2d$, *one has*

$$\int_{\underline{A}-M} |g(\xi)|^n \, d\xi \ll R^{n-d-\theta_2}$$

with $\theta_2 > 0$.

With the aid of the Poisson summation formula on $\underline{A}$, we can establish the

PROPOSITION 5. *If* $\delta < 1$, *and if* $n > 3D$, *then*

$$\int_M \{f(\xi)^n - g(\xi)^n\} \chi(-t\xi) d\xi \ll R^{n-d-\theta_3}$$

with $\theta_3 > 0$.

Since $R^d \sim t$, the Hardy-Littlewood theorem is then a consequence of relations (*) and (**) and of Propositions 3, 4 and 5.

The reader will find in the article [7] detailed proofs of the preceding propositions, together with a survey of the theory of the Gauss transform. What I would like to add, finally, is that I hope to have given an illustration of the fact that the adelic language can be useful in itself without any intervention of class field or algebraic group theory: like the theory of congruences, this topic, which appeared for rather elaborate purposes, can be used for current mathematics.

REFERENCES

[1] H. Davenport, Analytic methods for diophantine equations and approximations, Ann Arbor Publishers, Ann Arbor, 1962.

[2] P. Deligne, La conjecture de Weil I, Publ.Math. I.H.E.S. 43 (1974), 273-307.

[3] W.J. Ellison, Waring's problem, Amer.Math. Monthly 78 (1971), 10-36.

[4] G.H. Hardy and J.E. Littlewood, Some problems of 'Partitio Numerorum', I, II, IV, VI, in the Collected works of G.H. Hardy, vol. 1, 405-505, Oxford University Press, Oxford, 1966.

[5] J.I. Igusa, On a certain Poisson formula, Nagoya Math.J. 53 (1974) 211-233.

[6] J.I. Igusa, Lectures on forms of higher degree, Tata Institute of Fundamental Research Nr 59, Springer, Berlin, 1978.

[7] G. Lachaud, Une présentation adélique de la Série Singulière et du problème de Waring, to appear in 'L'Enseignement Mathématique'.

[8] T. Ono, Gauss transforms and Zeta functions, Ann. of Math. 91 (1970), 332-361.

[9] J.P. Serre, Majoration de Sommes exponentielles, Journées Arithmétiques de Caen, Astérisque 41-42 (1977), 111-126.

[10] A. Weil, Adeles and algebraic groups, I. A.S. Princeton, 1961.

[11] A. Weil, Sur la formule de Siegel dans la théorie des groupes classiques, Acta Math. 113 (1965), 1-87, in Oeuvres Scientifiques, vol. III, 71-157, Springer, Heidelberg, 1979.

[12] A. Weil, Basic number theory, Grundl.Math.Wiss.Bd. 144, Springer, Heidelberg, 1967, 3rd ed. 1974.

CLASS NUMBERS OF REAL ABELIAN NUMBER FIELDS OF SMALL CONDUCTOR

By F.J. van der Linden.

Introduction

For the class number of an abelian numberfield K we have a decomposition $h = h^+ . h^-$, where h^+ is the class number of the maximal real subfield K^+ of K, and h^- is a positive integer, for which there exists an explicit formula. In this lecture we are concerned with the determination of h^+ , i.e. the determination of the class number of a real abelian number field.

For an abelian number field K the *conductor* $f(K)$ is defined as the smallest positive integer f for which $K \subset \mathbb{Q}(\zeta_f)$, with ζ_f a primitive f-th root of unity. In this lecture we show how to compute the class numbers of most real abelian number fields of conductor ≤ 200 , in some cases assuming the generalized Riemann hypothesis. For more details see [3].

§ 1 The results

In theorems 1, 2, 3 and 4 we list our results. By GRH we denote the generalized Riemann hypothesis for the zeta-function of the Hilbert class field of $\mathbb{Q}(\zeta_{f(K)})$. The Euler function is denoted by ϕ.

Theorem 1 Suppose that $f(K) = q$ is a prime power. Then

$h(K) = 1$ if $\phi(q) \leq 66$.

Theorem 2 Suppose that $f(K) = q$ is a prime power, and assume GRH. Then

$h(K) = 4$ if $q = 163$

$h(K) = 1$ for all other K for which $\phi(q) \leq 162$.

Suppose next that $f(K) = n$ is not necessarily a prime power. The *genus field* $G(K)$ of K is defined as the maximal totally unramified extension of K which is abelian over $\mathbb{Q}$. It is contained in $\mathbb{Q}(\zeta_n)$, and it is equal to $G^*(K) \cap \mathbb{R}$, where $G^*(K)$ is the smallest field containing K which is a composite of abelian extensions of $\mathbb{Q}$ of prime power conductors. Because $g(K)$ is a subfield of the *Hilbert class field* $H(K)$ of K , it follows that the *genus factor* $g(K) = [G(K) : K]$ divides the class number $h(K)$. Clearly $g(K) = 1$ for fields of prime power conductor.

Theorem 3 Suppose that $f(K) = n$ is not a prime power. Then

$h(K) = 2.g(K) = 2$ for $K = \mathbb{Q}(\zeta_{136})^+$

$h(K) = g(K)$ for all other K for which $n \leq 200$, $\phi(n) \leq 72$, $n \neq 148$, $n \neq 152$.

$h(K) = g(K)$ for $n = 165$.

Theorem 4 Suppose that $f(K) = n$ is not a prime power. Assume GRH. Then

$h(K) = 2.g(K) = 2$ for $K = \mathbb{Q}(\zeta_{136})^+$

$h(K) = 2.g(K)$ if $n = 145$ and $\sqrt{145} \in K$,

$h(K) = 4.g(K) = 4$ if $n = 183$ and $12 \mid [K : \mathbb{Q}]$,

$h(K) = g(K)$ for all other K with $n \leq 200$.

§ 2 The method

The results of section 1 are derived by a method developed by J.M. Masley [4]. This method consists of two parts:

1) Determine an upper bound B for the class number h.

2) Test for each prime $p \leq B$ whether $p \mid h$.

1) The classical nethods from geometry of numbers lead, for most fields, to class number bounds that are too large to be useful. For fields with small conductors, however, there is a better technique, depending on the discriminant lower bounds proved by Odlyzko [5; cf. 6]. If K is a totally real field of degree n and discriminant Δ over $\mathbb{Q}$, then Odlyzko proved that

$$\Delta > A^n . e^{-E}$$

for several pairs (A, E), like

$$A = 28.668 \quad \text{and} \quad E = 8.0001$$

$$A = 60.704 \quad \text{and} \quad E = 200.01$$

$$A \to 60.840 \quad \text{for} \quad E \to \infty \quad .$$

If we assume the generalized Riemann hypothesis for the zeta-function of K we can also take

$$A = 30.338 \quad \text{and} \quad E = 8.0894$$

$$A = 213.626 \quad \text{and} \quad E = 5.7672 \times 10^{26}$$

$$A \to 215.333 \quad \text{for} \quad E \to \infty$$

These bounds lead to class number bounds in the following way. Apply the above inequality to the Hilbert class field $H(K)$ of K. Since $H(K)$ has discriminant Δ^h and degree $n.h$ over $\mathbb{Q}$, this yields

$$\Delta^h > A^{h.n} . e^{-E}$$

so

$$h < \frac{E}{n \log A - \log \Delta}$$

provided that the right hand side is positive.

In this way we can get class number upper bounds if the *root discriminant* $\Delta^{1/n}$ satisfies $\Delta^{1/n} < A$ for some A ; that is, $\Delta^{1/n} < 60.840$, or when we assume GRH: $\Delta^{1/n} < 215.333$. In the latter case practical upper bounds are obtained only when $\Delta^{1/n}$ is smaler then about 160. There are only finitly many abelian K satisfying this condition. Since the root discriminant and the conductor are roughly of the same size, the conductor is a good measure of how far we can go.

2) There are two main methods to decide whether a prime divides the class number.
The first ones uses the Galois action and works mostly for primes not dividing $[K : \mathbb{Q}]$.

<u>Definition 5</u> Let L/K be an abelian extension, and p a prime number not dividing $n = [L : K]$. Then we denote by $Cl_p(L)$ the p-primary part of the class group of L. If L/K is cyclic and σ is a generator of Gal (L/K) then we put

$$Cl_p^*(L/K) = \{\alpha \in Cl_p(L) \mid \alpha^{\Phi_n(\sigma)} = 1\}$$

where Φ_n is the n-th cyclotomic polynomial.

<u>Theorem 6</u> Let L/K be an abelain extension, and p a prime number not dividing $[L : K]$. For the p-primary part of the class group of L we have a decomposition:

$$Cl_p(L) = \oplus \, Cl_p^*(M/K) ,$$

where the summation is over the M with $K \subset M \subset L$ and M/K is *cyclic*.

This theorem and the following one are due to Fröhlich [1]. A proof of them is given in section 3.

Using theorem 6 gives us information about the p-part of the class number from the p-parts of the class number of its subfields, especially if L/K is non-cyclic. Even if L/K is cyclic we can use it in may cases because of the following theorem.

Theorem 7 For $Cl_p^*(L/K)$, as defined in definition 5 we have: $\# Cl_p^*(L/K)$ is a power of p^f, where f is the smallest positive integer for which $p^f \equiv 1 \bmod [L : K]$.

This is a very strong restriction because often we have $p^f > B$.

For p dividing $[K : \mathbb{Q}]$ we can make use of the following theorem

Theorem 8 Let K/K_0 be a p-extension, i.e. a Galois extension with $Gal(K/K_0)$ a p-group. Let P be a set of (finite or infinite) primes of K and q a prime of K. Suppose that K/K_0 is unramified outside $P \cup \{q\}$. If $p \mid h(K)$ then there exists a cyclic extension M/K_0 of degree p that is unramified outside P.

Proof See Masley [4] (2.6) □

If we take $P = \emptyset$ in theorem 8 we recover a result of Iwasawa [2]: if K/K_0 is a p-extension ramifying at at most one prime, then $p \mid h(K) \Rightarrow p \mid h(K_0)$.

If $p \mid [K : \mathbb{Q}]$ we choose a subfield K_0 of K for which K/K_0 is a p-extension. Suppose that $p \mid h(K)$. The extension of M, implied by theorem 8, is equivalent, by class field theory, to the exsistence of a certain quotient of a ray class group of K_0. In many cases it is possible to disprove the existence of this quotient group, by calculations with the units of K_0. If it turns out that M does exist we can look if MK is unramified over K. For the fields of conductors 136 and 183 the Hilbert class fields and class numbers were found in this way.

When we have used the above methods, we are sometimes left

with a few primes $p \leq B$. Usually these are primes of which a small power is $1 \bmod [K : \mathbb{Q}]$. For these primes we can try to use reflection principles or the relation between cyclotomic units and the class number. In some cases we can use the following theorem, which is a counter part to theorem 8.

<u>Theorem 9</u> Let L/K be an extension of number fields. Then

$h(K) \mid h(L).[L : K]$.

If no intermediate field $M \neq K$ of L/K is unramified over K, then $h(K) \mid h(L)$.

<u>Proof</u> See Masley. [4] (2.2). □

§ 3 The proofs of theorem 6 and 7

Proof of theorem 6

Let L/K be an abelian extension with Galois group G. For any intermediate field M of L/K we write $C(M) = Cl_p(M)$. We denote by $C(L)^{Gal(L/M)}$ the subgroup of $C(L)$ consisting of those elements which are left fixed by $Gal(L/M)$. We have maps:

$$\iota : C(M) \to C(L)^{Gal(L/M)} , \quad N : C(L)^{Gal(L/M)} \to C(M)$$

induced by the inclusion $M \subset L$ and the ideal norm map respectively. For elements α of $C(M)$ and β of $C(L)^{Gal(L/M)}$ we have

$$N\iota(\alpha) = \alpha^{[L : M]} \quad \text{and} \quad \iota N(\beta) = \beta^{[L : M]}.$$

since $p \nmid [L : M]$ this means that ι and N are isomorphisms. We identify $C(M)$ with $C(L)^{Gal(L/M)}$ by means of ι.

For a ring R and a cyclic group U of order m, with generator σ. We define

$$R(U) = R[U] / \Phi_m(\sigma). R[U]$$

where Φ_m is the m-th cyclotomic polynomial

Let $\mathcal{H}$ be the set of subgroups $H \subset G$ for which G/H is cyclic. For each $H \in \mathcal{H}$ we have an natural ring homomorphism

$$\phi_H : R[G] \to R[G/H] \to R(G/H).$$

The kernel of ϕ_H is generated by $\{\tau - 1 : \tau \in H\}$ and $\Phi_m(\sigma)$, where σ and H generate G and $M = \#G/H$.

Take $R = \mathbb{Q}$. Then each $\mathbb{Q}(G/H)$ is a *field*, so the ideals $\ker(\phi_H)$ are maximal, and it is easy to see that they are different. By the chinese remainder theorem it follows that the combined ring homomorphism

$$\phi : \mathbb{Q}[G] \to \Pi_{H \in \mathcal{H}} \mathbb{Q}(G/H)$$

Is surjective. An easy calculation shows that both sides have the same $\mathbb{Q}$-dimension so ϕ is an isomorphism. Tensoring with $\mathbb{Q}_p$ and restricting the coefficients to $\mathbb{Z}_p$ we obtain an inclusion of $\mathbb{Z}_p[G]$ as a subring of finite index in $\Pi_{H \in \mathcal{H}} \mathbb{Z}_p(G/H)$. This index divides the discriminant of $\mathbb{Z}_p[G]$ over $\mathbb{Z}_p$, which is a unit in

$\mathbb{Z}_p$. We conclude that

$$\mathbb{Z}_p[G] \simeq \Pi_{H \in \mathcal{H}} \mathbb{Z}_p(G/H) .$$

This isomorphism leads, for any $\mathbb{Z}_p[G]$-module A , to a decomposition

$$A \simeq \Pi_{H \in \mathcal{H}} A_H$$

Where A_H is the largest submodule of A annihalited by $\ker(\phi_H)$. With $A = C(L)$ we find

$$A_H = Cl_p^*(L^H/K),$$

where L^H is the fixed field of H . This proves theorem 6.

<u>Proof of theorem 7</u>

Let L/K be a cyclic extension of degree n , with Galois group G generated by σ . If we prove that for each $\alpha \neq 1$ in $Cl_p^*(L/K)$, the stabilizor of α in G is $\{1\}$, then by counting orbits we find that $\#Cl_p^*(L/K) \equiv 1 \bmod n$, and the theorem is proved.

So suppose that $\sigma^d\alpha = \alpha$, where $d \neq n$ divides n , and L' the fixed field of σ^d . Then

$$N_{L/L'}(\alpha) = \Pi_{i=1}^{n/d} \sigma^{di}(\alpha) = \alpha^{n/d}.$$

On the other hand, since $\Phi_n(\sigma)$ divides $\Sigma_{i=1}^{n/d} \sigma^{di}$ in the groupring and $\alpha^{\Phi_n(\sigma)} = 1$, we have

$$N_{L/L'}(\alpha) = 1$$

Hence $\alpha^{n/d} = 1$, and from $p \nmid n$ it follows that $\alpha = 1$ as required.

§ 4 Example

Let K be the field of conductor 95 and degree 12 over $\mathbb{Q}$. It is cyclic over $\mathbb{Q}$. Denote by K_i the subfield of K of degree i over $\mathbb{Q}$, for i = 2, 3, 4, 6. From [4] we get $h(K_i) = 1$. The class number bound gives $h(K) \leq 6$. Using theorem 7 for the extension K/K_3 we get $3 \nmid h(K)$, using it for the extension $K/\mathbb{Q}$ we get $5 \nmid h(K)$. So h(K) is a 2-power.

The extension K/K_3 ramifies only at the prime q over 5 and the prime p over 19. Using theorem 8 for the extension K/K_3, with $P = \{p\}$ we see that if $2 \mid h(K)$ then there exists an extension M/K_3, cyclic of degree 2 and only ramifying at p. By class field theory M belongs to a quotient group of order 2 of

$$(\mathcal{O}(K_3)/p)^* \ / \ (\mathcal{O}(K_3)^* \bmod p)$$

But this is a group of odd order, because $N(p) = 19 \equiv 3 \bmod 4$ and $-1 \in \mathcal{O}(K_3)^*$ is of order 2. So we have reached a contradiction and $2 \nmid h(K)$. This means that $h(K) = 1$.

Literature

[1] A. Fröhlich On the class group of relative abelian fields. *Quart. J. Oxf.* (2) 3 98-106 (1952)

[2] K. Iwasawa A note on class numbers of algebraic number fields. *Abh. Math. Sem. Univ. Hamburg* 20 257-258 (1956)

[3] F.J. van der Linden Class number computations of real abelian number fields. To appear.

[4] J.M. Masley Class numbers of real cyclic number fields with small conductor. *Compositio Math.* 37 297-319 (1978).

[5] A.M. Odlyzko Discriminant bounds. Unpublished tables (nov. 1976).

[6] G. Poitou Minorations des discriminants. *Sém. Bourbaki* 479 (1976).

ALGEBRAIC INDEPENDENCE PROPERTIES OF VALUES OF ELLIPTIC FUNCTIONS

D.W. Masser
Department of Mathematics
University of Nottingham
University Park, Nottingham, U.K.

G. Wüstholz
Gesamthochschule Wuppertal
Gaußstraße 20
D-5600 Wuppertal, W. Germany

1. INTRODUCTION

In 1949 A.O. Gelfond published a method of obtaining algebraic independence results for certain values of the exponential function. Since then a number of authors have developed this method, notably W.D. Brownawell, A.A. Smelev, R. Tijdeman, and M. Waldschmidt. A good exposition of the most recent theorems can be found in [6], together with references and historical remarks. The object of this note is to announce the natural elliptic analogues of these theorems.

For the convenience of the reader we state first the known results in the exponential case. For an integer $n \geq 1$ let $u_1, \ldots, u_n$ be complex numbers linearly independent over the rational field Q, and for an integer $m \geq 1$ let $v_1, \ldots, v_m$ also be complex numbers linearly independent over Q.

THEOREM 1. *If* $mn \geq 2m + 2n$ *then at least two of the numbers*

$$\exp(u_i v_j) \qquad (1 \leq i \leq n, \ 1 \leq j \leq m)$$

are algebraically independent over Q.

THEOREM 2. *If* $mn \geq 2m + n$ *then at least two of the numbers*

$$u_i, \exp(u_i v_j) \qquad (1 \leq i \leq n, \ 1 \leq j \leq m)$$

are algebraically independent over Q.

THEOREM 3. *If* $mn > m + n$ *then at least two of the numbers*

$$u_i, v_j, \exp(u_i v_j) \qquad (1 \leq i \leq n, \ 1 \leq j \leq m)$$

are algebraically independent over Q.

A corollary of Theorem 2, which was first proved by Gelfond [2], is the algebraic independence of α^β and α^{β^2} whenever $\alpha \neq 0,1$ is algebraic and β is a cubic irrationality.

2. THE RESULTS

For the elliptic analogues we take a Weierstrass elliptic function $\wp(z)$ whose invariants g_2, g_3 are algebraic numbers. If $\wp(z)$ has complex multiplication we denote by K the associated imaginary quadratic field; otherwise, if $\wp(z)$ has no complex multiplication, K denotes simply $\mathbb{Q}$. For an integer $n \geq 1$ let $u_1, \ldots, u_n$ be complex numbers linearly independent over K, and for an integer $m \geq 1$ let $v_1, \ldots, v_m$ be complex numbers linearly independent over $\mathbb{Q}$. The possible lack of symmetry in these conditions leads to the following four analogues of the above three theorems.

THEOREM 4. *If* $mn \geq 2m + 4n$ *then at least two of the numbers*

$$\wp(u_i v_j) \qquad (1 \leq i \leq n \,,\ 1 \leq j \leq m)$$

are defined and are algebraically independent over $\mathbb{Q}$.

THEOREM 5. *If* $mn \geq 2m + 2n$ *then at least two of the numbers*

$$u_i, \wp(u_i v_j) \qquad (1 \leq i \leq n \,,\ 1 \leq j \leq m)$$

are defined and are algebraically independent over $\mathbb{Q}$.

THEOREM 5'. *If* $mn \geq m + 4n$ *then at least two of the numbers*

$$v_j, \wp(u_i v_j) \qquad (1 \leq i \leq n \,,\ 1 \leq j \leq m)$$

are defined and are algebraically independent over $\mathbb{Q}$.

THEOREM 6. *If* $mn > m + 2n$ *then at least two of the numbers*

$$u_i, v_j, \wp(u_i v_j) \qquad (1 \leq i \leq n \,,\ 1 \leq j \leq m)$$

are defined and are algebraically independent over $\mathbb{Q}$.

From Theorem 5 or Theorem 5' we deduce that if $\wp(z)$ has complex multiplication over the imaginary quadratic field K, and β is cubic over K, then for any complex number u such that $\wp(u)$ is defined and algebraic, the numbers $\wp(\beta u)$, $\wp(\beta^2 u)$ are defined and are algebraically independent over $\mathbb{Q}$.

It seems likely that Theorems 4, 5, 5', and 6 are the best that can be obtained by Gelfond's method as it stands. We note that Theorem 5 improves upon a recent result of the second author [7].

3. ZERO-ESTIMATES

The essential tools for the proofs of the results of section 2 are suitable zero-estimates for polynomials in elliptic functions. The corresponding estimates in the exponential case were obtained in a very sharp form by Tijdeman [5] using analytic methods. But it seems difficult to generalize these methods in the correct way. By contrast, our arguments make use of techniques from commutative algebra. This approach was initiated by J.V. Nesterenko in [4], and developed further by Brownawell and the first author in [1].

For simplicity we state only the zero-estimate needed for the proof of Theorem 4. In the notation of section 2, let w be a complex number such that $w + u_i(s_1v_1 + \ldots + s_mv_m)$ does not lie in the period lattice of $\wp(z)$ for any integer i with $1 \le i \le n$ and any integers $s_1,\ldots,s_m$.

PROPOSITION. *Suppose* $mn \ge 2m + 4n$. *For real numbers* $D \ge 1$, $S \ge 0$ *let* $P(x_1,\ldots,x_n)$ *be a non-zero polynomial of total degree at most* D *such that the function*

$$f(z) = P(\wp(w + u_1z),\ldots,\wp(w + u_nz))$$

vanishes at the points

$$z = s_1v_1 + \ldots + s_mv_m$$

for all integers $s_1,\ldots,s_m$ *with* $0 \le s_1,\ldots,s_m \le S$. *Then* $D \ge 3^{-N}(S/n)^{m/n}$, *where* $N = 3^n - 1$.

For Theorem 5 we need a similar conclusion when $f(z)$ has zeroes of high multiplicity, and for Theorem 5' we require a result on the simple zeroes of functions of the form $P(z,\wp(w + u_1z),\ldots,\wp(w + u_nz))$ where $P(x_0,x_1,\ldots,x_n)$ is a polynomial whose degree in x_0 need not be the same as its total degree in $x_1,\ldots,x_n$. Finally for Theorem 6 we need a combination of all these estimates.

4. GENERALIZATIONS

There is no essential difficulty in extending all the results of section 2 to abelian functions, provided we have the appropriate zero-estimates. As an example, we give the estimate that yields generalizations of Theorem 4; it puts the above Proposition in the context of more general group varieties.

For integers $n \ge 1$, $N \ge 1$ let G be a quasi-projective commutative group variety of dimension n in complex projective space of N dimensions, and for an integer $\ell \geq 0$ let Γ be a finitely generated subgroup of G of rank ℓ. We define a number $\mu = \mu(\Gamma, G)$ in the following way. For an integer r with $1 \leq r \leq n$ put $\ell_r = 0$ if G has no algebraic subgroup of dimension $n - r$; otherwise let ℓ_r be the maximum rank of any subgroup of Γ which lies in some algebraic subgroup of G of dimension $n - r$. Then

$$\mu(\Gamma, G) = \min_{1 \leq r \leq n} (\ell - \ell_r)/r \quad .$$

For an integer $m \geq 1$ let $\gamma_1, \ldots, \gamma_m$ be generators of Γ, not necessarily linearly independent.

PROPOSITION. *There is a real number* $\lambda > 0$ *depending only on* G *with the following property. For real numbers* $D \geq 1$, $S \geq 0$ *let* $P(x_0, \ldots, x_N)$ *be a homogeneous polynomial of degree at most* D *which vanishes at the points*

$$s_1\gamma_1 + \ldots + s_m\gamma_m$$

for all integers $s_1, \ldots, s_m$ *with* $0 \leq s_1, \ldots, s_m \leq S$ *but does not vanish identically on* G. *Then* $D \geq \lambda(S/n)^{\mu}$.

This result is proved in [3], where it is also shown that the number μ is the natural exponent which leads to a best possible estimate.

REFERENCES

1. W.D. Brownawell and D.W. Masser, Multiplicity estimates for analytic functions II, Duke Math.J. 47 (1980), 273-295.

2. A.O. Gelfond, Transcendental and algebraic numbers, Dover, New York (1960).

3. D.W. Masser and G. Wüstholz, Zero-estimates on group varieties I, to appear in Inventiones Math.

4. J.V. Nesterenko, Bounds on the order of zeroes of a class of functions and their application to the theory of transcendental numbers, Izv.Akad.Nauk. SSSR, Ser.Mat. 41 (1977), 253-284.

5. R. Tijdeman, On the number of zeroes of general exponential polynomials, Indagationes Math. 33 (1971), 1-7.

6. M. Waldschmidt, Nombres transcendants, Lecture Notes in Math. 402, Springer-Verlag (1974).

7. G. Wüstholz, Algebraische Unabhängigkeit von Werten von Funktionen, die gewissen Differentialgleichungen genügen, J. reine angew Math. 317 (1980), 102-119.

ESTIMATIONS ÉLÉMENTAIRES EFFECTIVES SUR LES NOMBRES ALGÉBRIQUES

Maurice Mignotte

RÉSUMÉ

Si P est un polynôme à coefficients entiers, α un nombre algébrique qui n'est pas une racine de P, S un ensemble de places du corps $\mathbb{Q}(\alpha)$, nous obtenons une minoration de l'expression $\prod_{v\in S} |P(\alpha)|_v$. Cette minoration est parfois meilleure qu'une estimation de Liouville; c'est par exemple le cas lorsque P est fixé, que le degré de α tend vers l'infini tandis que la mesure de α reste bornée.

1. INTRODUCTION

Le présent travail constitue un prolongement de [3], où on considérait la quantité $|\alpha-\beta|$, α et β étant des nombres algébriques non conjugués. Cette fois, on cherche à minorer la quantité $\prod_{v\in S} |P(\alpha)|_v$, où P est un polynôme à coefficients entiers, α un nombre algébrique non racine de P et S un ensemble quelconque de places du corps $\mathbb{Q}(\alpha)$.

Comme annoncé dans le titre, la méthode est purement élémentaire, ell utilise tout au plus le principe du maximum pour les polynômes (exercice: en donner une preuve algébrique). Le début consiste en la construction d'un polynôme auxiliaire divisible par une puissance assez grande de P, ceci grâce à une variante convenable du lemme de Siegel.

L'estimation ainsi obtenue est raisonnable en fonction des mesures de P et α, et particulièrement bonne en fonction du degré de α.

Parmi les applications possibles de ce résultat, les deux suivantes me paraissent intéressantes:

- minoration du plus petit nombre premier non ramifié et totalement décomposé dans le corps $\mathbb{Q}(\alpha)$ (voir le corollaire 1 au théorème 4),
- majoration du nombre de conjugués réels d'un nombre de mesure assez petite (voir le corollaire 1 au théorème 5).

J'ai tenu à ne pas multiplier les corollaires curieux comme le suivant (non démontré dans le texte):

Si α est un nombre de Pisot ou de Salem, de degré D, pour tout $\varepsilon > 0$ on a l'inégalité

$$\|\alpha\| \geq |\alpha|^{-(1+\varepsilon)D} \quad \text{pour} \quad D \geq D_0 \ ,$$

(où $\| \ \|$ désigne la distance à l'entier le plus proche) meilleure, lorsque $|\alpha|$ reste borné, que l'estimation évidente

$$\|\alpha\| \geq (|\alpha| + 1 + \varepsilon)^{-D+1} \ .$$

Il n'est peut-être pas exclu que les estimations figurant ici, bien que particulières, puissent être utiles dans certaines démonstrations de la théorie des nombres transcendants.

Dans la suite nous utilisons les notations suivantes:

si α est un entier algébrique de degré D et si $\alpha_1,\ldots,\alpha_D$ sont les conjugués de α (dans le corps des complexes), la mesure de α est définie par la formule

$$M(\alpha) = \prod_{i=1}^{D} \max\{1, |\alpha_i|\} \ .$$

Si P est un polynôme à coefficients complexes

$$P(z) = a_0 \prod_{i=1}^{D} (z-z_i)$$

sa mesure vaut

$$M(P) = |a_0| \prod_{i=1}^{D} \max(1,|z_i|) \ ,$$

tandis que $L(P)$ désigne la somme des valeurs absolues des coefficients de P, c'est la longueur de P, la hauteur $H(P)$ étant le maximum des modules de ces coefficients.

Lorsque v est une place d'un corps de nombres et P un polynôme à coefficients entiers, il est commode de poser

$$L_v(P) = \begin{cases} L(P) & \text{si } v \text{ est archimédienne,} \\ 1 & \text{sinon.} \end{cases}$$

2. UNE VARIANTE DU LEMME DE SIEGEL

LEMME 1. *Soit* P *un polynôme à coefficients entiers, primitif (c'est-à-dire dont les coefficients sont premiers entre eux dans leur ensemble), de degré* d *, dont la décomposition en facteurs irréductibles sur* $\mathbb{Z}[X]$ *est de la forme* $P_1^{r_1} \ldots P_k^{r_k}$ *, les* P_j *étant distincts et de degré respectif* d_j *. Soient* N *et* T *des entiers positifs qui vérifient* $N > dT$ *. Alors, il existe un polynôme* F

non nul, à coefficients entiers, divisible par P^T *, de degré au plus* N *et qui vérifie*

$$H(F) \le \{4^{\Omega}(N+1)^{d^*(T+1)/2}\, M(P^N)\}^{\frac{T}{N+1-dT}} ,$$

où on a posé

$$\Omega := r_1 + \dots + r_k , \quad d^* := r_1^2 d_1 + \dots + r_k^2 d_k .$$

Cherchons F sous la forme $\sum_{i=0}^{N} a_i X^i$, les a_i étant les inconnues. Pour $j = 1,\dots,k$, soit α_j une racine de P_j dans $\mathbb{C}$. La condition P^T divise F est équivalente au système d'équations

$$\sum_{i=0}^{N} \binom{i}{t} \alpha_j^{i-t} a_i = 0 , \quad 0 \le t < r_j T , \ 1 \le j \le k .$$

Posons

$$\eta_j = \begin{cases} 1 & \text{si } \alpha_j \text{ est réel ,} \\ 2 & \text{sinon .} \end{cases}$$

Soient H , $\ell_{1,1},\dots,\ell_{1,r_1R},\dots,\ell_{k,1},\dots,{}_{k,r_kT}$ des entiers positifs qui vérifient l'inégalité

$$(\ell_{1,1} \cdots \ell_{1,r_1T})^{\eta_1} \cdots (\ell_{k,1} \cdots \ell_{k,r_kT})^{\eta_k} < (H+1)^{N+1} . \qquad (1)$$

Une utilisation classique du principe des tiroirs montre qu'il existe des entiers a_i , $0 \le i \le N$, non tous nuls, de valeur absolue au plus H , qui vérifient

$$\left|\sum_{i=0}^{N} a_i \binom{i}{t} \alpha_j^{i-t}\right| \le \sqrt{\eta_j}\; H\binom{N+1}{t+1} \max\{1, |\alpha_j|^N\}/\ell_{j,t+1} ,$$

pour $t = 0,\dots,r_jT-1$ et $j = 1,\dots,k$. Si n_j désigne la norme du corps $\mathbb{Q}(\alpha_j)$ sur $\mathbb{Q}$ et si b_j désigne le coefficient du terme de plus haut degré du polynôme P_j , ceci implique

$$\left|b_j^N n_j\left(\sum_{i=0}^{N} a_i \binom{i}{t} \alpha_j^{i-t}\right)\right| \le \eta_j \left(H\binom{N+1}{t+1}\right)^{d_j} M(\alpha_j)^N / \ell_{j,t+1}^{\eta_j} .$$

Le membre de gauche étant un entier naturel, la condition $F^{(t)}(\alpha_j) = 0$ sera réalisée pourvu que l'on ait

$$\ell_{j,t+1}^{\eta_j} > \eta_j \left(H\binom{N+1}{t+1}\right)^{d_j} M(\alpha_j)^N .$$

On choisit pour $\ell_{j,t+1}$ le plus petit entier tel que cette condition ait lieu, dans ce cas on a

$$\ell_{j,t+1} \leq 1 + \sqrt{n_j}\{(H(N+1)^{t+1})^{d_j} M(\alpha_j)^N\}^{1/n_j}$$

et

$$\ell_{j,t+1}^{n_j} < 4(H(N+1)^{t+1})^{d_j} M(\alpha_j)^N \quad .$$

La condition (1) est donc vérifiée si on a

$$\{4^{\Omega} H^d (N+1)^{d^*(T+1)/2} M(P)^N\}^T \leq (H+1)^{N+1} \quad ,$$

inégalité satisfaite pour

$$H = [\{4^{\Omega}(N+1)^{d^*(T+1)/2} M(P)^N\}^{\frac{T}{N+1-dT}}] \quad ,$$

d'où le résultat.

Le lemme 1 conduit aisément à l'assertion suivante, qui contient des résultats antérieurs de Ch. Pisot et M. Pathiaux.

Corollaire 1 Soit α un nombre algébrique non nul, il existe un polynôme R à coefficients entiers, non nul, qui admet α comme racine et qui vérifie

$$H(R) \leq M(\alpha) \quad .$$

De plus il existe un tel polynôme de degré au plus D , où D est une fonction effectivement calculable, qui ne dépend que du degré de α et de sa mesure.

3. ETUDE DE $P(\alpha)$

Dans toute la suite α désigne un nombre algébrique de degré $D \geq 2$, P désigne un polynôme à coefficients entiers de degré d qui ne s'annule pas en α . Grâce au lemme 1, on obtient le résultat général suivant.

Proposition 1 Soit S un ensemble de places du corps $\mathbb{Q}(\alpha)$ qui contient exactement s places archimédiennes. Soit P un polynôme à coefficients entiers qui ne s'annule pas en α . On a alors, pout tout entier positif T , la minoration

$$\prod_{v\in S} |P(\alpha)|_v \geq 4^{-\Omega} M(\alpha)^{-d} M(P)^{-D} \times$$

$$\exp\{-\frac{D}{T}(s \operatorname{Log} 2 + \mu + \operatorname{Log}(D+dR)) - dT\nu - \frac{d^*(T+1)}{2} \operatorname{Log}(D+dT)\} \quad ,$$

ou $\mu := \operatorname{Log} M(\alpha)$, $\nu := \operatorname{Log} M(P)$, Ω et d^* étant définis comme dans l'énoncé du lemme 1 .

Supposons P de la forme $a\tilde{P}$, a entier et $\tilde{P}$ primitif, alors on a

$$\prod_{v\in S} |P(\alpha)|_v = \prod_{v\in S} |a|_v \cdot \prod_{v\in S} |\tilde{P}(\alpha)|_v \quad ,$$

$$M(P) = |a| \; M(\tilde{P}) \quad ,$$

$$\prod_{v\in S} |a|_v \geq |a|^{-1} \quad ;$$

on peut donc se limiter au cas où P est primitif.

Lorsque P est primitif, considérons la fonction F construite au lemme 1 pour $N = dT + D - 1$; elle ne s'annule pas en α du fait qu'elle est de la forms $P^T Q$, $Q \in \mathbb{Z}[X]$ (car P est primitif), avec $P(\alpha) \neq 0$ par hypothèse et $\deg(Q) < D = \deg(\alpha)$.

Pour v en dehors de S on majore $|F(\alpha)|_v$ trivialement,

$$|F(\alpha)|_v \leq L_v(F) \max\{1, |\alpha|_v^{D+dt}\} \quad .$$

Pour v dans S on utilise la relation $F = P^T Q$, qui fournit la majoration évidente

$$|F(\alpha)|_v \leq |P(\alpha)|_v^T \, L_v(Q) \max\{1, |\alpha|_v^D\} \quad ,$$

et on sait (voir [2] par exemple; la démonstration qui figure en [2] est purement algébrique) que $L(Q)$ vérifie

$$L(Q) \leq 2^D \, L(F) \quad .$$

Grâce à la formule du produit

$$\prod_{v\in S} |F(\alpha)|_v \cdot \prod_{v\notin S} |F(\alpha)|_v = 1$$

et aux majorations précédentes on obtient l'inégalité

$$1 \leq (\prod_{v\in S} |P(\alpha)|_v^T) L(F)^D 2^{Ds} M(\alpha)^{D+dT} \quad .$$

Grâce à la majoration de $H(F)$ fournie par le lemme 1, on en déduit

$$- \operatorname{Log} \prod_{v\in S} |P(\alpha)|_v \le \frac{D \operatorname{Log}(L(F))}{T} + \frac{Ds \operatorname{Log} 2}{T} + \frac{\mu D}{T} + \mu d$$

$$\le \Omega \operatorname{Log} 4 + \mu d + D\nu + \frac{D}{T}(s \operatorname{Log} 2 + \mu + \operatorname{Log}(D+dT)) + \frac{d^*(T+1)}{2} \operatorname{Log}(D+dT) + dT\nu ,$$

ce qui équivaut au résultat annoncé.

En appliquant la proposition pour $T = 1$, on obtient le résultat très simple suivant.

THEOREME 1. *Soit* P *un polynôme quadratfrei à coefficients entiers qui ne s'annule pas en* α *, on a alors*

$$|P(\alpha)| \ge 4^{-\Omega}2^{-D}((D+d)M(\alpha)M(P))^{-(D+d)} ,$$

et

$$|\text{Norme}(P(\alpha))| \le 4^{\Omega}((D+d)M(\alpha)M(P))^{D+d} .$$

Pour obtenir la première inegalité on prend $S = \{v\}$, où $|\alpha|_v = |\alpha|$, alors $s = 1$. La seconde inégalité s'obtient en prenant pour S l'ensemble des places non archimédiennes du corps $\mathbb{Q}(\alpha)$, dans ce cas $s = 0$.

Un choix convenable de T conduit au résultat général suivant.

THEOREME 2. *Sous les hypothèses de la proposition 1, on a la minoration*

$$\prod_{v\in S} |P(\alpha)|_v \ge 4^{-\Omega}M(\alpha)^{-d}M(P)^{-(D+d)} \times$$

$$\times \exp\{-\tfrac{3}{2}\sqrt{(d^*D(\nu+\operatorname{Log}((\mu+3)dD))(s\operatorname{Log}2 + \mu+\operatorname{Log}((\mu+3)dD)))}$$

$$- d^*\operatorname{Log}((\mu+3)dD)\} .$$

On prend cette fois

$$T = \left[\frac{\sqrt{(D(s \operatorname{Log} 2 + \mu + \operatorname{Log}(D+d)))}}{d^*(\nu + \operatorname{Log}(D+d))}\right] + 1 .$$

On conclut en reportant cette valeur dans l'inégalité de la proposition 1 et en utilisant la majoration $D + dT \le (\mu+3)dD$.

Quitte à obtenir un résultat un peu moins précis, on peut déduire du théoremè 2 une inégalité plus simple.

Corollaire 1   Sous les hypothèses du théorème 2, on a

$$\prod_{v\in S} |P(\alpha)|_v \geq 4^{-\Omega} M(\alpha)^{-d} M(P)^{-(d+D)} \times$$

$$\times \exp\{-(\tfrac{3}{2}\sqrt{(d^*D(\nu+1)(s+\mu+1))}+d^*)\mathrm{Log}((\mu+3)dD)\} \quad .$$

Par le même raisonnement que dans la seconde partie du théorème 1, on obtient la majoration ci-dessous.

Corollaire 2   La norme de $P(\alpha)$ vérifie

$$|\mathrm{Norme}(P(\alpha))| \leq 4^{\Omega} M(\alpha)^{d} M(P)^{d+D} \times$$

$$\times \exp\{\tfrac{3}{2}\sqrt{(d^*D(\nu+\mathrm{Log}((\mu+3)dD))(\mu+\mathrm{Log}((\mu+3)dD)))}+d^*\mathrm{Log}((\mu+3)dD)\} \quad .$$

Afin d'y voir plus clair, il n'est peut-etre pas inutile de considérer un cas particulier simple du corollaire 2. On a par exemple.

Corollaire 3   Soit P un polynôme cyclotomique et ζ une racine de l'unité de degré D, on a alors

$$|\mathrm{Norme}(P(\zeta))| \leq 4.(3dD)^{d+2\sqrt{dD}} \quad .$$

En effet, ici μ et ν sont nuls; de plus, comme P est irréductible, on a $\Omega = 1$ et $d^* = d$.

Lorsque d et D ont le même ordre de grandeur, on en déduit la majoration

$$|\mathrm{Norme}(P(\zeta))| \leq c_1^{D\ \mathrm{Log}\ D} \quad ,$$

tandis que l'estimation évidente

$$|\mathrm{Norme}(P(\zeta))| \leq L(P)^D$$

fournit seulement

$$|\mathrm{Norme}(P(\zeta))| \leq 2^{Dd} \leq c_2^{D^2} \quad .$$

REFERENCES

[1]   T. Callahan, M. Newman and M. Sheingorn, Fields with a large Kronecker constant, J. of Number Th. 9 (1977), 182-186.

[2]   M. Mignotte, An inequality about factors of polynomials, Math. of Comp. 28 (1974), 1153-1157.

[3] M. Mignotte, Approximation des nombres algébriques par des nombres algébriques de grand degré, Annales de la Faculté des Sciences de Toulouse,

CONTINUED FRACTIONS AND RELATED ALGORITHMS

G.J. Rieger
Universität
D-3000 Hannover

In this short lecture, let us first speak about 3 types of infinite continued fractions.

Write $\xi \in X := [0,1] \setminus \mathbf{Q}$ as an ordinary continued fraction,

$$\xi = \cfrac{1}{a_1 + \cfrac{1}{a_2 + \ddots}} \; .$$

Define the corresponding operator $S : X \to X$ by

$$\xi = \frac{1}{a_1 + S(\xi)} \; .$$

Denote by λ the Lebesgue measure on $\mathbf{R}$. For $\alpha \in [0,1]$, $0 \le n \in \mathbf{Z}$ let

$$F_n(\alpha) := \lambda\{\xi \in X : S^n(\xi) \le \alpha\}, \quad R_n(\alpha) := F_n(\alpha) - \frac{\log(1+\alpha)}{\log 2} \; .$$

In a letter to Laplace of 1812, Gauss [3] mentions

$$\lim_{n\to\infty} R_n(\alpha) = 0 \; ;$$

the first proofs were published by Kusmin [7] in 1928, showing

$$R_n(\alpha) \ll C_1^{-\sqrt{n}} \; ,$$

and by Lévy [8] in 1929, showing

$$R_n(\alpha) \ll C_2^{-n} \quad (1 < C_{1,2} \in \mathbf{R}) \; .$$

Knopp [6] proved already in 1926 that S is ergodic. Furthermore, S is mixing. The asymptotic expansion of $R_n(\alpha)$ was given by Babenko [1] in 1976.

Write now $\xi \in Y := [-\frac{1}{2}, \frac{1}{2}] \setminus \mathbf{Q}$ as a continued fraction by nearest integers,

$$\xi = \cfrac{\varepsilon_1}{b_1 + \cfrac{\varepsilon_2}{b_2 + \ddots}} ,$$

$\varepsilon_n \in \{-1,1\}$, $2 \le b_n \in \mathbf{Z}$, $b_n + \varepsilon_{n+1} \ge 2$ $(n \ge 1)$. Define the corresponding operator $T : Y \to Y$ by

$$\xi = \frac{\varepsilon_1}{b_1 + T(\xi)} .$$

For $\alpha \in [0,1]$, $0 \le n \in \mathbf{Z}$ let

$$M_n(\alpha) := \begin{cases} \lambda\{\xi \in Y : 0 \le T^n(\xi) \le \alpha\} & \text{for } \alpha \le \tfrac{1}{2} , \\ \lambda\{\xi \in Y : 0 \le T^n(\xi) \le \tfrac{1}{2} \vee -\tfrac{1}{2} \le T^n(\xi) \le \alpha - 1\} & \text{for } \alpha > \tfrac{1}{2} . \end{cases}$$

Let $G := \frac{1+\sqrt{5}}{2}$. By [12], we have

$$M_n(\alpha) = \frac{\log \frac{G+\alpha}{G}}{\log G} + O(C_3^{-n}) \qquad (1)$$

with a certain $C_3 > 1$. T is mixing by [15] and especially ergodic.

Closely related are the singular continued fractions of Hurwitz (see e.g. [9], §44). The corresponding operator is ergodic, by [14], and an analogue of (1) holds, by [13].

We now turn to finite continued fractions.

Write a/b with $a \in \mathbf{Z}$, $b \in \mathbf{Z}$, $0 < a < b$, $(a,b) = 1$ as an ordinary continued fraction,

$$\frac{a}{b} = \cfrac{1}{a_1 + \cfrac{1}{a_2 + \ddots + \cfrac{1}{a_n}}} ;$$

let $E(a,b) := n$. Let $\sigma(b) := \sum_{d|b} d^{-1}$. Denote by ϕ the function of Euler. Heilbronn [5] proved

$$\sum_{\substack{0<a<b \\ (a,b)=1}} E(a,b) = 12(\log 2)\pi^{-2}\phi(b)\log b + O(b(\sigma(b))^3) . \qquad (2)$$

Write now a/b with $a \in \mathbf{Z}$, $b \in \mathbf{Z}$, $0 < 2a < b$, $(a,b) = 1$ as a continued fraction by nearest integers (see e.g. [9], §39),

$$\frac{a}{b} = \cfrac{1}{a_1 + \cfrac{\varepsilon_1}{a_2 + \ddots + \cfrac{\varepsilon_{n-1}}{a_n}}} \tag{3}$$

where $2 \le a_j \in \mathbb{Z}$ $(0 < j \le n)$, $\varepsilon_j \in \{-1,1\}$, $a_j + \varepsilon_j \ge 2$ $(0 < j < n)$; let $N(a,b) := n$. It is known $N(a,b) \le E(a,b)$ always (see e.g. [9], §39). By [10], we have

$$\sum_{0 < a < \frac{b}{2},\, (a,b) = 1} N(a,b) = 6 (\log G)\pi^{-2}\phi(b) \log b + O(b(\sigma(b))^3).$$

For $a \in \mathbb{Z}$, $b \in \mathbb{Z}$, $0 < a \le b$ the conditions

$$b = ga + r\,,\ 2 \nmid g\,,\ -a < r \le a$$

determine $g \in \mathbb{Z}$, $r \in \mathbb{Z}$ uniquely. Repetition leads to an expansion like (3), where now $a_j \in \mathbb{Z}$, $2 \nmid a_j$ $(0 < j \le n)$, $\varepsilon_j \in \{-1,1\}$, $a_j + \varepsilon_j > 0$ $(0 < j < n)$ and a/b has been written as a continued fraction with odd partial quotients; let

$$U(a,b) := n\ . \tag{4}$$

By [17], we have

$$\sum_{\substack{0<a<b\\(a,b)=1}} U(a,b) = 18 (\log G)\pi^{-2}\phi(b) \log b + O(b(\sigma(b))^3)\ .$$

Harris [4] introduces the following algorithm. For $a \in \mathbb{Z}$, $b \in \mathbb{Z}$, $0 < 2a < b$, $2 \nmid ab$, the conditions

$$b = ga + r\,,\ |r| < a\,,\ 2 \mid r$$

determine $g \in \mathbb{Z}$, $r \in \mathbb{Z}$ uniquely; then $2 \nmid g$, $g \ge 3$. In case $r = 0$, we are done. In case $r \ne 0$, let c be the largest odd divisor of r, and we continue with c, a instead of a, b. This stops after $H(a,b)$ steps, say.

Example $141 = 3.41 + 2.9$, $41 = 5.9 - 4.1$, $9 = 9.1$ gives $H(41,141) = 3$.

By [11], for every $k \in \mathbb{Z}$, $k > 0$ there exist odd integers $b_k > a_k > 0$, $d_k > c_k > 0$ with

$$E(a_k,b_k) \le 5 \ , \quad H(a_k,b_k) = k \ ,$$

$$E(c_k,d_k) \ge N(c_k,d_k) > k \ , \quad H(c_k,d_k) = 2 \ .$$

An analogue of (2) for $H(a,b)$ is not known.

For every $x > 2$, the number of pairs $a \in \mathbb{Z}$, $b \in \mathbb{Z}$ with $0 < a < b \le x$ which do not satisfy

$$\frac{1}{2} \log b \le E(a,b) \le \frac{1 + \log b}{\log G} \tag{5}$$

is at most $5x^{1.99}$, by [2].

For every $x > 2$, the number of pairs $a \in \mathbb{Z}$, $b \in \mathbb{Z}$ with $0 < 2a < b \le x$ which do not satisfy

$$\frac{1}{8} \log b < N(a,b) < \frac{\log (3b)}{\log(1+\sqrt{2})} - 1 \tag{6}$$

is at most $36x^{7/4}$, by [16].

For every $x > 2$, the number of pairs $a \in \mathbb{Z}$, $b \in \mathbb{Z}$ with $0 < a < b \le x$ which do not satisfy

$$\frac{1}{8} \log b < U(a,b) < \frac{\log(b\sqrt{5})}{\log G} \tag{7}$$

is at most $36x^{7/4}$, by [18].

For every $x > 2$, the number of pairs $a \in \mathbb{Z}$, $b \in \mathbb{Z}$ with $0 < 2a < b$, $2 \nmid ab$ which do not satisfy

$$\frac{1}{12} \log b < H(a,b) < \frac{\log(b+1)}{\log 2} - 1 \tag{8}$$

is at most $18x^{7/4}$, by [16].

In connection with (5), (6), (7), (8) one shows that there are no exceptions to the upper estimate and not too many exceptions to the lower estimate. The treatment of the exceptions to the lower estimate of (6) and (7) is, by the way, literally the same. So far we have only reported. To make this lecture more satisfactory, we prove

$$U(a,b) < \frac{\log(b\sqrt{5})}{\log G} \qquad (0 < a < b) \ . \tag{9}$$

For the infinite sequence

$$A = (a_1,\varepsilon_1,a_2,\varepsilon_2,a_3,\varepsilon_3,\dots)$$

with $a_j \in \mathbb{Z}$, $2 \nmid a_j$, $\varepsilon_j \in \{-1,1\}$, $a_j + \varepsilon_j > 0$ $(j > 0)$ we define the derived sequence $(q_0, q_1, q_2, \ldots)$ by $q_0 := 1$, $q_1 := a_1$, $q_n := a_n q_{n-1} + \varepsilon_{n-1} q_{n-2}$ $(n > 1)$. Comparison with (3) gives

$$b = q_n \quad . \qquad (10)$$

Following Fibonacci, define $f : \mathbb{Z} \to \mathbb{Z}$ by $f_0 := 1$, $f_1 := 1$, $f_n - f_{n-1} - f_{n-2} := 0$ $(n \in \mathbb{Z})$. Induction gives

$$f_n = \frac{G^{n+1} + (-1)^n G^{-n-1}}{G + G^{-1}} > \frac{G^n}{\sqrt{5}} \qquad (n \geq 0) , \qquad (11)$$

$$f_n = f_k f_{n-k} + f_{k-1} f_{n-k-1} \qquad (n \in \mathbb{Z} , k \in \mathbb{Z}) ,$$

$$f_j \leq f_{2j-2} , f_{2j} \geq 0 \ (j \in \mathbb{Z}) , f_j \geq 0 \ (j > -3) ,$$

and consequently

$$f_n \leq f_{2k-2} f_{n-k} + f_{2k-4} f_{n-k-1} \qquad (k \leq n + 1) . \qquad (12)$$

LEMMA 1. *The sequence* A *satisfies* $q_n \geq f_n$ $(n \geq 0)$.

Proof This is clear for $n = 0$ and $n = 1$. Let $n > 1$ and take $q_j \geq f_j$ $(0 \leq j < n)$ as induction hypothesis. Case 1: there exists a $k \in \mathbb{Z}$ with $0 < k < n$ such that $e_{n-j} = -1$ $(0 < j < k)$, $e_{n-k} = 1$. We know

$$a_n \geq 1 , a_{n-j} \geq 3 \quad (0 < j < k) . \qquad (13)$$

We have the $k-1$ inequalities

$$q_n \geq q_{n-1} - q_{n-2} , \ q_{n-j} \geq 3q_{n-j-1} - q_{n-j-2} \ (0 < j < k - 1) \qquad (14)$$

and the inequality

$$q_{n-k+1} \geq 3q_{n-k} + q_{n-k-1} \ . \qquad (15)$$

Since $3f_j - f_{j-2} = f_{j+2}$ $(j \in \mathbb{Z})$, (14) gives

$$q_n \geq f_{2j-2} q_{n-j} - f_{2j-4} q_{n-j-1} \qquad (0 < j < k) . \qquad (16)$$

But (16) for $j = k - 1$ and (15) give

$$q_n \geq f_{2k-2} q_{n-k} + f_{2k-4} q_{n-k-1} \ .$$

From this, the induction hypothesis, and (12) we deduce $q_n \geq f_n$.
Case 2: we have $e_{n-j} = -1$ $(0 < j < n)$. Then (13), (14) and (16) hold for $k = n$. But (16) for $k = n$, $j = k-1$ reads

$$q_n \geq f_{2n-4}q_1 - f_{2n-6}q_0 .$$

Using $q_1 = a_1 \geq 3$ (instead of (15)), we obtain

$$q_n \geq 3f_{2n-4} - f_{2n-6} = f_{2n-2} \geq f_n .$$

(10), (4), Lemma 1, and (11) imply (9).

A different proof of Lemma 1 can be found in [18].

Unfortunately, many beautiful related results had to be omitted in this short lecture.

REFERENCES

1. K.I. Babenko, A problem of Gauss, (Russian) Dokl.Akad. Nauk SSSR 238 (1978), 1021-1024 (English translation: Soviet Math.Dokl. 19 (1978), 136-140).

2. J.D. Dixon, A simple estimate for the number of steps in the euclidean algorithm, Amer.Math. Monthly 1971, 374-376.

3. C.F. Gauss, Werke X/1, 371-374.

4. V.C. Harris, An algorithm for finding the greatest common divisor, Fibonacci Quarterly 8 (1970), 102-103.

5. H. Heilbronn, On the average length of a class of finite continued fractions, Abhandlungen aus Zahlentheorie und Analysis (zur Erinnerung an Edmund Landau), Berlin 1968.

6. K. Knopp, Mengentheoretische Behandlung einiger Probleme der diophantischen Approximationen und der transfiniten Wahrscheinlichkeiten, Math.Ann. 95 (1926), 409-426.

7. R.O. Kusmin, Sur une problème de Gauss, Atti Congr.Intern. Bologne 6 (1928), 83-89.

8. P. Lévy, Sur le loi de probabilité dont dépendent des quotients complèts et incomplèts d'une fraction continue, Bull.Soc.Math. France 57 (1929), 178-194.

9. O. Perron, Die Lehre von den Kettenbrüchen, 3. Auflage, Stuttgart 1954.

10. G.J. Rieger, Über die mittlere Schrittanzahl bei Divisionsalgorithmen, Math.Nachr. 82 (1978), 157-180.

11. G.J. Rieger, On the Harris modification of the Euclidean algorithm, Fibonacci Quarterly 14 (1976), 196.

12. G.J. Rieger, Ein Gauss-Kusmin-Lévy-Satz für Kettenbrüche nach nächsten Ganzen, Manuscripta Math. 24 (1978), 437-448.

13. G.J. Rieger, Ein Gauss-Kusmin-Lévy-Satz für die singulären Kettenbrüche im Sinn von Hurwitz, Abh. Braunschweigische Wiss.Ges. 28 (1977), 81-88.

14. G.J. Rieger, Die metrische Theorie der Kettenbrüche seit Gauss, Abh. Braunschweigische Wiss.Ges. 27 (1977), 103-117.

15. G.J. Rieger, Mischung und Ergodizität bei Kettenbrüchen nach nächsten Ganzen, J. reine angew.Math. 310 (1979), 171-181.

16. G.J. Rieger, Über die Schrittanzahl beim Algorithmus von Harris und dem nach nächsten Ganzen, Arch.d.Math. (in print).

17. G.J. Rieger, Ein Heilbronn-Satz für Kettenbrüche nach ungeraden Teilnennern, Math.Nachr. (in print).

18. G.J. Rieger, Über die Länge von Kettenbrüchen mit ungeraden Teilnennern, Abh. Braunschweigische Wiss.Ges. 32 (1981), 51-53.

IWASAWA THEORY AND ELLIPTIC CURVES: SUPERSINGULAR PRIMES

Karl Rubin
Department of Mathematics
Harvard University

1. PRELIMINARIES

We are interested in the following situation. E is an elliptic curve defined over the number field F, and $K \subset F$ is an imaginary quadratic field. Let $\mathcal{O}$ denote the ring of integers of K, and suppose that E has complex multiplication by an order $\mathcal{O}'$ in $\mathcal{O}$. Fix a prime $\mathfrak{p}$ of $\mathcal{O}$ such that E has good reduction at all primes of F above $\mathfrak{p}$ and $\mathfrak{p}$ is prime to 6.

Write $[\tau]$ for the endomorphism of E corresponding to $\tau \in \mathcal{O}'$. If $\mathfrak{A}$ is any ideal of $\mathcal{O}$, let $E_{\mathfrak{A}}$ denote the points of E annihilated by all endomorphisms $\{[\tau] \mid \tau \in \mathcal{O}' \cap \mathfrak{A}\}$. Consider the tower of number fields

$$F_n = F(E_{\mathfrak{p}^{n+1}}), \qquad F_\infty = \bigcup_n F_n .$$

<u>Definition</u> Let $\kappa : \mathrm{Gal}(F_\infty/F) \to \mathcal{O}_{\mathfrak{p}}^*$ be the map obtained by considering the action on $E_{\mathfrak{p}^\infty} = \bigcup_n E_{\mathfrak{p}^n}$.

The character κ is an isomorphism between $\mathrm{Gal}(F_\infty/F)$ and an open subgroup of $\mathcal{O}_{\mathfrak{p}}^*$, so $\mathrm{Gal}(F_\infty/F)$ breaks up into $\mathrm{Gal}(F_\infty/F_0) \times \Delta$ where $\Delta \cong \mathrm{Gal}(F_0/F) \hookrightarrow (\mathcal{O}/\mathfrak{p})^*$. Further, $\mathrm{Gal}(F_\infty/F_0)$ is isomorphic to an open subgroup of $\mathcal{O}_{\mathfrak{p}}$, so $\mathrm{Gal}(F_\infty/F_0) \cong \mathbb{Z}_p^{\deg(\mathfrak{p})}$, where p is the rational prime below $\mathfrak{p}$.

In [1,2] Coates and Wiles studied the tower F_∞/F in the special case $F = K$, $\deg(\mathfrak{p}) = 1$ and used their results to prove an important theorem in the direction of the Birch and Swinnerton-Dyer conjecture. The remainder of this paper is concerned with an investigation of the case $\deg(\mathfrak{p}) = 2$. So from now on we assume $\mathfrak{p} = (p)$, p a rational prime which remains inert in $\mathcal{O}$.

We will study the $\mathbb{Z}_p^2$-extension F_∞/F by introducing the Iwasawa algebra Λ and two modules over Λ.

<u>Definition</u>

$$\Lambda = \varprojlim \mathbb{Z}_p[\mathrm{Gal}(F_n/F_0)] = \mathbb{Z}_p[[\mathrm{Gal}(F_\infty/F_0)]] \cong \mathbb{Z}_p[[T_1,T_2]]$$

where the last isomorphism depends on the choice of two topological generators of $\mathrm{Gal}(F_\infty/F_0)$.

<u>Definition</u> $X_\infty = \mathrm{Gal}(M_\infty/F_\infty)$ where M_∞ is the maximal abelian p-extension of F_∞ which is unramified outside the primes above p .

<u>Definition</u> $Y_\infty = U_\infty/C_\infty$ where $U_\infty = \varprojlim U_n$ and U_n is the group of local units congruent to 1 in the completion $F_n \otimes_K K_p$; and $C_\infty = \varprojlim \bar{C}_n$, $\bar{C}_n$ being the closure of the elliptic units of F_n (see [1], section 5) in U_n .

As we will show in §2, X_∞ is closely connected with the arithmetic of the curve E . Also, X_∞ and Y_∞ are related by class field theory. In [2], with $\mathfrak{p}$ of degree 1 , Coates and Wiles connected Y_∞ with the $\mathfrak{p}$-adic L-function attached to E . Combining these facts gives interesting results relating the arithmetic of the curve with its complex L-function. This is the motivation for the present investigation.

2. X_∞ AND GALOIS COHOMOLOGY

We first introduce the Tate-Shafarevich group Ш , and a slight variant Ш' .

<u>Definition</u>

$$Ш_n = \ker[H^1(E/F_n) \to \oplus H^1(E/(F_n)_\mathfrak{q})] \qquad Ш_\infty = \varinjlim Ш_n$$

$$Ш'_n = \ker[H^1(E/F_n) \to \oplus H^1(E/(F_n)_\mathfrak{q})] \qquad Ш'_\infty = \varinjlim Ш'_n$$

where the first direct sum is over all primes $\mathfrak{q}$ of F_n and the second is over all primes not above p .

Galois cohomology gives a Kummer theory exact sequence

$$0 \to E(F_n)/p^{n+1}\, E(F_n) \to \mathrm{Hom}(\mathrm{Gal}(M_n/F_n), E_{p^{n+1}}) \to (Ш'_n)_{p^{n+1}} \to 0$$

where M_n is the maximal abelian p-extension of F_n unramified outside primes above p . Taking the direct limit over n yields

$$0 \to E(F_\infty) \otimes K_p/\mathcal{O}_p \to \mathrm{Hom}(X_\infty, E_{p^\infty}) \to Ш'_\infty(p) \to 0 \qquad (2.1)$$

where $Ш'_\infty(p)$ denotes the p-primary part of $Ш'_\infty$.

Our main result here is

THEOREM. $Ш'_\infty(p) = Ш_\infty(p)$.

<u>Sketch of proof</u> The theorem follows from the following local

statement. Suppose L is a finite extension of K_p and $L' = \bigcup L_n$ is a $\mathbf{Z}_p$-extension of L. Then $\varinjlim H^1(E/L_n) = 0$. This implies that any element of $H^1(E/F_n)$ vanishes in $H^1(E/(F_m)_{\mathfrak{q}})$ for every $\mathfrak{q}$ above p, if m is taken large enough.

This local statement is proved by Tate duality: $\varinjlim H^1(E/L_n)$ is dual to $\varprojlim E(L_n)$ where the inverse limit is taken with respect to the norm map on E. We are therefore reduced to showing that there are no sequences of points $P_n \in E(L_n)$ such that $P_m = N_{n/m}P_n$ for $n \geq m$. This is just a statement about formal groups, and follows because the formal group here has height 2.

Combining the theorem with (2.1) yields

$$0 \to E(F_\infty) \otimes K_p/\mathcal{O}_p \to \mathrm{Hom}(X_\infty, E_{p^\infty}) \to Ш_\infty(p) \to 0 . \tag{2.2}$$

Greenberg [4] showed that X_∞ as a Λ-module has free rank $r_2(F_0) > 0$. It follows from this and (2.2) that either $E(F_\infty) \times K_p/\mathcal{O}_p$ or $Ш_\infty(p)$ is 'very large' -- both cannot be co-torsion Λ-modules. Unfortunately it seems very difficult to decide which one is large, even assuming the standard conjectures about the Tate-Shafarevich group.

3. THE STRUCTURE OF Y_∞

For this section we make the additional assumption that E is defined over K, so $F = K$. Then the character κ defined in §1 maps onto $\mathcal{O}_p^*$, so $\Delta \cong (\mathcal{O}/p)^* \cong \mathbb{F}_{p^2}$. Since the characters of Δ are $\mathcal{O}_p$-valued, it is useful to introduce the following notation.

Definition $\tilde{\Lambda} = \Lambda \otimes_{\mathbb{Z}_p} \mathcal{O}_p \cong \mathcal{O}_p[[T_1,T_2]]$; and if Z is any Λ-module, $\tilde{Z} = Z \otimes_\Lambda \tilde{\Lambda}$.

Definition If $\chi : \Delta \to \mathcal{O}_p^*$ is any character and Z any $\tilde{\Lambda}$-module on which Δ acts, then $Z(\chi)$ is the submodule of Z on which Δ acts via χ.

With this notation we have $Z = \bigoplus_\chi Z(\chi)$.

THEOREM. *Let χ be the restriction of κ to Δ. Then if $i \not\equiv 0, p+1$ modulo p^2-1, we have $\tilde{U}_\infty(\chi^i) \cong \tilde{\Lambda}^2$, and $\tilde{C}_\infty(\chi^i) \cong \tilde{\Lambda}$. Further, we can choose a basis α, β for $\tilde{U}_\infty(\chi^i)$ so that $\tilde{C}_\infty(\chi^i) \subset \tilde{\Lambda}\cdot\beta$. The significance of this last statement is that if $\tilde{C}_\infty(\chi^i) = \tilde{\Lambda}\cdot\varepsilon$, say, then there is an F in $\tilde{\Lambda}$ so $\varepsilon = F\cdot\beta$, and then $\tilde{Y}_\infty(\chi^i) = (\widetilde{U_\infty/C_\infty})(\chi^i) \cong \tilde{\Lambda} \oplus \tilde{\Lambda}/F\tilde{\Lambda}$.*

<u>Remarks</u> If $i \equiv 0$ or $p+1$ modulo p^2-1, then we get pseudo-isomorphisms $U_\infty(\chi^i) \hookrightarrow \Lambda^2$, $C_\infty(\chi^i) \hookrightarrow \Lambda$. χ^{p+1} is exceptional because it is the eigenspace containing the p-power roots of unity.

Unfortunately, the splitting of $Y_\infty(\chi^i)$ is not canonical, and although the ideal $F\tilde{\Lambda}$ is uniquely determined, this only specifies F up to a unit in $\tilde{\Lambda}$.

So what can we say about the power series F? Here we make use of the 'logarithmic differentiation' homomorphisms of Coates-Wiles. Briefly, Coleman [3] has defined an injection $\rho : U_\infty \hookrightarrow (\mathcal{O}_p[[X]])^*$, so for each $k > 0$ we have the homomorphism $\phi_k(u) = D^k(\log(\rho(u))$ on U_∞, where D is a derivation associated to the formal group of the kernel of reduction mod p on E.

<u>Proposition</u> Suppose $i \not\equiv 0$, $p+1$ modulo p^2-1. Then $\tilde{C}_\infty(\chi^i)$ has a generator ε such that $\phi_k(\varepsilon) = 12\,\Omega^{-k}f^{-k}L(\bar{\psi}^k,k)$ for every $k \equiv i \pmod{p^2-1}$, where Ω is a period of E such that $E \cong \mathbb{C}/\Omega\mathcal{O}'$; f is the conductor of E; and ψ is the Grossencharacter attached to E.

Then the properties of the ϕ_k give us the following corollary:

<u>Corollary</u> With notation as in the theorem and the proposition, we have $F(\kappa(1+T_1)^k-1, \kappa(1+T_2)^k-1) = 12\,\Omega^{-k}f^{-k}L(\bar{\psi}^k,k)/\phi_k(\beta)$ for every $k \equiv i \pmod{p^2-1}$.

Unfortunately, the values $\phi_k(\beta)$ are still a mystery. However, I should remark that the power series F is uniquely determined by the special values above.

<u>Proposition</u> For any u in U_∞, p^{n_k} divides $\phi_k(u)$, where $n_k = \left[\frac{(k-1)p-1}{p^2-1}\right]$.

This follows because the formal group that the ϕ_k come from has height two. This result is a slight improvement on a similar result of Katz [5]; Katz also obtained congruences among the numbers $\phi_k(u)/p^{n_k}$. Clearly, to understand fully the significance of F, and the interpolation properties of the numbers $L(\bar{\psi}^k,k)$, we must understand the numbers $\phi_k(\beta)$. In the case where $\mathfrak{p}$ is a prime of degree 1, this is the role of the Γ-transform.

BIBLIOGRAPHY

1. Coates, J., Wiles, A.: On the Conjecture of Birch and Swinnerton-Dyer. Inventiones math. 39 (1977), 223-251.

2. Coates, J., Wiles, A.: On p-adic L-functions and Elliptic Units. J.Austral.Math.Soc. (A) 26 (1978), 1-25.

3. Coleman, R.: Division Values in Local Fields. Inventiones math. 53 (1979), 91-116.

4. Greenberg, R.: On the Structure of Certain Galois Groups. Inventiones math. 47 (1978), 85-99.

5. Katz, N.: Formal Groups and p-adic Interpolation. Société Math. de France Astérisque 41-42 (1977), 55-65.

ON RELATIONS BETWEEN GAUSS SUMS AND CYCLOTOMIC UNITS

C.-G. Schmidt
Fachbereich 9 Mathematik
Universität des Saarlandes
D-6600 Saarbrücken

1. INTRODUCTION

For a natural number $m \not\equiv 2(4)$ let $K := \mathbb{Q}(\zeta_m)$ be the cyclotomic field generated by a primitive m-th root of unity ζ_m. The arithmetic of K is closely connected to certain Gauss sums and cyclotomic units as is well known. For a prime divisor k of the rational prime $p \nmid m$ in K we define the Gauss sums

$$\tau_x(k) := - \sum_{\alpha \bmod k} (\alpha/k)_m{}^x \cdot \zeta_p{}^{\mathrm{tr}\,\alpha} \qquad (x \bmod m),$$

where $(\alpha/k)_m$ denotes the m-th power residue symbol and tr the absolute trace of the residue class field of K mod k. The sum is taken over a complete set of representatives of K mod k. In general the numbers

$$\eta(x) := 1 - \zeta_m{}^x \qquad (x \bmod m)$$

are not units in K but only suitable quotients are units. Nevertheless we call them cyclotomic units.

We consider the multiplicative relations of Gauss sums and cyclotomic units or in more detail the relation modules

$$R_{id}(k) := \{\underline{a} = (a_1,\dots,a_{m-1}) \in \mathbb{Z}^{m-1} \;;\; \prod_{x=1}^{m-1} \tau_x(k)^{a_x} \simeq 1\}$$

('specific ideal relations'),

$$R_{el} := \{\underline{a} \in \mathbb{Z}^{m-1} \;;\; \prod_{x=1}^{m-1} \tau_x(k)^{a_x} = 1 \quad \text{for all } k \nmid m\}$$

('simultaneous element relations'),

$$R' := \{\underline{a} \in \mathbb{Z}^{m-1} \;;\; \prod_{x=1}^{m-1} |\eta(x)|^{a_x} = 1\}$$

('relations of cyclotomic units mod torsion').

2. THE GAP GROUPS OF THE IDEAL RELATIONS AND THE UNIT RELATIONS

For y mod m set $\delta_y := (0,\dots,0,1,0,\dots,0) \in \mathbb{Z}^{m-1}$ with 1 in the y-th component. There are well known explicit relations:

a) Davenport-Hasse (or distribution) relations [2]:

$$r_{d,x} := \delta_{dx} - \sum_{j=0}^{d-1} \delta_{x+jm/d} + \sum_{j=1}^{d-1} \delta_{jm/d}$$

for $d \mid m$, $x = 1,\dots,(m/d) - 1$, where $r_{d,x} \in R_{id}(k)$ for all $k \nmid m$. As an analog we have for the same d,x

$$r_{d,x}' := \delta_{dx} - \sum_{j=0}^{d-1} \delta_{x+jm/d} \in R' .$$

b) Norm relations: For $y \bmod m$ there is

$$n_y := \delta_y + \delta_{m-y} - \delta_1 - \delta_{m-1} \in R_{id}(k) \quad \text{for all} \quad k \nmid m$$

and

$$n_y' := \delta_y - \delta_{m-y} \in R' .$$

Hasse [4], p.465 respectively Milnor [1] asked: Do these relations generate $R_{id}(k)$ respectively R'? Let R_1 be the submodule of $R_{id}(k)$ generated by the $r_{d,x}$ and n_y, and R_1' the submodule of R' generated by the $r_{d,x}'$ and n_y'. Bass has partly answered Milnor's question. He proved in [1]:

$$R_1' \otimes \mathbb{Q} = R' \otimes \mathbb{Q} .$$

Yamamoto [8] showed the following result by combinatorical methods: Let r be the number of rational primes dividing m and take $p \equiv 1(m)$. Then we have an isomorphism of 2-elementary abelian groups:

$$R_{id}(k)/R_1 \cong (\mathbb{Z}/2\mathbb{Z})^{2^{r-1}-1} .$$

But there is a totally different, more structural approach which offers in addition an analogous result for R'. We set

$$A := \mathbb{Z}^{m-1} / \langle r_{d,x}' \; ; \; d \mid m , x = 1,\dots,(m/d) - 1\rangle$$

and for $J := \{\pm 1 \bmod m\}$ we consider A as a J-module (J operates on A by: $(-1) \circ \delta_y := \delta_{m-y}$). This provides us with a cohomological description of the quotient groups in question. (For details see [5].)

THEOREM 1. $R_{id}(k)/R_1 \cong H^0(J,A)$ *for* $p \equiv 1(m)$.

$$R'/R'_1 \cong H^1(J,A) .$$

Now using a formula of Sinnott [7] we can compute these cohomology groups (see [5]).

THEOREM 2. $H^0(J,A) \cong (\mathbb{Z}/2\mathbb{Z})^{2^{r-1}-1}$,

$$H^1(J,A) \cong (\mathbb{Z}/2\mathbb{Z})^{2^{r-1}-r} .$$

In this way we obtain again Yamamoto's result and also give a complete answer to Milnor's question.

3. THE GAP GROUP OF ELEMENT RELATIONS

We now ask for the simultaneous element relations of the Gauss sums. For $\underline{a} \in \mathbb{Z}^{m-1}$ and $k \nmid m$ set

$$\chi_{\underline{a}}(k) := \prod_{x=1}^{m-1} \tau_x(k)^{a_x}$$

and further set $R := \cap R_{id}(k)$ where the intersection is taken over all $k \nmid m$. All $\chi_{\underline{a}}$ for $\underline{a} \in R$ extend to characters of the ray class group $\bmod m^2$ which we denote by S_{m^2} . The values of these characters are roots of unity of K . Thus we get a pairing

$$R \times S_{m^2} \to \mathrm{Tor}\, K^\times , \ (\underline{a},k) \mapsto \chi_{\underline{a}}(k) .$$

By class field theory there is a Kummer extension

$$L := \mathrm{Fix}(\bigcap_{\underline{a}\in R} \mathrm{Ker}\, \chi_{\underline{a}})$$

over K and we have

$$R/R_{el} \simeq \mathrm{Gal}(L/K)^*$$

where the star denotes the corresponding character group. So for all $\underline{a} \in R$ the Kummer extension $L_{\underline{a}} := \mathrm{Fix}(\mathrm{Ker}\, \chi_{\underline{a}})$ measures the order of the class of $\underline{a}$ considered as an element of R/R_{el} . By a theorem of Deligne [3] the extension $L_{\underline{a}}/K$ is given explicitly by adjunction of a certain product of values of the classical Γ-function.

THEOREM (Deligne). *For* $\underline{a} \in R$ *set*

$$\Omega_{\underline{a}} := \prod_{x=1}^{m-1} \Gamma(x/m)^{a_x} .$$

Then $L_{\underline{a}} = K(\Omega_{\underline{a}})$ *and the Frobenius automorphism operates like*

$$\sigma_{k}(\Omega_{\underline{a}}) = \chi_{\underline{a}}(k).\Omega_{\underline{a}} .$$

Deligne's proof is not yet published. The result is reported in [3] where it is also stated that the proof involves Hodge-theory of Fermat hypersurfaces. In the case $\underline{a} \in R_1$ the theorem was independently proved by Gross, Koblitz [3] (using a result of Katz on some p-adic cohomology of the Fermat curve) and the author [6] (using only arguments from class field theory and Kummer theory). To prove the theorem for all $\underline{a} \in R$ it would be sufficient to show

$$\prod_{x=1}^{m-1} (\Gamma(x/m)^{p-1}/\Gamma(x(p-1)/m + 1))^{a_x} \equiv 1 \bmod k$$

for $p \equiv 1(m)$ and $\underline{a} \in R$ but it is not clear how to attack this directly.

REFERENCES

1. H. Bass, Generators and relations for cyclotomic units, Nagoya Math.J. 27 (1966), 401-407.

2. H. Davenport and H. Hasse, Die Nullstellen der Kongruenzzetafunktion in gewissen zyklischen Fällen, J. reine angew.Math. 172 (1934), 151-182.

3. B.H. Gross and N. Koblitz, Gauss sums and the p-adic Γ-function, Ann.Math. 109 (1979), 569-581.

4. H. Hasse, Vorlesungen über Zahlentheorie (Berlin 1964).

5. C.-G. Schmidt, Die Relationenfaktorgruppen von Stickelberger-Elementen und Kreiszahlen, J. reine angew.Math. 315 (1980), 60-72.

6. C.-G. Schmidt, Gauss sums and the classical Γ-function, Bull. London Math.Soc. 12 (1980), 344-346.

7. W. Sinnott, On the Stickelberger ideal and the circular units of a cyclotomic field, Ann.Math. 108 (1978), 107-134.

8. K. Yamamoto, The gap group of multiplicative relationship of Gaussian sums, Symp.Math. Vol.XV (1975), 427-440.

SUR LA PROXIMITÉ DES DIVISEURS

Gérald Tenenbaum
Résidence Saint Sébastien,
tour A, F-500 Nancy

Soit n un nombre entier, et, $1 = d_1 < d_2 < \ldots < d_{\tau(n)} = n$ la suite croissante de ses diviseurs. Nous nous proposons ici de présenter certains résultats liés à l'étude de la fonction arithmétique

$$E(n) := \min_{i=1}^{\tau(n)-1} \frac{d_{i+1}}{d_i} .$$

Erdös a conjecturé il y a plus de quarante ans que l'on a pour presque tout entier n

$$1 < E(n) < 2 \quad ; \tag{1}$$

il a montré en 1948 [1] que la suite des entiers n satisfaisant (1) possède effectivement une densité asymptotique mais sa méthode ne permet pas d'établir que cette densité a pour valeur 1.

Une justification heuristique de (1) peut être formulée ainsi: comme le nombre des valeurs distinctes des quantités $\log d'/d$, d et d' parcourant les diviseurs de n, est égal à

$$U(n) := \mathrm{card}\{d|n, d'|n : (d,d') = 1\} = \prod_{\substack{p^\nu || n \\ p \text{ premier}}} (2\nu+1) ,$$

on a pour tout n

$$3^{\omega(n)} \leq U(n) \leq 3^{\Omega(n)}$$

où $\omega(n)$ (resp. $\Omega(n)$) désigne le nombre des facteurs premiers de n comptés sans (resp. avec) leur ordre de multiplicité; le théorème de Hardy et Ramanujan implique donc

$$U(n) = (\log n)^{\log 3 + o(1)}$$

pour presque tout n, et, l'on peut s'attendre à ce qu'un intervalle I inclus dans $[-\log n, \log n]$ et de longueur λ contienne usuellement $\lambda \cdot (\log n)^{\log 3 - 1 + o(1)}$ points $\log d'/d$ distincts - la conjecture d'Erdös exprime que le nombre de ces points est au moins égal à 1 dans le cas $I =]0, \log 2[$. Cet argument a conduit

Erdös à conjecturer qu'en fait la formule asymptotique

$$E(n) = 1 + (\log n)^{1 - \log 3 + o(1)} \tag{2}$$

a lieu pour presque tout n . Il a même annoncé pouvoir établir (2) en 1964 [2] mais son argument s'est malheureusement avéré inexact. Erdös et Hall ont récomment prouvé un théorème qui implique l'inégalité

$$E(n) \geq 1 + (\log n)^{1 - \log 3 + o(1)}$$

pour presque tout n , mais le problème de la majoration de l'ordre normal de $E(n)$ reste entier.

L'étude de $E(n)$ a suscité la définition de plusieurs fonctions arithmétiques:

$$T(n,\alpha) := \mathrm{card}\{d|n, d'|n : |\log d'/d| \leq (\log n)^{\alpha}\} \quad (\alpha < 1)$$

$$U(n,\alpha) := \mathrm{card}\{d|n, d'|n : (d,d') = 1, |\log d'/d| \leq (\log n)^{\alpha}\} \quad (\alpha < 1)$$

$$\tau^{+}(n) := \mathrm{card}\{k \in \mathbf{N} : \exists\, d|n, 2^k \leq d < 2^{k+1}\}$$

$$g(n) := \mathrm{card}\{i (1 \leq i \leq \tau(n)-1) : d_i | d_{i+1}\}$$

Les deux premières définitions sont dues à Hall, les deux autres à Erdös. L'inégalité (1) est impliquée par

$$\tau^{+}(n) < \tau(n) \tag{3}$$

et, si l'on y remplace 2 par e (ce qui représente une modification sans importance), par

$$T(n,0) > \tau(n) \tag{4}$$

ou encore par

$$U(n,0) > 1 \quad . \tag{5}$$

Enfin, notons que, comme chaque intervalle $[2^k, 2^{k+1}[$ contient au plus un diviseur di tel que $d_i | d_{i+1}$, on a pour tout entier n

$$g(n) \leq \tau^{+}(n) \quad . \tag{6}$$

Hall a étudié [6] la valeur moyenne de $T(n,\alpha)$ pour α dans $[0,1]$:

THÉORÈME 1 (Hall). *Pour tout réel* α *de* $[0,1]$ *il existe une constante positive* $C(\alpha)$ *telle que l'on ait pour* κ *infini*

$$\sum_{n<\kappa} \frac{T(n,\alpha)}{(2\alpha+2)^{\Omega(n)}} \sim C(\alpha)\kappa(\log\log \kappa)^{\gamma(\alpha)} ,$$

où l'on a posé $\gamma(\alpha) = 1$, *si* $\alpha = 0$, *et* $\gamma(\alpha) = 0$, *si* $\alpha \neq 0$.

Pour $\alpha = 0$, la formule n'est pas explicitement prouvée dans [6] mais on peut constater facilement que la méthode de transformation de Fourier utilisée pour traiter le cas $\alpha > 0$ est encore applicable, au prix de certaines précautions supplémentaires. La valeur explicite de $C(\alpha)$ est donnee dans [6]; elle présente une discontinuité en $\alpha = 1$.

On déduit sans peine du théorème 1 que presque tout entier n satisfait à $\log T(n,\alpha) \leq \{\log(2\alpha+2) + o(1)\} \log\log n$. On trouvera dans [6] ou dans [7] la démonstration, due à Erdös, de l'inégalité $\log T(n,\alpha) \geq \{(\alpha+1)\log 2 + o(1)\} \log\log n$ pour presque tout n . Dans [7], Hall et l'auteur établissent l'existence d'un ordre normal pour $\log T(n,\alpha)$, $0 \leq \alpha \leq 1$.

THÉORÈME 2 (Hall-Tenenbaum). *Pour presque tout entier* n *on a*

$$\frac{\log T(n,\alpha)}{\log\log n} = (\alpha+1)\log 2 + O\left(\sqrt{\frac{\log\log\log\log n}{\log\log n}}\right) ,$$

uniformement par rapport a α *dans* $[0,1]$.

Ainsi, $T(n,0)$ est usuellement de l'ordre de $\tau(n)$ et l'inégalité (4), si elle est vérifiée pour presque tout n , est pratiquement optimale et risque d'être difficile à établir. En revanche, l'argument heuristique présenté plus haut semble indiquer que $U(n,0)$ est une fonction mieux adaptée au problème. En effet, la conjecture naturelle concernant l'ordre normal de $\log U(n,\alpha)$ est

$$\log U(n,\alpha) = \{\log 3 - 1 + \alpha + o(1)\} \log\log n \qquad (7)$$

pour presque tout n ; en particulier, on s'attend donc à ce que $U(n,0)$ soit usuellement de l'ordre d'une puissance positive de $\log n$ et, partant, que l'inégalité (5) soit une forme très affaiblie de la réalité. Dans [7], nous avons étudié l'ordre moyen et l'ordre normal de $U(n,\alpha)$. Nous avons obtenu les résultats suivants:

THÉORÈME 3 (Hall-Tenenbaum). *Pour tout réel* α *de* $[0,1]$ *il existe une constante positive* $D(\alpha)$ *telle que l'on ait pour* κ *infini*

$$\sum_{n<\kappa} y(\alpha)^{\Omega(n)} U(n,\alpha) \sim D(\alpha)\kappa(\log\log \kappa)^{\delta(\alpha)}$$

oū l'on a posé

$$y(\alpha) = \begin{cases} (2-\alpha)/3 & \text{si} \quad 0 \le \alpha \le \frac{1}{2} \\ 1/(2\alpha+1) & \text{si} \quad \frac{1}{2} \le \alpha \le 1 \end{cases}, \text{ et, } \delta(\alpha) = \begin{cases} 1 \text{ si } \alpha = \frac{1}{2} \\ 0 \text{ si } \alpha \neq \frac{1}{2} \end{cases}.$$

La valeur de $D(\alpha)$ est explicitée dans [7], comme dans le cas de $T(n,\alpha)$, on observe une discontinuite pour $\alpha = 1$.

THÉORÈME 4 (Hall-Tenenbaum). *Pour presque tout entier* n *on a*

$$\alpha \log 3 + o(1) \le \frac{\log U(n,\alpha)}{\log\log n} \le \log 3 - 1 + \alpha + o(1)$$

La majoration du théorème 4 est en accord avec (7). Il suffirait donc pour établir (7), et par conséquent (5), de montrer que l'on a asymptotiquement

$$\sum_{n<\kappa} \log U(n,0) \ge \{\log 3 - 1 + o(1)\}\kappa \log\log \kappa \quad .$$

Erdös a souvent conjecturé (voir [3] par exemple) que le rapport $\tau^+(n)/\tau(n)$ tend vers 0 sur une suite de densité 1 . A l'opposé, Montgomery a conjecturé, lors du Symposium de Théorie Analytique des Nombres de Durham, en 1979, que non seulement $\tau^+(n)/\tau(n)$ mais même $g(n)/\tau(n)$ peut être minoré par une constante positive sur une suite d'entiers n de densité positive. Son argument heuristique peut être résumé ainsi: soit n un entier et p son plus petit facteur premier; quitte à négliger une suite de n de densité nulle, on peut supposer $\log p \le (\log n)^{o(1)}$; de plus, la moitié au moins des diviseurs de n divisent n/p , et, si $d_i | (n/p)$ mais $d_i \nmid d_{i+1}$, on a $d_{i+1} < pd_i$, or la mesure de $\bigcup_{d_i|(n/p)}]\log d_i, \log pd_i[$ ne dépasse pas $\tau(n) \log p$, elle est donc usuellement $\le (\log n)^{\log 2 + o(1)}$ et l'on peut s'attendre à ce que la proportion des $\log d_{i+1}$ contenus dans cette réunion tende vers 0 comme $(\log n)^{\log 2 - 1 + o(1)}$; la conjecture nécessite seulement qu'elle soit majorée par une constante $< \frac{1}{2}$.

Dans [5], Erdös et l'auteur établissent la conjecture de Montgomery. Plus précisément, on a

THÉORÈME 5 (Erdös-Tenenbaum). *Pour tout* α *de* $[0,1]$ *désignons par* $\Delta_1(\alpha)$ *(resp.* $\Delta_2(\alpha)$*) la densité supérieure de la suite des entiers* n *satisfaisant à* $\tau^+(n) \leq \alpha\tau(n)$ *(resp.* $g(n) \leq \alpha\tau(n)$*). Alors*

(i) pour tout reel positif ε *, il existe une constante* $c(\varepsilon)$ *telle que l'on ait*

$$\Delta_1(\alpha) \leq c(\varepsilon)\alpha^{1-\varepsilon}$$

pour tout α *de* $[0,1]$.

(ii) on a $\lim_{\alpha \to 0} \Delta_2(\alpha) = 0$.

Cela implique que non seulement $\tau^+(n)/\tau(n)$ (resp. $g(n)/\tau(n)$) ne tend pas vers 0 pour presque tout n mais même que toute suite A telle que $\lim_{n \in A} \frac{\tau^+(n)}{\tau(n)} = 0$ (resp. $\lim_{n \in A} \frac{g(n)}{\tau(n)} = 0$) est de densité nulle. Il est également à noter que le résultat (ii) est optimal en ce sens que $\Delta_2(\alpha)$ est positif lorsque α est positif [5].

L'ensemble de ces résultats suggère que, même s'il est usuellement non nul, le nombre des rapports d_{i+1}/d_i qui sont proches de 1 est relativement faible. Cela explique peut être, au moins partiellement, pourquoi la conjecture initiale d'Erdös reste encore à prouver.

BIBLIOGRAPHIE

[1] P. Erdös, On the density of some sequences of integers, Bull. Amer.Math.Soc. 54 (1948), 685-692.

[2] P. Erdös, On some applications of probability to analysis and number theory, J. London Math.Soc. 39 (1964), 692-696.

[3] P. Erdös, Some unconventional problems in number theory, Astérisque 61 (1979), 73-82.

[4] P. Erdös and R.R. Hall, The propinquity of divisors, Bull. London Math.Soc. 11 (1979), 304-307.

[5] P. Erdös et G. Tenenbaum, Sur la structure de la suite des diviseurs d'un entier, Ann.Inst. Fourier, (1) 31 (1981), 17-37.

[6] R.R. Hall, The propinquity of divisors, J. London Math.Soc. (2) 19 (1979), 35-40.

[7] R.R. Hall et G. Tenenbaum, Sur la proximité des diviseurs, Recent Progress in Analytic Number Theory, vol. 1 (1981), 103-113 [H. Halberstam and C. Hooley ed.], Academic Press.